A Multidisciplinary Approach to Myelin Diseases II

NATO ASI Series

Advanced Science Institutes Series

A series presenting the results of activities sponsored by the NATO Science Committee, which aims at the dissemination of advanced scientific and technological knowledge, with a view to strengthening links between scientific communities.

The series is published by an international board of publishers in conjunction with the NATO Scientific Affairs Division

A	**Life Sciences**	Plenum Publishing Corporation
B	**Physics**	New York and London
C	**Mathematical and Physical Sciences**	Kluwer Academic Publishers
D	**Behavioral and Social Sciences**	Dordrecht, Boston, and London
E	**Applied Sciences**	
F	**Computer and Systems Sciences**	Springer-Verlag
G	**Ecological Sciences**	Berlin, Heidelberg, New York, London,
H	**Cell Biology**	Paris, Tokyo, Hong Kong, and Barcelona
I	**Global Environmental Change**	

Recent Volumes in this Series

Volume 254 —Slow Potential Changes in the Human Brain
edited by W. C. McCallum and S. H. Curry

Volume 255 —Quantitative Assessment in Epilepsy Care
edited by Harry Meinardi, Joyce A. Cramer, Gus A. Baker, and Antonio Martins da Silva

Volume 256 —Advances in the Biomechanics of the Hand and Wrist
edited by F. Schuind, K. N. An, W. P. Cooney III, and M. Garcia-Elias

Volume 257 —Vascular Endothelium: Physiological Basis of Clinical Problems II
edited by John D. Catravas, Allan D. Callow, and Una S. Ryan

Volume 258 —A Multidisciplinary Approach to Myelin Diseases II
edited by S. Salvati

Volume 259 —Experimental and Theoretical Advances in Biological Pattern Formation
edited by Hans G. Othmer, Philip K. Maini, and James D. Murray

Volume 260 —Lyme Borreliosis
edited by John S. Axford and David H. E. Rees

Series A: Life Sciences

A Multidisciplinary Approach to Myelin Diseases II

Edited by

S. Salvati

Istituto Superiore di Sanità
Rome, Italy

Springer Science+Business Media, LLC

Proceedings of a NATO Advanced Research Workshop on
A Multidisciplinary Approach to Myelin Diseases II,
held March 1–4, 1993,
in Rome, Italy

NATO-PCO-DATA BASE

The electronic index to the NATO ASI Series provides full bibliographical references (with keywords and/or abstracts) to more than 30,000 contributions from international scientists published in all sections of the NATO ASI Series. Access to the NATO-PCO-DATA BASE is possible in two ways:

—via online FILE 128 (NATO-PCO-DATA BASE) hosted by ESRIN, Via Galileo Galilei, I-00044 Frascati, Italy

—via CD-ROM "NATO Science and Technology Disk" with user-friendly retrieval software in English, French, and German (©WTV GmbH and DATAWARE Technologies, Inc. 1989). The CD-ROM also contains the AGARD Aerospace Database.

The CD-ROM can be ordered through any member of the Board of Publishers or through NATO-PCO, Overijse, Belgium.

Library of Congress Cataloging-in-Publication Data

A Multidisciplinary approach to myelin diseases II / edited by S. Salvati.
p. cm. -- (NATO ASI series. Series A. Life sciences ; v. 258)
"Proceedings of a NATO Advanced Research Workshop on a Multidisciplinary Approach to Myelin Diseases II, held March 1-4, 1993, in Rome, Italy"--T.p. verso.
"Published in cooperation with NATO Scientific Affairs Division."
Includes bibliographical references and index.
ISBN 978-94-010-4391-5
1. Demyelination--Congresses. 2. Myelination--Congresses. 3. Myelin sheath--Pathophysiology--Congresses. I. Salvati, S. II. North Atlantic Treaty Organization. Scientific Affairs Division. III. NATO Advanced Research Workshop on a Multidisciplinary Approach to Myelin Diseases (2nd : 1993 : Rome, Italy) IV. Series.
[DNLM: 1. Demyelinating Diseases--congresses. 2. Myelin Proteins--genetics--congresses. 3. Myelin Sheath--physiology--congresses. WL 100 M961 1993]
RC366.M85 1993
616.8'3--dc20
DNLM/DLC
for Library of Congress 93-48655
CIP

ISBN 978-1-4613-6034-6 ISBN 978-1-4615-2435-9 (eBook)
DOI 10.1007/978-1-4615-2435-9

Originally published by Plenum Press,New York in 1994
Softcover reprint of the hardcover 1st edition 1994

PREFACE

The diseases that fall under the generalized group of demyelinating diseases - Multiple Sclerosis, Leukodystrophyes, Encephalomyelitis- are the focus of worldwide concern.

This volume contains papers presented by leading scientists who attended the NATO Advanced Research Workshop held at the Istituto Superiore di Sanità, Rome, March 1-4, 1993. This book is an update of the previous one published in 1987 of the research discussed at a similar meeting held in 1986.

It was decided to hold this 2nd meeting since there has been great progress in the advances in understanding the myelinogenesis process in the last five years.

The workshop gathered together scientists from many fields such as cellular and molecular biology, immunology, pathology, virology and of course clinical neurology. Stimulating ideas were exchanged in the hope that more knowledge of demyelinating diseases can lead to new theraupetic approaches.

Although the workshop was on the whole similar to the previous one, this time there was more emphasis on experimental models and clinical aspects. In the former the use of animal and cellular models as tools for understanding the pathological mechanisms linked to human disease were discussed; in the latter the clinicians described the filtering down of basic research to clinical treatment.

The publication of this interdisciplinary exchange is to make known the results of the most recent research among the investigators from all over the world involved in these studies.

I am very grateful to the participants who with their interesting and stimulating discussion contributed to the success of the meeting. I would like to thank the NATO Scientific Affairs Division, Associazione Italiana Sclerosi Multipla, Associazione Italiana Leucodistrofie Unite and the Banca di Roma for their financial support.

Acknowledgments are also given to L. Attorri, A. Confaloni, A. Di Biase , L. Malvezzi Campeggi and F. Pieroni for their highly appreciated contribution to the success of the meeting.

Last but not least the contribution of Elizabeth Duncan in her help with the publication of this volume is gratefully acknowledged.

S. Salvati

CONTENTS

MYELIN GENES AND THEIR REGULATION

MYELIN ASSEMBLY

EXPERIMENTAL MODELS TO STUDY DEMYELINATING DISEASES

IMMUNOLOGICAL ASPECTS OF MYELIN DISEASES

REMYELINATION AND MYELIN REPAIR

ROUND TABLE: CLINICAL ASPECTS OF NEW DISCOVERIES

THE *PROTEOLIPID PROTEIN GENE FAMILY*: THREE NEW MEMBERS AND EVIDENCE FOR THEIR COMMON ORIGIN FROM A PRIMITIVE MEMBRANE CHANNEL

K. Kitagawa and D.R. Colman

Brookdale Center for Molecular Biology
The Mount Sinai School of Medicine
One Gustave L. Levy Place
Box 1126, Room 25-22
New York, New York 10029

INTRODUCTION

It has been presumed that the primary role for the proteolipid polypeptides (PLPs: PLP and DM-20) and P_0 in the generation of the compact myelin sheath in the CNS (PLPs), and PNS (P_0) is essentially the same: to mediate via homophilic interactions the self-adhesion of the plasma membrane that yields compact myelin. The molecular basis for adhesive functions at the intraperiod line in the respective nervous systems in which these proteins operate must however be completely different; curiously, these most abundant integral membrane proteins in tetrapods share absolutely no biochemical, genetic or structural features in common with each other.

P_0-ordered CNS myelin in the CNS and PNS in lower vertebrates (fish) yields morphologically classic, perfectly regular and periodic myelin sheaths. Apparently, P_0 performs the appropriate membrane adhesion functions at both the major dense and intraperiod lines, and so mediates by itself the assembly of a well-ordered sheath. The emergence and dominance of the PLPs in the brains and spinal cords of tetrapods - "higher" organisms - is therefore mystifying. It should be stressed that there is no experimental evidence that the PLPs and P_0 are *actually* functionally equivalent in the generation of the myelin sheath. Although descriptively and perhaps accurately termed "adhesive struts" (1), convincing experimental evidence that directly demonstrates adhesive properties for the PLPs has not been forthcoming, and so an understanding of the precise functional role the PLPs play in CNS myelinogenesis remains elusive.

A Multidisciplinary Approach To Myelin Diseases II
Edited by S. Salvati, Plenum Press, New York, 1994

The appreciation of the existence of an immunoglobulin (Ig)-related domain in the extracellular segment of P_0 and therefore it's membership in the immunoglobulin gene superfamily (2, 3) has led to a variety of studies designed to directly and precisely examine the role this domain plays in membrane adhesion in general. Transfection studies have demonstrated unequivocally the strong effects that P_0-mediated adhesion brings about in non-adhesive cell lines (4, 5). PLP and its isoform DM-20, by contrast have eluded assignment to a known gene family; these unusually hydrophobic proteins, presumed to be derived from a unique gene that emerged *de novo* in tetrapods, as we have stated are not known to be adhesive; in fact; studies on a variety of dysmyelinating mutants (6), and on the expression of the PLP gene during development (7, 8, 9) have intimated pleiotropic (i.e. not only "structural") roles for these proteins in oligodendrocyte maturation and survival.

We have for a while been intrigued by the finding of an immunochemically-reactive PLP in the CNS of the "living fossils" coelacanth and lungfish (for review, see 10), considered to be the closest living relatives of contemporary amphibians (11). These data implied for the first time that the traditionally accepted notion that PLPs are limited to tetrapod nervous systems was not entirely accurate, and suggested to us that multiple forms of proteolipid proteins so far undetected by current technologies may exist in other myelinated organisms, perhaps even in elasmobranchs (sharks), in which myelin first appeared, roughly 400 Myr BP (million years before the present). We therefore set out to search for PLPs in the brain of *Torpedo californica* (the electric ray), an elasmobranch of the suborder *Batoidea*, which separated from other sharks about 200 Myr BP. For our initial studies in this cartilaginous fish we used a molecular biological approach to ascertain whether PLP-like mRNAs were detectable in *Torpedo* CNS.

Several remarkable and unexpected findings emerged from our studies. We identified in *Torpedo* three distinct DM-20 related nucleotide sequences, and conclude that the proteolipids, specifically a DM-20 related subset, form a family of contemporary proteins that originated at least as far back as the first, ancient myelin-forming nervous systems that arose in cartilaginous fish. We term the new proteolipid proteins $\mathbf{DM}_\alpha$, $\mathbf{DM}_\beta$, and $\mathbf{DM}_\gamma$, and include in this newly discovered gene family the well-studied mammalian DM-20, which is most homologous with, but not identical to the *Torpedo* DM_α polypeptide. The contemporary set of *Torpedo* DMs are derived from different although very closely related genes. The expressed molecules in brain tissue are not only highly homologous to each other but as a family bear substantial amino acid sequence identities and similarities to segments of channel-forming regions of the subunit proteins that assemble in *Torpedo* to form the nicotinic acetylcholine receptor, and in the rat the glutamate receptor macromolecular complexes, ligand-gated channels proposed to have diverged at least 750 million years ago from a single ancestral gene encoding a pore-forming protein that contained four putative membrane-spanning transmembrane segments (12). It is likely that the DM polypeptide family, whose distribution in brain may not be limited to myelinating cells, originated as well from this ancestral precursor.

RESULTS AND DISCUSSION

A family of DM polypeptides exists in cartilaginous fish

We synthesized degenerate oligonucleotides representing segments of the first, [plp 1: TT(T/C)TT(T/C)GGIGTIGCI(T/C)TITT(T/C)TG(T/C)GGITG(T/C)GG] third [plp

3: TG(A/G)CAIGTIGTCCAIGT(A/G)TT(A/G)AA(A/G)TAIAT(A/G)TA] and fourth [plp 4: GC(A/G)AA(A/G)TT(A/G)TAIGTIGCIGCIATCAT(A/G)AA] hydrophobic domains of the rat PLP sequence, and used them as primers in polymerase chain reactions (PCR) where first strand cDNAs transcribed from brain mRNAs from *Torpedo californica* were present as templates. PCR reactions yielded single bands on agarose gels with each pair of primers. In *Torpedo californica* a band of about 370 bp was obtained with the plp 1 - plp 3 primers, and 610 bp with the plp1/plp 4 primers (Fig.1):

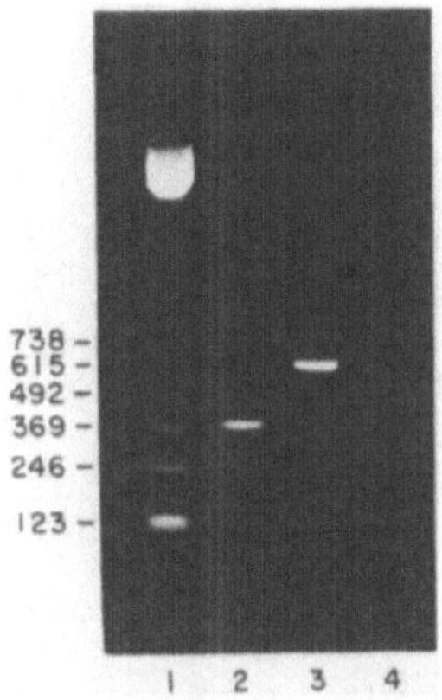

Figure 1: *Torpedo californica* brain contains DM-20 protein-related mRNAs:
Brain cDNA template was synthesized by reverse transcription of poly $(A)^+$ RNA. After forty cycles, aliquots of each PCR reaction were analyzed by agarose gel electrophoresis. Lanes: **1**, markers; **2**, plp1/plp3 primers; **3**, plp1/plp4 primers; **4**, all primers but no cDNA template was added. In lanes 2 and 3, single bands were obtained that we now know to contain at least three distinct DM sequences.

PCR-synthesized cDNAs were cloned into the vector pGEM1, and after transformation, single colonies were isolated and analyzed. The deduced amino acid sequences of individual cDNA inserts were found to be assignable to three groups of DM-20 related polypeptides, that we term **DM$_\alpha$**, **DM$_\beta$**, and **DM$_\gamma$**, with DM$_\alpha$ having the highest sequence identities and similarities (64.1% identity and 82.5% similarity to mammalian DM-20 across a span of 103 amino acids, or 42% of the entire rat DM-20 polypeptide length; see Fig 2).

It is of great interest that, surprisingly, no cDNA was found that contained a nucleotide sequence corresponding to the unique region in PLP, since it has been assumed that in tetrapods, the DM-20 polypeptide arose from the introduction of a new splice site in a pre-existing gene that encoded PLP (13, 14). Implicit in this hypothesis is acceptance of the notion that PLP *predates* DM-20 in evolution. By contrast, our results (a) provide the first evidence that the DM family of polypeptides existed prior to the acquisition of the nucleotide sequence that encodes the peptide segment unique to PLP in tetrapods, and (b) illustrate the ubiquitous distribution throughout vertebrate evolution (at least as far back as the emergence of cartilaginous fish), of a DM family of proteins in brain.

In *Torpedo californica,* we detected on brain RNA blots a single band of 1.5 kb with the DM$_\alpha$ probes, a single band of 3.4kb with DM$_\beta$, and two bands of 2.5 kb and 3.4kb with the DM$_\gamma$ probe. Clearly, then, the DM polypeptides are products of

```
           1                                                                                                     103
Rat  DM20  HEALTGTEKLIETYFSKNYQDYEYLINVIHAFQYVIYGTASFFFLYGALLLAEGFYTTGAVRQIFGDYKTTICGKGLSATFVGITYALTVVWLLVFACSAVPV
           ||||||||||: .||.:: ||..| .:::.|||||||||||||||.||||||||||:||: :||::||:||:.:||||: :|. |.|:|: ||| |||||
Torp DMα   HEALTGTEKLIGQHFSQHFVDYALLATLVQVFQYVIYGTASFFFLYGVLLLAEGFYTTSAVKSVFGEFRTTMCGRCVSATFIFLTHILSVIWMGVFAFSAVPV

           1                                                                                                       103
Rat  DM20  HEALTGTEKLIETYFS..KNYQDYEYLINVIHAFQYVIYGTASFFFLYGALLLAEGFYTTGAVRQIFGDYKTTICGKGLSATFVGITYALTVVWLLVFACSAVPV
           ||||.|| .:::. |.  :. .|   ::.:|. |.|||||.|. ||:|| ||:.|||:||||::::||:|.|.||:.:|: |: :|| : :.|| | | ..:||
Torp DMβ   HEALSGTVTILQNNFEVVRGAGDTLDVFTMIDIFKYVIYGVAAAFFVYGILLMVEGFFTTGAIKDLYGDFKITTCGRCVSGWFIMLTYIFMLAWLGVTAFTSLPV

           1                                                                                                  100
Rat  DM20  HEALTGTEKLIETYFSKNYQDYEYLINVIHAFQYVIYGTASFFFLYGALLLAEGFYTTGAVRQIFGDYKTTICGKGLSATFVGITYALTVVWLLVFACSA
           | ||| .|:::  |||.|  |. .|.:||: :||||||.|||||||| :||||||||.:||::| |::|||:||:.:|: || :|| |.:.|| ||: ||
Torp DMγ   HVALTKVERIVLMYFSNNPSDHVLLTDVIQIMQYVIYGVASFFFLYGIILLAEGFYTSSAVKEIHGEFKTTLCGRCISGMFVFLTYLLGIAWLGVFGFSA

           101                                                                         185
Rat  DM20  VPVYIYFNTWTTCQSIAFPS.KTSASIGSLCADARMYGVLPWNAFPGKVCGSNLLSICKTAEFQMTFHLFIAAFVGAAATLVSLLT
           |||:||:| |..||.|. |. . ...|:.:|.|.| :|::|||| ||:.|||:|  ||.|.||::.:||||.| .||:||::.||.
Torp DMγ   VPVFIYYNMWSACQTISSPTVNLTTAIEEICVDVRQFGIIPWNASPGRACGSDLTIICNTSEFDLSYHLFIVACAGAGATVIALLI
```

Figure 2

```
           1          -                                 50     -            +   -  +   +                                100
Rat  DM20  HEALTGTEKL IETYFS..KN YQDYEYLINV IHAFQYVIYG TASFFFLYGA LLLAEGFYTT GAVRQIFGDY KTTICGKGLS ATFVGITYAL TVVWLLVFAC
Torp DMα   HEALTGTEKL IGQHFS..QH FVDYALLATL VQVFQYVIYG TASFFFLYGV LLLAEGFYTT SAVKSVFGEF RTTMCGRCVS ATFIFLTHIL SVIWMGVFAF
Torp DMβ   HEALSGTVTI LQNNFEVVRG AGDTLDVFTM IDIFKYVIYG VAAAFFVYGI LLMVEGFFTT GAIKDLYGDF KITTCGRCVS GWFIMLTYIF MLAWLGVTAF
Torp DMγ   HVALTKVERI VLMYFS..NN PSDHVLLTDV IQIMQYVIYG VASFFFLYGI ILLAEGFYTS SAVKEIHGEF KTTLCGRCIS GMFVFLTYLL GIAWLGVFGF
           ↑                                         ↑                                                    ↑          ↑
           P(shaking pup)                            P(md rat)                                            I(P-M4)    R(P-M1)

           101                           - +         150                    -                  189
Rat  DM20  SAVPVYIYFN TWTTCQSIAF PS.KTSASIG SLCADARMYG VLPWNAFPGK VCG.SNLLSI CKTAEFQMTF HLFIAAFVGA AATLVSLLT
Torp DMα   SAVPVYVYFT FWSSCQTVRH VT.ENGTGFD DVCVDARQYG ILPWNASPGK ICG.LNLANV CNTSDLELTY HLFIATFAGA AATAIALLT
Torp DMβ   TSLPVFMYFN IWTLCQNITI VD.ST..... DLCLDLRKFG IVPIHEQKTV CTLNENFSKL CQSNDLNMTF HLFIVALAGA GAAVIAMVH
Torp DMγ   SAVPVFIYYN MWSACQTISS PTVNLTTAIE EICVDVRQFG IIPWNASPGR ACG.SDLTII CNTSEFDLSY HLFIVACAGA GATVIALLI
                            ↑                                    ↑  ↑                        ↑
                            T(rumpshaker)                        S  F(P-M5)                   V(jimpy msd)
                                                              (P-M2)
```

Figure 3

```
Torp  DMγ            IQIMQYVIYGVASFFFLYGIIL--LAEGFYTSSAV-KEIHGEFKTTLCGRCISGMFVFLTYLLGIAWLGVFGFSAVPVFIYYNMWSACQTISS
                     ::|  ..| .  .::|: .:|:  |  :|.|: ..  ...:: | ||   :||.:: : .:||.|. | . . ||||::  | ::. . .|||
Torp nACHRα     (223)LDITYHFIMQRIPLYFVVNVIIPCLLFSFLTGLVFYLPTDSGEKMTL---SISVLLSLTVFLLVIVELIPSTSSAVPLIGKYMLFTMIFVISS(312)

Torp  DMα            YGTASFFF--LYGVLLLAEGFYTTSAVKSVFGEFRTTMCGRCVSATFIFLTHILSVIWMG-VFAFSAVPVYVYFTFWSSCQTVRHVTENGTGFDDVCVD-ARQYGIL
                     || |.. .  |  .|: .|.:  ..:..|:  .:......:: .:.| ||..:    |||. |||: :|.| :::. : | .  :| .| :.| |:.. | ..::||:
Rat   GluR1     (483)YGRADVAVAPLTITLVREEVIDFSKPFMSLGISIMIKKPQKSKPGVFSFLDPLAYEIWMCIVFAYIGVSVVLFLVSRFS-PYEWHSEEFEEGRDQTTSDQSNEFGIF(588)
```

Figure 4

different sized mRNAs, at least one of which, DMγ, is alternatively spliced, or contains different polyA addition sites. Genomic DNA blotting experiments revealed that each DM cDNA we found in *Torpedo* is derived from a different, although closely related gene.

Sequence analysis of the DM family

We identified a number of common and intriguing features (Fig. 3) between the deduced amino acid sequences of the lengthy PCR-generated *Torpedo* brain cDNAs, and the homologous region of mammalian DM-20 (exemplified by the rat sequence). We anticipate from the data reported here and ongoing studies (Kitagawa et al., in preparation) that full-length shark DMs contain, like tetrapod DM-20 and PLP, uncleaved amino terminal insertion signals and four membrane-spanning hydrophobic segments.

Comparison of the deduced amino acid sequences revealed a generally higher degree of amino acid conservation in the hydrophobic, presumably membrane-spanning domains, than in the hydrophilic regions. Interestingly however, there is extensive charge conservation, especially in the third hydrophilic domain, although there are substitutions (D/E and K/R). Tryptophan (W), which has the lowest relative mutation index of any amino acid, cysteine (C), which can exist in a variety of molecular states, and proline (P), an amino acid that as a helix-breaker confers certain structural constraints on a polypeptide chain, were also found to be conserved. It is of substantial interest that amino acid positions that are the subject of point mutations leading to dysmyelination (for review, see 6) in the *shaking pup*, *md* rat, Pelizaeus-Merzbacher (P-M) disease 4, and P-M 1, are absolutely conserved between *Torpedo* and rat. Apparently, these positions are crucial for the proper function of these proteolipid proteins in these very different organisms. Conversely, amino acid positions corresponding to mutated sites in *rumpshaker* (15), P-M 2 (6), P-M 5 (16), and *jimpy*^{msd} (6) can apparently tolerate substantial flexibility, since amino acids in these positions are not conserved within the DM family.

In light of the finding of a *glycosylated* proteolipid protein in lungfish (17), we noted with interest the presence of the sequence N-X-S/T, the consensus sequence for N-linked glycosylation in the *Torpedo* DMs in hydrophilic and membrane-bordering regions, at least one of which is presumably exposed at the extracellular surface of the plasma membrane bilayer, according to current topological models for tetrapod DM-20 and PLP (15, 18, 19, 20).

Regional homology between the DM family and ligand-gated channels

A search of the Genebank database for related sequences revealed only the expected high homologies with PLP cloned from a variety of tetrapods. However, when we directly compared each of the *Torpedo* DM sequences with those of polypeptides believed to possess four transmembrane segments [for example, the connexins (21), PMP-22 (22), the *rds* protein (23), and several ligand-gated channels (24, 25)] we were surprised to find substantial identities and similarities (Fig. 4) between DMγ and the α subunit of the *Torpedo* nicotinic acetylcholine receptor (27% identity and 49.4% similarity across 92 amino acids), and between *Torpedo* DM_{α} and the rat glutamate receptor (25.5% identity and 46.1% similarity across 103 amino acids).

The similar regions were located in and between the membrane-spanning segments of all of these molecules; that is to say that similarity was with the *channel* or pore-forming regions of the receptors, and *not* with the N-terminal ligand-binding domains. It is important to stress that no such substantial, or even minimal, sequence resemblance exists between the channel-forming region in the *Torpedo* nicotinic receptor and rat glutamate receptor; in this respect the DM family is closer in sequence similarity to each of these receptors than the receptors are to each other.

Proteolipid polypeptide origin

It is clear from these studies that contrary to popular belief, there are at least three proteolipid proteins that co-exist in *Torpedo* CNS, forming a gene family that probably arose by duplication from a single ancestral gene, which is likely to have functioned even in organisms that greatly preceded the invention of the vertebrate phylum. We can infer that the original DM gene encoded a polypeptide that possessed an uncleaved amino terminal hydrophobic segment which functioned as a signal for membrane insertion into the rough endoplasmic reticulum, as is still the case with DM-20 and PLP in tetrapods (26). This ancestral polypeptide probably possessed as well four membrane-spanning domains, an evolutionarily conserved structural motif of co-assembling subunits of several different macromolecular complexes that function as pores or channels. The detection of sequence similarity between the *Torpedo* DM family and certain ligand-gated channels lends further support to the developing notion of possible channel-like properties for myelin proteolipid proteins (15, 27, 28, 29, 30, 31).

It is thought-provoking to speculate that the *Torpedo* DM polypeptides *may actually be* channel subunits. Further, these proteins may be expressed in brain in other cells besides oligodendrocytes, and in fact may eventually be found in tissues throughout the organism. It is also entirely possible, and indeed even likely, that these or other new members of the DM gene family may be discovered in tetrapods, or at the other end of the vertebrate spectrum, possibly in the unmyelinated vertebrate genus *Agnatha*, as well.

Finally, it is important to emphasize that these studies support the notion of pleiotropic properties for proteolipid proteins in the myelin sheath, and lend credence to the concept that P_0 and the PLPs are not functionally equivalent in myelin sheath biogenesis. Speculatively, it may be suggested that the oligodendrocyte may have co-opted a primitive channel protein or proteins, harnessing the tendency for self-assembly to aid in the generation of the sheath, and perhaps even retaining the pore-forming functional capabilities of the channel polypeptide(s).

Acknowledgments

This work was supported by a National Institute of Health Grant (#NS20147).

This is manuscript #148 from the Brookdale Center for Molecular Biology, Mt. Sinai School of Medicine.

Bibliography

1. Kirschner, D., Ganser A., and Caspar, D. (1984) Diffraction studies of molecular organization and membrane interactions in myelin, in Myelin (Morell , P., ed), pp 51-95. Plenum Press, New York.

2. Lai, C. et al. (1987) Neural protein 1B236/myelin associated glycoprotein (MAG) defines a subgroup of the immunoglobulin gene superfamily. Immunol. Rev. 100: 129-151.
3. Lemke, G., Lamar, E. and Patterson, J. (1988) Isolation and characterization of the gene encoding the major structural protein of peripheral myelin. Neuron 1: 73-83.
4. D'Urso D., Brophy, P.J., Staugaitis, S.M., Gillespie, C.S., Frey A.B., Stempak, J.G. and Colman, D.R. (1990) Protein zero of peripheral nerve myelin: biosynthesis, membrane insertion, and evidence for homotypic interaction. Neuron 4: 449-460.
5. Filbin, M.T., Walsh, F.K., Trapp, B.D. , Pizzey, J.A. and Tennekoon, G.I. (1990) Role of myelin P0 protein as homophilic adhesion molecule. Nature 344: 871-872.
6. Hudson, L. (1990) Molecular genetics of X-linked mutants. Ann. New York Acad Sci. 605: 155-165.
7. Ikenaka, K., Kagawa, T. and Mikoshiba, K. (1992) Selective expression of DM-20, an alternavely spliced myelin proteolipid protein gene product, in developing nervous system and in nonglial cells. J. Neurochem. 58: 2248-2253.
8. Timsit, S. G., Bally-Cuif, L., Colman, D.R. and Zalc, B. (1992) DM-20 mRNA is expressed during the embyonic development of the nervous system of the mouse. J. Neurochem. 58: 1172-1175.
9. LeVine, S.M., Wong, D. and Macklin, W.B. (1990) Developmental expression of proteolipid protein and DM20 mRNA and proteins in the rat brain. Dev. Neurosci. 12: 235-250.
10. Waehneldt, T. (1990) Phylogeny of myelin proteins. Ann. New York Acad. Sci. (605) 15-28.
11. Gorr, T., and Kleinschmidt, T. (1993) Evolutionary relationships of the coelacanth. American Scientist 81: 72-82.
12. Hille, B. (1992) Ionic channels of excitable membranes. (Sinauer Asoociates, Inc., pub., Sunderland, Massachusetts).
13. Nave, K.A., Lai, C., Bloom, F.E. and Milner R.J. (1987) Splice site selection in the proteolipid protein (PLP) gene transcript and primary structure of the DM20 protein of central nervous system myelin. Proc. Natl. Acad. Sci. USA 84: 5665-5669.
14. Diehl, H.J., Schaich, M., Budzinski, R.M. and Stoffel, W. (1986) Individual exons encode the integral membrane domains of human myelin proteolipid protein. Proc. Natl. Acad. Sci. USA 83, 9807-9811.
15. Schneider et al. (1992) Uncoupling of hypomyelination and glial cell death by a mutation in the proteolipid gene. Nature 238: 758-761.
16. Pham-Dinh et al. (!991) Pelizaeus-Merzbacher disease: A valine to phenylalanine point mutation in a putative extracellular loop of myelin proteolipid. Proc. Natl. Acad. Sci. USA 88: 7562-7566.
17. Waehneldt, T.V., Matthieu, J.M. and Jeserich, G. (1986) Major central nervous system myelin glycoprotein of the african lungfish (Protopterus dolli) cross-reacts with myelin proteolipid protein antibodies, indicating a close phylogenetic relationship with amphibians. J. Neurochem. 46: 1387-1391.
18. Hudson, L.D., Friedrich, V.L., Behar, T., Dubois-Dalcq, M. and Lazzarini, R.A. (1989) The initial events in myelin synthesis: Orientation of proteolipid protein in the plasma membrane of cultured oligodendrocytes. J. Cell Biol. 109: 717-727.

19. Popot, J. L., Dinh, D.P. and Dautigny, A. (1991) Major myelin proteolipid: the 4-α-helix topology. J. Memb. Biol. 120: 233-246.
20. Schliess, F. and Stoffel, W. (1991) Evolution of the myelin integral membrane proteins of the central nervous system. Biol. Chem. Hoppe-Seyler 372, 865-874.
21. Beyer, E.C., Paul, D.L., and Goodenough, D.A., (1990) Connexin family of gap junction proteins. J. Memb. Biol. 116: 187-194.
22. Suter et al. (1992) Trembler mouse carries a point mutation in a myelin gene. Nature 356: 241-244.
23. Travis, G.H., Sutcliffe, J.G., and Bok, D. (1991) The retinal dgeneration slow (rds) gene product is a photoreceptor disc membrane associated glycoprotein. Neuron 6: 61-70.
24. Hollmann, M., O'Shea-Greenfield, A., Rogers, S.W. and Heinemann, S. (1989) Cloning by functional expression of a member of the glutamate receptor family. Nature 342, 643-648.
25. Noda et al. (1983) Primary structure of α-subunit precursor of *Torpedo californica* acetylcholine receptor deduced from a cDNA sequence. Nature 301: 251-255.
26. Colman, D.R., Kreibich, G., Frey, A.B. and Sabatini, D.D. (1982) Synthesis and incorporation of myelin polypeptides into CNS myelin. J. Cell Biol. 95:598-608.
27. Lin, L.F.H. and Lees, M.B. (1982) Interaction of dicyclohexylcarbodiimide with myelin proteolipid. Proc. Natl. Acad, Sci. USA 79: 941-945.
28. Ting-Beall, H.P., Lees, M.B. and Robertson, J.D. (1979) Interaction of Folch-Lees proteolipid apoprotein with planar lipid bilayers. J. Memb. Biol. 51: 33-46.
29. Diaz, R.S., Monreal, J., and Lucas, M. (1990) Calcium movements mediated by proteolipid protein and nucleotides in liposomes prepared with the endogenous lipids from brain white matter. J. Neurochem. 55: 1304-1309.
30. Helynck, G. et al. (1983) Brain proteolipids. Isolation, purification, and effect on ionic permeability of membranes. J. Biochem. 133: 689-695.
31. de Cozar, M., Lucas, M., and Monreal, J. (1987) Ionophoric properties of the proteolipid apoprotein from bovine brain myelin. Biochem. Int. 14: 833-841.

TRANSGENIC MODELS FOR INVESTIGATING OLIGODENDROCYTE DIFFERENTIATION AND MYELIN FORMATION

Wendy B. Macklin[1], Patricia A. Wight[1]*,
Cynthia S. Duchala[1] and Carol Readhead[2]**

[1]Mental Retardation Research Center
UCLA Medical Center
Los Angeles, CA 90024
[2]Department of Biology
California Institute of Technology
Pasadena, CA 91125

INTRODUCTION

The myelin membrane is a highly specialized extension of the oligodendroglial plasma membrane, which surrounds axons and forms a tightly compacted multilamellar structure[1]. Myelin has a relatively simple protein composition, with the proteolipid (PLP) and DM20 proteins comprising approximately 50% of adult CNS myelin protein[2,3]. These proteins are translated from alternatively spliced mRNAs encoded by the PLP gene[4,5,6]. Thus, a major portion of the oligodendrocyte differentiation program is dedicated to the expression of the PLP gene. Mutations in this gene are devastating, and animals such as the *jimpy* mouse or the *md* rat are severely compromised, dying within the first postnatal month. CNS changes have been noted in these animals, at ages prior to the time of oligodendrocyte differentiation[7,8,9,10,11]. In earlier studies, we and others demonstrated that the DM20 mRNA and protein appear prior to PLP in the developing nervous system[12,13,14,15]. In addition, it has now been shown that the DM20 mRNA is expressed in the developing embryonic nervous system[16,17]. Thus, a number of studies have suggested that in addition to the production of the most abundant proteins of the myelin membrane, the PLP gene may encode a protein(s), perhaps the DM20 protein, which is expressed in cells in the developing nervous system prior to oligodendrocyte differentiation.

* Current address: Department of Physiology, University of Arkansas for Medical Sciences, Little Rock, AR 72205
** Current address: Department of Medicine, Cedars Sinai Medical Center, D-3067, Los Angeles, CA 90048

A Multidisciplinary Approach To Myelin Diseases II
Edited by S. Salvati, Plenum Press, New York, 1994

Transgenic technology is an invaluable tool in investigating brain development and function. In such studies, foreign DNA is introduced into the germline and its expression can be examined under *in vivo* conditions throughout development. Expression of the transgene can be controlled by a constitutive promoter such as the actin promoter, an inducible promoter such as the metallotheionine promoter, or a tissue specific promoter such as the neurofilament promoter. The *E. coli* LacZ reporter gene has been used as a transgene to track expression of many constitutive, inducible or cell-specific promoters in the brain, such as neuron-specific enolase[18], amyloid precursor protein[19], glial fibrillary acidic protein[20] and c-*fos*[21]. Expression of this reporter gene is easy to detect with a simple histological assay for ß-galactosidase activity. The current studies were undertaken to investigate oligodendrocyte differentiation, using this reporter gene and a cell-specific promoter, the mouse myelin PLP promoter. The transgene effectively tracked PLP gene expression in the developing nervous system. Thus, transgenic mice expressed the PLP-LacZ transgene at developmental ages and in cell types that are consistent with normal PLP gene expression. Interestingly, the data support potentially new features of PLP gene regulation. For example, transgene expression was seen in both the central and peripheral nervous system of developing embryos, providing support for the hypothesis that the PLP gene is expressed in cells other than differentiating oligodendrocytes. An important aspect of transgene expression is the observation that the fusion protein is localized in the myelin membrane itself, due apparently to the inclusion of PLP N-terminal amino acids. These studies are extensions of studies reported earlier[22,23,24].

METHODS AND MATERIALS

Generation of Transgenic DNA Construct

The transgene was generated as described[24]. The final transgene, PLP(+)Z, contained approximately 2.4 kb of 5'-flanking PLP DNA, exon 1, intron 1, and the first 37 bp of PLP exon 2 ligated to a trpS-LacZ fusion gene and SV40 polyadenylation signals (nucleotides 209-3736 of pCH110, Promega) (Figure 1). Thus, PLP intron 1 splice sites were maintained for accurate splicing and protein synthesis initiates at the proper PLP AUG.

Transfection of N20.1 Cells

N20.1 cells, a cell line derived by immortalization of primary oligodendrocytes with SV40 large T antigen[25], were grown in at 34°C in Ham's F12/DME supplemented with 10% fetal bovine serum (Irvine Scientific). Cells were transfected with either pPLP(+)Z or pRSVZ[26] by calcium phosphate-mediated gene transfer as described by the methods of Chen and Okayama[27]. Briefly, 1.5×10^5 N20.1 cells were seeded in each well of a 6-well plate containing 2 ml of growth medium and incubated at 34°C/5% CO_2 overnight. The next day, a calcium phosphate-DNA mixture was added to the medium with a final DNA concentration of 2.4 µg/ml. The plate was incubated at 34°C in a 3% CO_2 atmosphere. The medium was removed after 24h and the cells were washed two times with growth medium, refed with medium and incubated at 34°C in an atmosphere of 5% CO_2. After 48 h, transfected cells were fixed in 1% glutaraldehyde and stained as described by Lim and Chae[28].

Generation of transgenic mice

The transgene was purified away from vector sequences and approximately 200 copies/pl were microinjected into the pronuclei of BDF2 fertilized mouse eggs[29,30]. After incubation, 2-cell stage embryos were transferred into BDF1 pseudopregnant mice. Pups were screened at three weeks of age to identify those carrying the transgene, by Southern blot analysis of tail DNA probed with random primed LacZ DNA.

Histochemistry

Animals were perfused with 0.5% paraformaldehyde-1.0% glutaraldehyde in PBS, pH 7.3 under halothane anesthesia. Tissue was dissected out after one hour and immersed in fixative containing 10% sucrose. After several hours, tissue was immersed in fixative containing 25% sucrose and allowed to sink overnight at 4°C. Tissue was then stored at -80°C and subsequently cryostat sectioned. Sections (20-30 µm) were stained with 1 mg/ml X-gal (5-bromo-4-chloro-3-indolyl-ß-galactopyranoside; Promega) in 5mM potassium ferricyanide, 5 mM potassium ferrocyanide, 2 mM $MgCl_2$, 0.02% NP-40, 0.01% sodium deoxycholate and PBS[24,31] .

Electron microscopy

Animals (5 day) were perfused with 0.5% paraformaldehyde, 1% glutaraldehyde, and tissue was isolated. Samples were dissected for staining with Bluo-gal (halogenated indolyl-ß- D-galactoside; Gibco BRL) as described by Loewy et al.[32] The Bluo-gal stain contained 10 mM Bluo-gal, 16 mM potassium ferricyanide, 16 mM potassium ferrocyanide and 2 mM $MgCl_2$ in PBS. Vibratome sections (500 µm) were stained for 14-18 h at 37°C, embedded and processed for electron microscopy.

RESULTS

Generation of Transgenic Mice

The PLP-LacZ transgene contained PLP sequences encompassing approximately 2.4 kb 5'-flanking DNA, exon 1, intron 1 and part of exon 2 fused to the LacZ gene (Figure 1). To determine whether this transgene could be appropriately expressed in oligodendrocytes, it was transfected into an immortalized oligodendrocyte cell line N20.1.

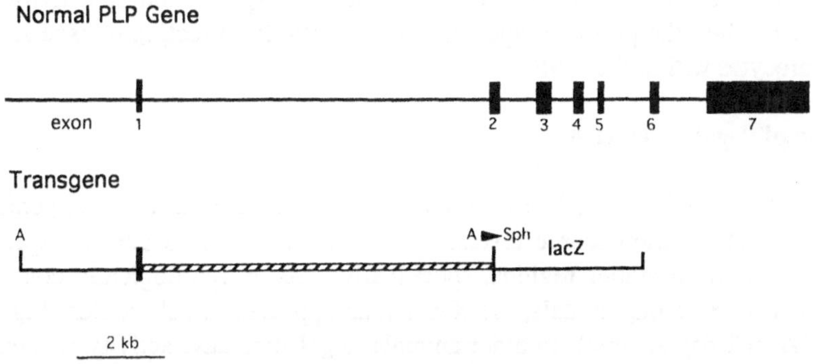

Figure 1. Structure of PLP-LacZ transgene. The normal PLP gene is shown with black bars representing the seven exons of the mouse PLP gene. The transgenic construct is represented below, which contains PLP genomic DNA from an *Apa*I (A) site in the 5'-flanking DNA to an *Apa*I site in exon 2, which has been converted to a *Sph*I site, at which the LacZ is fused.

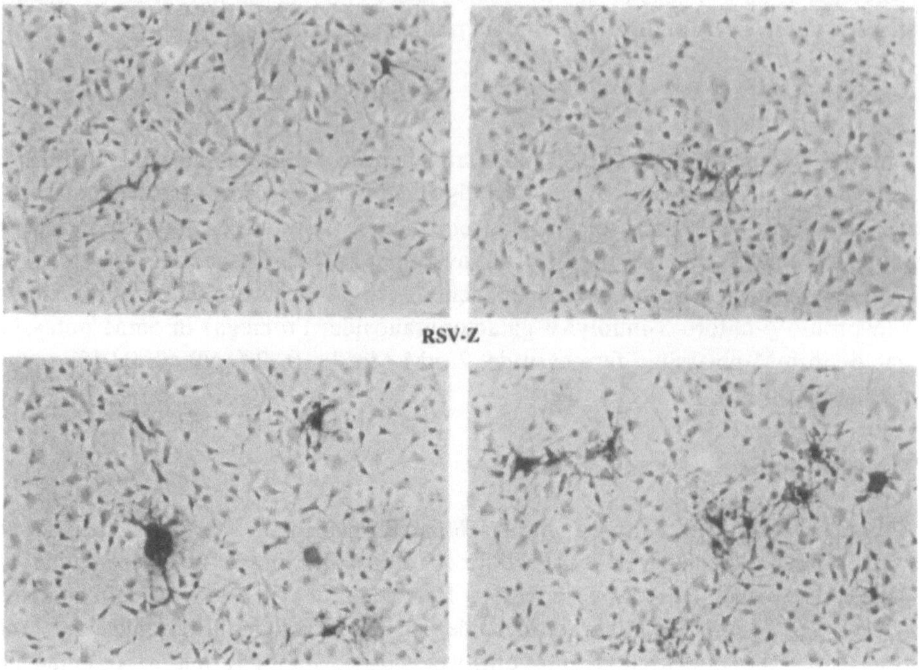

Figure 2. Transfection of N20.1 cells with PLP-LacZ transgene and RSV-Z. Two representative fields of N20.1 cells transfected with either plasmid, and stained with X-gal to identify cells expressing ß-galactosidase are shown.

While it was not expressed as well as a control plasmid, RSV-LacZ, which is controlled by a strong viral promoter, the PLP-LacZ transgene was expressed in the N20.1 cells (Figure 2).

Three transgenic lines were generated, all of which express the transgene in white matter regions of the central nervous system. Lines 26 and 27 expressed high levels of enzyme activity (Figure 3), while line 20 showed much lower levels (Figure 4), and no ß-galactosidase activity was observed in nontransgenic brains (Figure 4). All three lines demonstrated essentially the same cellular distribution of ß-galactosidase activity within the nervous system. Thus, the promoter appears to be specific for directing transgene expression to oligodendrocytes within the CNS.

Distribution of ß-galactosidase

The distribution of the ß-galactosidase activity was studied in the transgenic brains at the light and electron microscopic levels. As shown in Figures 3 and 4, expression was observed only in white matter regions, specifically cells morphologically consistent with oligodendrocytes. In young animals, the X-gal stain appeared in cell bodies (Figure 5, note cell at asterisk in 2 day sample). In older animals, ß-galactosidase activity was extensive in myelinated regions, and appeared to be associated with the myelin sheath itself. It

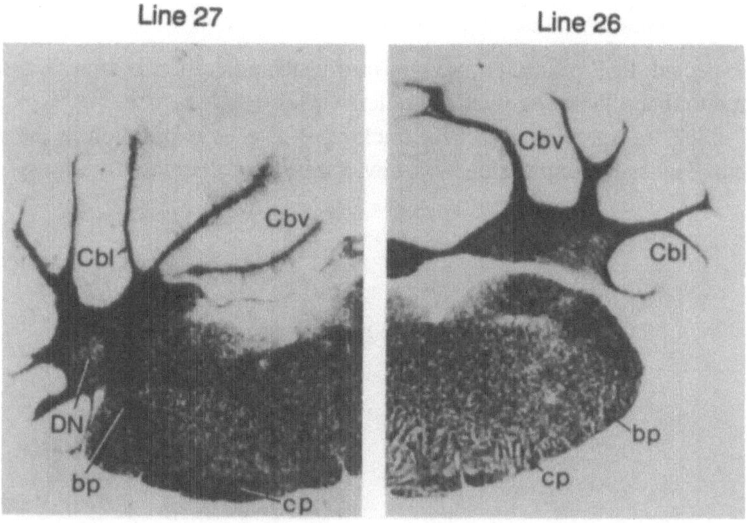

Figure 3. X-gal histochemistry of transgenic mice. Coronal sections of cerebellum from two high expressing transgenic lines, 26 and 27 at day 19 and 17 respectively. Tissue was stained for two hours. Staining seen in lateral cerebellum (Cbl), ventral cerebellum (Cbv), dentate nucleus (DN), brachium pontis (bp), cerebral peduncle (cp)

appeared less in cell bodies, and more in the cable-like network of myelinating axons (note arrows in 19 and 25 day samples, Figure 5). This localization in the myelin sheath was confirmed at the electron microscopic level (Figure 6). The enzyme activity was predominantly found in the multilamellar structures of the myelin membrane, never in gray matter areas, and occasionally associated with the nuclear membrane. This unique membrane localization for ß-galactosidase likely results from the addition of the N-terminal 13 amino acids of the proteolipid protein to the fusion protein.

Developmental Expression of Transgene

The PLP-LacZ transgene was developmentally controlled. Very low levels of expression were seen in neonatal animals (Figure 7), while enzyme levels increased as animals developed (Figure 8). Significant enzyme activity was present in adult animals. At all ages studied, ß -galactosidase activity was localized to white matter areas of the brain.

DISCUSSION

Expression of the PLP-LacZ transgene was controlled by the myelin PLP promoter and most probably sequences within intron 1. PLP sequences contained within the transgene were sufficient to promote transgene expression in both transfected cells and transgenic animals. As expected, PLP promoter strength in transfected cells was significantly reduced when compared to that of a strong viral promoter.

The PLP-LacZ transgene is useful for tracking PLP gene expression in the developing nervous system. Transgene expression was low at birth and increased with age, appearing

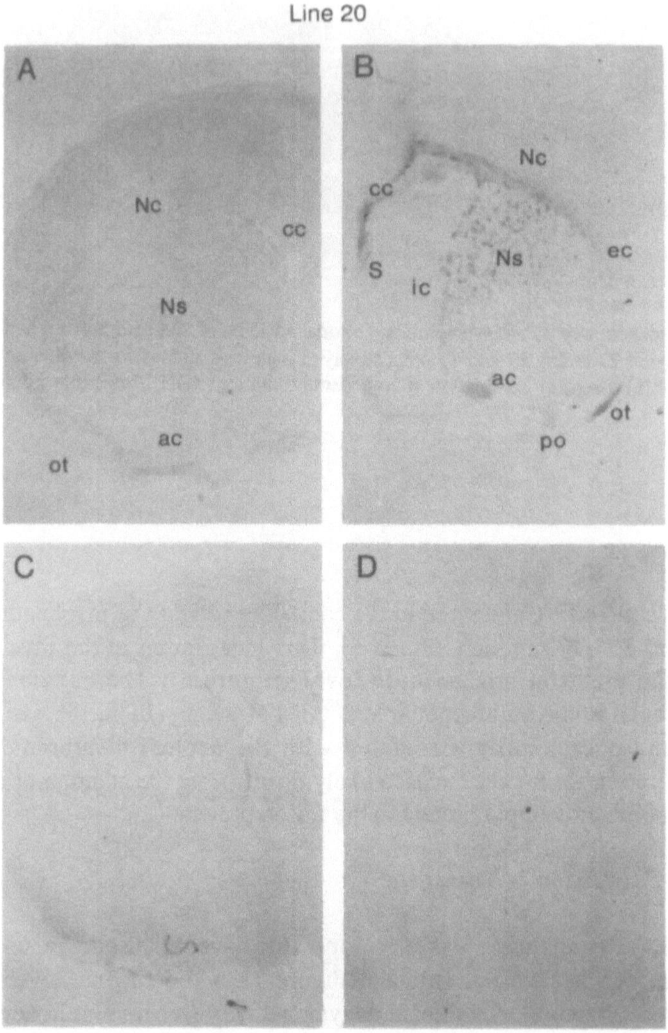

Figure 4. X-gal histochemistry of transgenic mice. Coronal section of low expressing line 20 at day 25. Tissue was stained with X-gal for 17 hours (B,D). Comparable section stained with cresyl violet (A,C). A and B are sections from a transgenic animal; C and D are from a non-transgenic littermate. Staining seen in anterior commisure (ac), corpus callosum (cc), internal capsule (ic) neostriatum (Ns), olfactory tract (ot), primary olfactory cortex (po), and septum (S). Less was seen in the external capsule (ec) and neocortex (Nc).

only in white matter areas. Transgene expression was strong in adult animals, indicating that either the gene was still expressed at that age, or the enzyme was stabilized.

In other studies, expression of the transgene was demonstrated in embryos[24]. In these studies, expression was quite extensive throughout the peripheral nervous system and in selected regions of the central nervous system. Specifically, ß-galactosidase activity was noted in the spinal cord, dorsal root ganglia, and olfactory bulb, furthermore, activity was observed in peripheral nerves of the forearm, and in the trigeminal nerve. This distribution is similar to that recently reported for DM20 expression in the developing embryo. DM20 is an alternatively spliced protein produced by the PLP gene, which has been found in

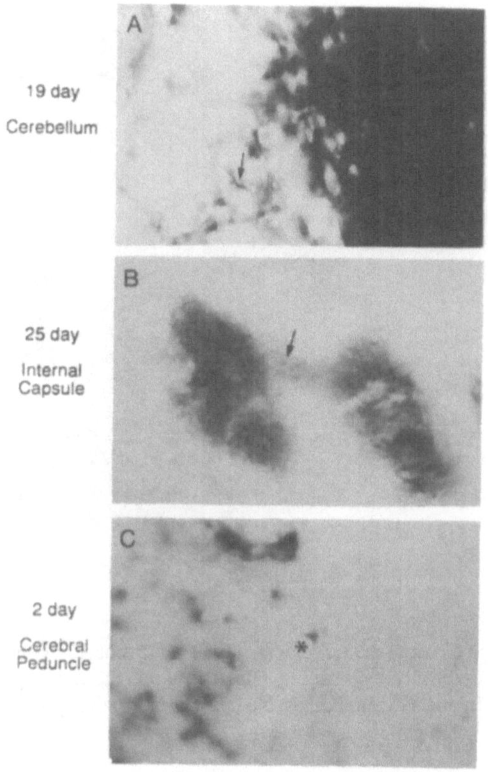

Figure 5. X-gal histochemistry of transgenic mice. High magnification (100X objective) of ß-galactosidase staining in A) 19 day cerebellum, B) 25 day internal capsule and C) 2 day cerebral peduncle. Arrows identify fibers that were stained in older animals; asterisk denotes stained cell body in 2 day animal.

embryos[16,17], in heart[33] and in adult Schwann cells[34,35,36,37,38,39]. Thus the elements of this gene that control DM20 (and/or PLP) expression in embryos and in postnatal brain appear to be contained in the 2.4 kb of 5'-flanking DNA or intron 1 of the mouse myelin PLP gene.

Expression of the transgene was quite strong in two of the transgenic lines (26 and 27), and significantly weaker in line 20. Expression of the transgene in line 20 was quite low, and only major white matter areas of the brain were detected. This may reflect a positional effect of transgene integration in line 20. It does not appear to be due to a gene dosage effect, since both lines 20 and 27 contained comparable amounts of DNA, but had dramatically different levels of expression. In contrast, line 26 contained substantially higher copies of the transgene, but both lines 26 and 27 expressed the transgene at the same relative level[24]. Therefore, the low expression observed in line 20 may be due to a *cis*-acting element(s) near the site of integration. Alternatively the transgene may be maintained in the chromatin in a conformation less favorable for transcription.

The electron microscopic analysis of the expression of the transgene in the CNS indicated that the ß-galactosidase was present in the myelin membrane itself. This is quite intriguing since this enzyme is normally a soluble protein. This results presumably from the presence of the first 13 amino acids of the PLP protein in the fusion protein. The

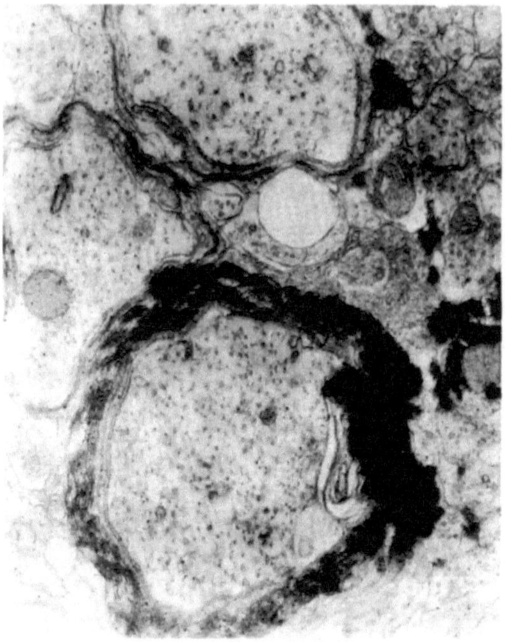

Figure 6. Electron micrograph of white matter region of transgenic mouse. Tissue from a five day mouse was analyzed for ß-galactosidase activity by staining with Bluo-gal. Staining appeared primarily in myelin membranes and occasionally in nuclear membranes. No cytoplasmic staining was observed.

presence of the transgene product in the myelin membrane has had no obvious clinical effect on these animals, since animals have survived longer than one year with no apparent dysmyelination. Additionally, when myelin was analyzed by electron microscopy from older animals, it appeared normal and fully compacted. When the ß-galactosidase enzyme assay was conducted on vibratome sections, the tissue preservation of the myelin was poor (see Figure 6). In contrast, when identical samples were incubated with all reagents except the Bluo-gal, or when nontransgenic samples were incubated with all reagents including the Bluo-gal, no tissue disruption was observed (data not shown). Thus, the disruption of the compact myelin in these samples appears to result solely from conversion of the Bluo-gal substrate to product. How this enzymatic reaction disrupts the compact myelin is unknown. Interestingly, when myelin was purified from adult transgenic animals, it had some biochemical differences from nontransgenic myelin. The myelin fraction did not form as compact a fraction during purification as nontransgenic myelin. When purified myelin from transgenic animals was studied at the electron microscopic level, very little multilamellar myelin membrane was found (data not shown). It may be that the osmotic shock used to purify myelin was sufficient to open up the multilamellar structure of the myelin membrane when it contains this large (>100,000 kD) protein.

It is not known whether this transgene protein is localized at the major dense line or the intraperiod line. Several models exist for the tertiary structure of the myelin PLP in the membrane. Most of these models place the N-terminal sequence on the extracellular surface[41,42,43], but the most recent model[44], which proposes a four alpha-helix domain protein and which is consistent with the tertiary structure of a number of channel proteins, places the N-terminal sequence on the cytoplasmic surface. Several of these models have had experimental data supporting their specific model, but most of these data were obtained

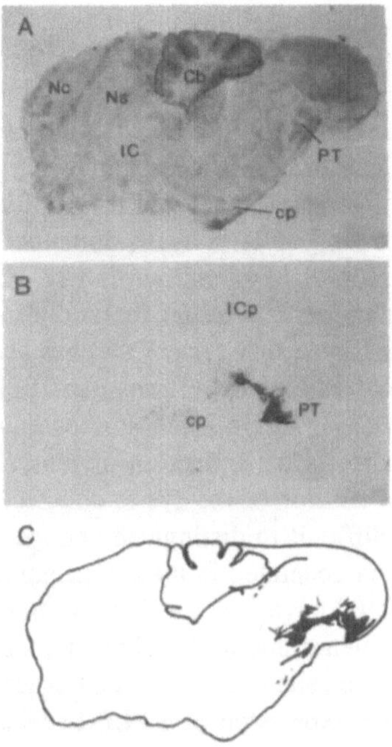

Figure 7. X-gal histochemistry of transgenic mice. Sagittal sections of 2 day old transgenic mouse, stained with A) cresyl violet or B) X-gal. Panel C is an overlay of the X-gal stain on the cresyl violet stain from panels A and B. Staining primarily seen in cerebral peduncle (cp), pyramidal tracts (PT) and inferior cerebellar peduncle (ICp).

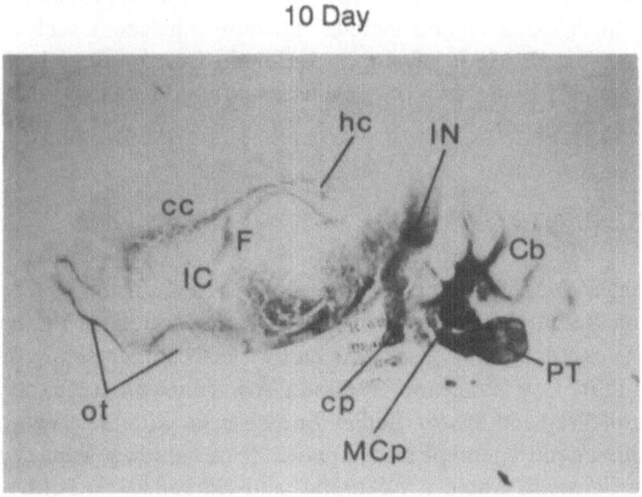

Figure 8. X-gal histochemistry of transgenic mice. Sagittal section of 10 day old transgenic mouse, stained overnight with X-gal. Staining primarily seen in cerebellum (Cb), corpus callosum (cc), cerebral peduncle (cp), fornix (F), hippocampal commisure (hc), internal capsule (IC), inferior colliculus (IN), olfactory tract (ot), middle cerebellar peduncle (MCp) and pyramidal tract (PT).

from mapping PLP epitopes that are expressed in cultured oligodendrocytes. Identification of the localization of the ß-galactosidase within this multilamellar membrane isolated from tissue will be extremely valuable for validation of these models.

Oligodendrocytes from these transgenic animals are now tagged with the ß-galactosidase enzyme. They appear to produce normal myelin and should be ideal for studies on oligodendrocyte transplantation. In general, two kinds of oligodendrocyte transplantation studies have been undertaken. The first studies have focused on transplanting normal oligodendrocytes into mutant tissue[45,46]. Normal cells that express both the PLP and myelin basic protein (MBP) genes are transplanted into *jimpy* or *shiverer* mice, which express either no PLP or no MBP respectively[47]. It is then possible to track the migration of the transplanted oligodendrocytes and their ability to myelinate by identifying PLP-expressing or MBP-expressing cells in these mutants. These studies have provided some valuable new information on this system, but they have the inherent problem that the migration and/or myelination patterns may differ markedly in these mutants, which do not have normal oligodendrocytes. The second system that has been used is the generation of a lesion in a normal adult animal followed by transplantation of fresh oligodendrocytes, which can be tagged with dyes[48,49]. Again, such studies have been quite valuable, but a potential problem in these studies is the induction of remyelination by endogenous cells near the lesion and the eventual dilution of the dye to undetectable levels with each cycle of cell division. Often it is quite difficult to distinguish endogenous remyelinating cells from transplanted cells. Thus, the cells generated in these transgenic mice would be an excellent source of transplantable oligodendrocytes, since they are already tagged, the tag does not dilute, and it marks the myelin membrane itself. Thus, it should be possible to follow both migration and myelination of these cells in either mutant or lesioned systems.

This overview of studies on expression of a PLP-LacZ transgene in oligodendrocytes demonstrates that the myelin PLP sequences are sufficient to direct cell-specificity. The PLP gene is normally expressed in the white matter of the developing central nervous system, which is consistent with the expression of the transgene. A more extensive investigation of these transgenic animals shows that the transgene is not expressed in non-neural tissues, providing further support for the fact that this promoter is extremely specific[24]. This promoter should be quite effective for directing expression of other genes to oligodendrocytes. Such studies could include antisense transgenes such as were used by Katsuki et al.[50] to block MBP gene expression in transgenic mice. Additionally, overexpression or ectopic expression of proteins could be directed by the PLP sequences contained within this transgene.

ACKNOWLEDGEMENTS

The authors thank Dr. Robin Fisher, Director of the Neurohistology Core of the UCLA Mental Retardation Research Center for extensive consultation on the neuroanatomical aspects of transgene expression; Michael Kremen for electron microscopy; Janet Miyashiro, Kristine Engelbrecht, Lee Zane and Svetlana Arutyunovna for excellent technical assistance; and Carol Gray and Sharon Belkin for assistance with the artwork. We thank Dr. Leroy Hood for his encouragement and support. These studies were supported by the National Multiple Sclerosis Society, NS25304, and HD25831 (W.B.M.) and AG07687 (C.R.) P.A.W. was supported by a fellowship from the NIH (NS8774) and C.D. by a fellowship from the National Multiple Sclerosis Society.

REFERENCES

1. Raine, C.S.. Morphology of myelin and myelination, *in* "Myelin," P. Morell, ed., Plenum Press, New York, pp 1-50 (1984).
2. Eng, L.F., Chao, F.C., Gerstl, B., Pratt, D., and Tavaststjerna, M.G. The maturation of human white matter myelin: fractionation of the myelin membrane proteins. *Biochem.* 7:4455-4465 (1968).
3. Norton, W.T., and Poduslo, S.E. Myelination in rat brain: changes in myelin composition during brain maturation. *J. Neurochem.* 21:759-773 (1973).
4. Macklin, W.B., Campagnoni, C.W., Deininger, P.L., and Gardinier, M.V. Structure and expression of the mouse myelin proteolipid protein gene. *J. Neurosci. Res.* 18:383-394 (1987).
5. Nave, K.-A., Lai, C., Bloom, F.E., and Milner, R.J. Splice site selection in the proteolipid protein (PLP) gene transcript and primary structure of the DM-20 protein of CNS myelin. *Proc. Natl. Acad. Sci.* 84:5666-5669 (1987).
6. Simons, R., Alon, N., and Riordan, J.R. Human myelin DM-20 proteolipid protein deletion defined by cDNA sequence. *Biochem. Biophys. Res. Commun.* 146:666-671 (1987).
7. Keen, P., Osborne, R.H., and Pehrson, U.M.M. Respiration and metabolic compartmentation in brain slices from a glia-deficient mutant, the *jimpy* mouse. *J. Physiol.* (Lond.) 245:22-33 (1976).
8. Skoff, R.P. Myelin deficit in jimpy mouse may be due to cellular abnormalities in astroglia. *Nature* 264:560-562 (1976).
9. Hertz, L. Chaban, G., and Hertz, E. Abnormal metabolic response to excess potassium in astrocytes from the *jimpy* mouse, a convulsing neurological mutant. *Brain Res.* 181:482-487.
10. Omlin, F.X., and Anders, J.J. Abnormal cell relationships in *jimpy* mice: electron microscopic and immunocytochemical findings. *J. Neurocyt.* 12: 767-784 (1983).
11. Vermeesch, M.K., Knapp, P.E., Skoff, R.P., Studzinski, D.M., and Benjamins, J.A. Death of individual oligodendrocytes in *jimpy* brain precedes expression of proteolipid protein. *Dev. Neurosci.* 12: 303-315 (1990).
12. Kronquist, K.E., Crandall, B.F., Macklin, W.B., and Campagnoni, A.T. Expression of myelin proteins in the developing human spinal cord: cloning and sequencing of human proteolipid protein cDNA. *J. Neurosci. Res.* 18:395-401 (1987).
13. Gardinier, M.V., and Macklin, W.B. Myelin proteolipid protein gene expression in *jimpy* and *jimpy^{msd}* mice. *J. Neurochem.* 51:360-369 (1988).
14. LeVine, S.M., Wong, D., and Macklin, W.B. Developmental expression of proteolipid protein and DM-20 mRNAs and proteins in the rat brain. *Dev. Neurosci.* 12:235-250 (1990).
15. Schindler, P., Luu, B., Sorokine, O., Trifilieff, T., and Van Dorsselaer, A. Developmental study of proteolipids in bovine brain: a novel proteolipid and DM-20 appear before proteolipid protein during myelination. *J. Neurochem.* 55:2079-2085 (1990).
16. Ikenaka, K., Kagawa, T., and Mikoshiba, K. Selective expression of DM-20, an alternatively spliced myelin proteolipid protein gene product, in developing nervous system and in nonglial cells. *J. Neurochem.* 58:2248-2253 (1992).
17. Timsit, S.G., Bally-Cuif, L., Colman, D.R., and Zalc, B. DM-20 mRNA is expressed during the embryonic development of the nervous system of the mouse. *J. Neurochem.* 58:1172-1175 (1992)
18. Forss-Petter, S., Danielson, P. E., Catsicas, S., Battenberg, E., Price, J., Nerenberg, M., and Sutcliffe, J.G. Transgenic mice expressing β-galactosidase in mature neurons under neuron-specific enolase promoter control. *Neuron* 5:187-197 (1990)
19. Wirak, D.O., Bayney, R., Kundel, C.A., Lee, A., Scangos, G.A., Trapp, B.D. and Unterbeck, A.J. Regulatory region of human amyloid precursor protein (APP) gene promotes neuron-specific gene expression in the CNS of transgenic mice. *EMBO J.* 10:289-296 (1991).
20. Mucke, L., Oldstone, M.B., Morris, J.C., and Nerenberg, M.I. Rapid activation of astrocyte-specific expression of GFAP-lacZ transgene by focal injury. *New Biologist* 3:465-474 (1991).
21. Smeyne, R.J., Schilling, K., Robertson, L., Luk, D., Oberdick, J., Curran, R., and Morgan, J.I. Fos-lacZ transgenic mice: Mapping sites of gene induction in the central nervous system. *Neuron* 8:13-23 (1992).
22. Wight, P.A., Readhead, C., Duchala, C.S., and Macklin, W.B. Myelin proteolipid protein gene expression in transgenic mice. *Trans. Amer. Soc. Neurochem.* 23:182, 1992.
23. Wight, PA., Duchala, C.S., Readhead, C., and Macklin, W.B., Beta-galactosidase expression in oligodendrocytes in transgenic mice. *Trans. Soc. Neurosci.* 18:1089 1992.
24. Wight, P.A., Duchala, C.S., Readhead, C., and Macklin, W.B. Expression of a myelin proteolipid protein-LacZ fusion gene is spatially and temporally regulated in transgenic mice.J.Cell.Biol.In press 1993.
25. Verity, A.N., Bredesen, D., Voderscher, C., Handley, V.W. and Campagnoni, A.T. Expression of myelin protein genes and other myelin components in an ologodendrocytic cell line conditionally immortalized with a temperature-sensitive retrovirus. *J. Neurochem.* 60:577-587 (1993).

26. MacGregor, G.R. Histochemical staining of clonal mammalian cell lines expressing E. coli ß-galactosidase indicates heterogeneous expression of the bacterial gene. *Somatic Cell Mol. Genet.* 13:253-265 (1987).

27. Chen, C.A., and Okayama, H. Calcium phosphate-mediated gene transfer: A highly efficient transfection system for stably transforming cells with plasmid DNA. *BioTechniques* 6:632-638 (1988).

28. Lim, K., and Chae, C.-B. A simple assay for DNA transfection by incubation of the cells in culture dishes with substrates for beta-galactosidase. *BioTechniques* 7:576-579 (1989).

29. Hogan, B.L., Costantini, F., and Lacy, E. "Manipulation of the Mouse Embryo: A Laboratory Manual," Cold Spring Harbor, Cold Spring Harbor Lab, New York (1986).

30. Readhead, C., Popko, B., Takahashi, N., Shine, H.D., Saavedra, R.A., Sigman, R.L., and Hood, L. Expression of a myelin basic protein gene in transgenic *shiverer* mice: correction of the dysmyelinating phenotype. *Cell* 48:703-712 (1987).

31. Sanes, J.R., Rubenstein, J.L.R., and Nicolas, J.-F. Use of a recombinant retrovirus to study post-implantation cell lineage in mouse embryos. *EMBO J.* 5:3133-3142 (1986).

32. Loewy, A.D., Bridgman, P.C., and Mettenleiter, T.C. ß-galactosidase expressing recombinant pseudorabies virus for light and electron microscopic study of transneuronally labeled CNS neurons. *Brain Res.* 555:346-352 (1991).

33. Campagnoni, C.W., Garbay, B., Micevych, P., Pribyl, T., Kampf, K., Handley, V.W., and Campagnoni, A.T. DM-20 mRNA splice product of the myelin proteolipid protein gene is expressed in the murine heart. *J Neurosci. Res.* 33:148-155 (1992).

34. Puckett, C., Hudson, L., Ono, K., Friedrich, V., Benecke, J., Dubois-Dalcq, M., and Lazzarini, R.A. Myelin-specific proteolipid protein is expressed in myelinating Schwann cells but is not incorporated into myelin sheaths. *J. Neurosci. Res.* 18:511-518 (1987).

35. Griffiths, I.R., Mitchell, L.S., McPhilemy, K., Morrison, S., Kyriakides, E., and Barrie, J.A. Expression of myelin protein genes in Schwann cells. *J. Neurocytol.* 18:345-352 (1989).

36. Stahl, N., Harry, J., and Popko, B. Quantitative analysis of myelin protein gene expression during development in the rat sciatic nerve. *Mol. Brain Res.* 8:209-212 (1990).

37. Agrawal, H.C., and Agrawal, D. Proteolipid protein and DM-20 are synthesized by Schwann cell, present in myelin membrane, but they are not fatty acylated. *Neurochem. Res.* 16:855-858 (1991).

38. Gupta, S.K., Pringle, J., Poduslo, J.F., and Mezei, C. Levels of proteolipid protein mRNAs in peripheral nerve are not under stringent axonal control. *J. Neurochem.* 56:1754-1762 (1991).

39. Kamholz, K., Sessa, M., Scherer, S., Volgelbacker, H., Mokuno, K., Baron, P., Wrabetz, L., Shy, M., and Pleasure, D. Structure and expression of proteolipid protein in the peripheral nervous system. *J. Neurosci. Res.* 31:231-244 (1992).

40. Pham-Dinh, D., Birling, M-C., Roussel, G., Dautigny, A., and Nussbaum, J-L. Proteolipid DM-20 predominates over PLP in PNS. *Neuroreport* 2:89-92 (1991).

41. Laursen, R.A., Samiullah, M., and Lees, M.B. The structure of bovine brain myelin proteolipid and its organization in myelin. *Proc. Natl. Acad. Sci.* 81:2912-2916 (1984).

42. Stoffel, W., Hillen, H. and Giersiefen, H. Structure and molecular arrangement of proteolipid protein of central nervous system myelin. *Proc. Natl. Acad. Sci.* 81:5012-5016 (1984).

43. Hudson, L.D., Friedrich, V.L., Jr., Behar, T., Dubois-Dalcq, M., and Lazzarini, R.A. The initial events in myelin synthesis: orientation of proteolipid protein in the plasma membrane of cultured oligodendrocytes. *J. Cell Biol.* 109:717-727 (1989).

44. Popot, J.L., Pham Dinh, D., and Dautigny, A. Major myelin proteolipid: the 4-alpha-helix topology. *J. Memb. Biol.* 120: 233-246 (1991).

45. Gumpel, M., Gout, O., Lubetzki, C, Gansmuller, A, and Baumann, N. Myelination and remyelination in the central nervous system by transplanted oligodendrocytes using the shiverer model. Discussion on the remyelinating cell population in adult mammals. *Dev. Neurosci.* 11:132-139 (1989).

46. Lachapelle, F., Lapie, P., Gansmuller, A., Villarroya, H., Baumann, N., and Gumpel, M. What have we learned about the jimpy phenotype expression by intracerebral transplantations? *Ann. N.Y. Acad. Sci.* 605:332-345 (1990).

47. Campagnoni, A.T., and Macklin, W.B. Cellular and molecular aspects of myelin protein gene expression. *Mol. Neurobiol.* 2:41-89 (1988).

48. Blakemore, W.F., and Crang, A.J. Extensive oligodendrocyte remyelination following injection of cultured central nervous system cells into demyelinating lesions in adult central nervous system. *Dev. Neurosci.* 10:1-11 (1988).

49. Crang, A.J., Franklin, R.J., Blakemore, W.F., Noble, M., Barnett, S.C., Groves, A., Trotter, J., and Schachner, M. The differentiation of glial cell progenitor populations following transplantation into non-repairing central nervous system glial lesions in adult animals. *J. Neuroimmunol.* 40:243-253 (1992).

50. Katsuki, M., Sato, M., Kimura, M., Yokoyama, M., Kobayashi, K., and Nomura, T. Conversion of normal behavior to shiverer by myelin basic protein antisense cDNA in transgenic mice. *Science* 241:593-595 (1988).

AXONAL CONTACT INDUCES OLIGODENDROGLIAL SPECIFIC TRANSCRIPTION OF THE *MBP* GENE FROM THE PROXIMAL PROMOTER REGION

C. Goujet-Zalc[1], C. Lubetzki[1], Ch. Babinet[2], M. Monge[1], S. Timsit[1], C. Demerens[1], M. Miura[3], A. Gansmüller[1], M. Sanchez[1], S. Pournin[2], K. Mikoshiba[3], and B. Zalc[1]

[1]Laboratoire de Neurobiologie Cellulaire, Moléculaire et Clinique, INSERM U-134, Hôpital de la Salpêtrière, Université Pierre et Marie Curie, 75651, Paris Cedex 13, France
[2]Laboratoire de Génétique des Mammifères, Institut Pasteur, 28, Rue du Docteur Roux, 75015 Paris, France
[3]Department of Molecular Neurobiology, The Institute of Medical Science, The University of Tokyo, 4-6-1 Shirokane-dai, Minatoku, Tokyo 108, Japan

INTRODUCTION

In the CNS, oligodendrocyte produces and maintains the myelin sheath. Although much is known from earlier studies about the morphological changes that occur during oligodendroglial differentiation, relatively little is known about the sequential molecular events that occur during the acquisition of oligodendroglial characteristics by immature cells. Both *in vitro* and *in vivo* experiments have shown that the sequence of expression of the myelin constituents is not continuous, but occurs in two waves separated by a 3-5 day time lag: the early wave (GalC and CNP) corresponds to the period when the progenitor cell stops dividing and the late wave (MBP, PLP, MAG) occurs shortly before the oligodendrocyte processes start to enwrap the axon. There is evidence to suggest that MBP is the first marker of the late wave to be expressed, thus signing the entry of the oligodendrocyte into the stage of maturation[1,2].

MBP represents 30% of total adult myelin protein[3]. The mouse *MBP* gene is composed of at least 8 exons which, due to alternative splicing, can produce at least 16 potential isoforms [4-6]. Exons 1 to 7, span a 32 kb region of chromosome 18. Exon 0, which encodes an untranslated sequence, is located at least 20 kb upstream from exon 1 [ref.7]. Exon 1 is subdivided in two parts: exon 1b which starts from position +1 and exon 1a which extends 210nt upstream. Transcripts containing exon 0 are rare and also contain exon 1a [ref.7]. Transfections and foot-printing experiments have shown that the sequence immediately 5' to exon 1b can behave as a promoter for the transcription of the portion of the *MBP* gene

A Multidisciplinary Approach To Myelin Diseases II
Edited by S. Salvati, Plenum Press, New York, 1994

corresponding to exons 1b to 7 ref.8-10. The importance of the proximal 256bp of the *MBP* promoter region for brain and oligodendroglial specific expression has been confirmed by an *in vitro* transcription[11] and, more recently, by a retroviral mediated assay[12].

As a step towards understanding the mechanisms regulating oligodendrocyte-specific gene expression, it is necessary to identify and analyze the *cis*-acting DNA sequences and their corresponding transcription factors. We have chosen to analyze the regulatory sequences of the *MBP* gene using an *in vivo* approach through the generation of transgenic mice. We have previously reported expression in transgenic mice of a fusion gene (*MBP-lacZ*) containing the 256 bp 5' to exon 1b of the *MBP* gene linked to the E. coli *lacZ* gene[13]. Of four transgenic families, two (*MBP-lacZ* line 2 and 4) expressed ß-galactosidase activity in the nervous system but not in most other tissues. Histochemical and immunohistochemical analysis of adult brain from these two lines showed oligodendroglial-specific expression of the transgene. This was confirmed by assaying the ß-galactosidase activity in an highly enriched oligodendrocyte population purified from the *MBP-lacZ* animals. These data demonstrated that the 256bp proximal region of the *MBP* gene promoter is sufficient to confere oligodendroglial specific transcription. Although the transgene construct contains, in addition to the 256bp sequence upstream from the *MBP* major transcription site, the 90 first bp from exon 1b, *in situ* hybridization experiments with a *lacZ* cRNA probe showed that the transgene RNA was strictly located at the oligodendrocyte cell bodies, suggesting that the 5' region of exon 1b of the *MBP* gene is not responsible for the migration of the *MBP* transcripts. Developmental *in situ* analysis showed that, while immunodetectable MBP was seen in oligodendrocytes shortly before deposition of myelin, expression of the transgene occured only after the cells had started to myelinate. In primary cultures from the *MBP-lacZ* transgenic brain, only oligodendrocytes that had established contact with an axon express the transgene. These results suggest that *cis*-acting regulatory elements, activated in oligodendrocytes by an axonal signal, are located within 256 bp upstream from the *MBP* gene.

RESULTS

Tissue Expression of the *MBP-lacZ* Transgene is oligodendrocyte-specific

To determine the extent of transgene expression, various tissues from F1 animals from *MBP-lacZ* line 4 were analyzed for expression of ß-galactosidase, by histochemical staining with the chromogenic X-Gal substrate. With the exception of the testes, enzymatic ß-galactosidase activity was detected solely in the CNS. Although MBP is known to be synthesized by Schwann cells, no expression of the transgene was detected in the spinal nerves. The appearance of the X-Gal staining was diffused throughout the cytoplasm of the ß-galactosidase$^+$ cells, allowing their processes to be seen, as well as all the myelinated fibers either in tracts, or isolated. Blue stained cells in the brain were found essentially in white matter, strongly suggesting that the transgene is expressed in oligodendrocytes. To unambiguously identify the phenotype of the ß-galactosidase$^+$ cells, a series of immunolabeling experiments were undertaken. Expression of ß-galactosidase in the brain of transgenic *MBP-lacZ* line 4 mice was very extensive, and the X-Gal staining intense. In order to overcome the problem of quenching of immunofluorescence on these intensely X-Gal stained cells, immunotyping of the ß-galactosidase$^+$ cells in the brains of these animals was performed by double immunolabeling with a rabbit anti-ß-galactosidase antibody and the RIP monoclonal antibody which specifically recognizes oligodendrocytes[14]. All ß-galactosidase$^+$ cells were RIP$^+$ and most RIP$^+$ cells were ß-galactosidase$^+$. Double labeling with antibodies against GFAP failed to reveal any ß-galactosidase⁺/GFAP⁺ cells.

Ultrastructural analysis of the ß-galactosidase⁺ cells in brains of line 4 mice was undertaken using Bluo-gal as substrate. Oligodendroglial cell bodies and processes, vizualized on semi-thin sections examined under the light microscope, were clearly labeled with this Bluo-gal substrate, but not the myelin sheaths, unlike the staining with the X-Gal substrate. This was confirmed by electron microscopy on ultra-thin sections. More than 90% of the oligodendrocytes observed contained electron-dense Bluo-gal precipitates. In cell bodies, a very intense precipitate was bound to the nuclear membrane and accumulated in the cytoplasm as well as in the oligodendrocytes cell processes, but no precipitate could be detected in the myelin sheath (see Fig. 8 in ref.[13]).

In cultures of oligodendrocytes (> 80% GalC⁺ cells) from 4 week old transgenic mice from line 4, purified on a Percoll density gradient[15], double labeling with X-Gal and anti-GalC antibodies showed that more than 80% of the GalC⁺ cells were ß-galactosidase⁺, but none of the GFAP⁺ cells (5 to 10% of the cells in the preparation) were ß-galactosidase⁺.

Localization of the Transgene mRNA

Migration of MBP mRNA along oligodendrocyte processes and in the myelin sheath has been well documented since the pionneer work by D. Colman[16]. It is still not known which sequence of the *MBP* transcript is responsible for this migration. It has been postulated that either the 5' or the 3' untranslated sequence could be involved. The *MBP-lacZ* transgene used in the present study included, in addition to the 256nt upstream of the initiation of transcription site, the 47nt 5' untranslated and 43nt of the coding region of the corresponding *MBP* transcript. This construction made it possible to determine by *in situ* hybridization whether the 5' end of the *MBP* message is involved in migration. [^{35}S] labeled sense and antisense cRNAs were generated from a HpaI-HpaI fragment of the *lacZ* sequence subcloned into the plasmid Bluescript. Hybridization of the antisense cRNAs with paraffin sections of brains from 18 day-old line 4 transgenic mice showed that the transgene RNA was confined to the cell bodies. The intensity of the hybridization signal over cell processes and myelin sheaths was within the background range. In this respect, the pattern of hybridization observed with the *lacZ* probe was superimposable on the one observed on adjacent sections hybrized with a PLP probe, but not with an *MBP* probe (see Fig. 9 ref.[13]). No signals were observed on adjacent sections hybridized with sense *lacZ*, PLP or *MBP* riboprobes. This shows that, in contrast to *MBP* transcripts, mRNA transcribed from the trangene used in the present study does not migrate.

The *MBP-lacZ* Transgene is Expressed Later in Development than the *MBP* Gene

The temporal regulation of transgene expression was analyzed during postnatal development by X-Gal staining of sections from brains of animals of different ages. The temporal pattern of ß-galactosidase expression followed a caudo-rostral gradient. At birth, ß-galactosidase activity was detected only in the medulla oblongata. At 3 days of age, ß-galactosidase⁺ cell bodies could first be seen in the brain stem and in the central part of the cerebellum. On day 5, the first X-Gal⁺ myelinated fibers appeared in the brain stem, while in the cerebellum the number of ß-galactosidase⁺ cells increased. At postnatal day 8, ß-galactosidase⁺ cells had migrated in all the cerebellar folia and ß-galactosidase⁺ cells could be detected in more rostral parts of the brain.

Co-expression of the *MBP* gene and *MBP-lacZ* transgene was analyzed at postnatal days 3 and 5, either on adjacent serial cryostat sections or on the same tissue sections. In the latter case, to circumvent the frequent quenching of immunostaining by the product of the

X-Gal reaction, sections were first treated with anti-MBP antibodies for immunofluorescent labeling and fields showing fluorescent MBP^+ staining were photographed. Then the sections were unmounted, treated for enzymatic detection of the ß-galactosidase activity and the same fields were photographed. There was a good, but not complete, superimposition of the MBP^+ areas and the presence of ß-galactosidase$^+$ cells. Interestingly, ß-galactosidase$^+$ cells were detected in areas where myelination was already well advanced. In these areas, as already described[17], the anti-MBP antibodies stained the myelin sheaths, but oligodendrocytes cell bodies were no longer MBP^+, mostly because the MBP had migrated out of the cell body towards the cell processes and the myelin sheaths. In contrast, oligodendrocytes that had not yet started to myelinate, or were just starting to enwrap axonal processes were clearly labeled with the anti-MBP antibodies. These MBP^+ oligodendrocytes were not ß-galactosidase$^+$, suggesting a delay in the expression of the transgene as compared to the expression of the MBP itself.

To explore this observation in more detail, we analyzed the expression of the transgene in mixed primary cultures. Brains from E15 transgenic embryos were dissociated, seeded on class coverslips and maintained in culture. After 3-4 weeks in vitro, corresponding approximately to P15-21 in vivo, ß-galactosidase activity was detected by enzyme histochemistry, and several cell type specific antigens were labeled immunocytochemically. Oligodendrocytes were labeled with anti-GalC and anti-MBP antibodies, and neurons with the TuJ1 monoclonal antibody which recognizes the class III ß-tubulin neuronal specific isotype[18]. In these cultures, the ß-galactosidase$^+$ cells were not uniformely distributed on the coverslips, but were localized in bundle-like formations. These cells had the typical morphology of myelinating oligodendrocytes, i.e., they had a reduced number of processes and were $GalC^+$ and MBP^+ as described elswhere[19]. The bundles of fibers in which, or in the vicinity of which, the ß-galactosidase$^+$ cells localized were clearly labeled with the TuJ1 antibody, indicating they were neuritic processes. All ß-galactosidase$^+$ cells localized at a short distance from a TuJ1 positive neurite, had extended at least one process towards, and established a contact, with the neurite. It should be noted that, in the same cultures, $GalC^+$/MBP^+/ß-galactosidase$^-$ cells were also observed. These cells were more often seen in the periphery of the coverslip, had typical oligodendrocyte sun-like morphology, but had no contact with ß-tubulin positive ($TuJ1^+$) processes.

DISCUSSION

The 256bp Proximal Region of the *MBP* Gene Promoter is Sufficient to Confer Oligodendroglial Specific Transcription

A construct containing putative upstream regulatory elements from the mouse *MBP* gene fused to an E. coli *lacZ* reporter gene was introduced into the germline of mice[13]. Several groups have generated transgenic mice with construct containing 5' flanking sequences of the *MBP* gene, varying from 1,9 to 3.2 kb, fused to a bacterial reporter gene and shown oligodendrocyte-specific expression[20-21].

Our data suggest that *cis*-acting regulatory elements, capable of appropriately targeting oligodendrocyte-specific gene expression, are located within 256nt upstream of the cap site of the *MBP* gene. The diffuse intracellular distribution of the X-Gal reaction product in mouse *MBP-lacZ* line 4 might be interpreted as evidence that transcripts of the transgene migrate within the cell, as do native *MBP* transcripts. This is not the case, however. Diffusion of the X-Gal product has been reported to occur even when higher concentrations

(20mM instead of 5mM) of ferro - and ferricyanide are used[22]. When Bluo-gal was used instead of X-Gal the reaction product remained localized in the cell body and processes, leaving the myelin unstained. ß-galactosidase would therefore not seem to be synthesized in myelin from transgene transcripts which had migrated. The differences between the in situ hybridization patterns obtained with labeled *lacZ, MBP* and *PLP* cRNAs (see Fig. 9 in ref.[13]) supports this conclusion. Since the *MBP-lacZ* transcript includes the 5' *MBP* mRNA non coding sequence proposed to be responsible for migration of the *MBP* mRNA, this hypothesis can probably be excluded.

The 256bp Proximal Region of the *MBP* Gene Promoter Mediates Transcriptional Activation Following Axonal Contact

Although the temporal and spatial expression of the transgene during development closely paralleled that of myelination, the MBP protein could consistently be detected by immunocytochemical methods before activity of the reporter transgene. It is possible that immunofluorescence is more sensitive than enzymatic staining with X-Gal, but this has not been our experience previously[23], and there are, to our knowledge, no reports in the literature that support this hypothesis. An alternative explanation is that, during development, transcription of the *MBP* gene in newly differentiated oligodendrocytes is initiated from another promoter located upstream from the 256bp proximal sequence. After the oligodendrocytes have started to wrap around the axons, the 256bp proximal promoter would then be activated. This is consistent with developmental observations in vivo and in vitro. In the developing brain, oligodendrocytes that had not yet started to deposit myelin, or cells that had just started to myelinate, were MBP^{+}/ß-galactosidase^{-}, while ß-galactosidase^{+} oligodendrocytes were detected only in brain areas where myelination was already well advanced. In dissociated cell culture from transgenic brains, the oligodendrocytes segregated into two sub-populations, only one of which expressed the transgene. The ß-galactosidase^{-} oligodendrocytes were non-myelinating, and had an extremely dense network of processes. The ß-galactosidase^{+} cells had fewer processes, they were localized in bundles of TuJ1 positive fibers, and in most of the cases, we could see that at least one oligodendrocyte process was in close contact with a neurite. This strongly suggests that transcription of the *MBP* gene from the proximal promoter is induced by contact of an oligodendrocyte process with an axon. Recently, de Waegh et al.[24], showed that close intercellular contacts between myelinating Schwann cells and axons modulate a kinase-phosphatase system acting on neurofilaments. Our observation illustrates a reverse cross-talk between a myelinating cell and the axon. The mechanism by which axons activate the proximal portion of the *MBP* proximal promoter in the oligodendrocyte remains to be discovered.

Acknowledgements

We are greatly indebted to Drs. B. Rantsch, D. Colman, M. Dent, A. Frankfurter, B. Friedman, and J. Seeley for the generous gift of valuable antibodies. C.L. is recipient of a "Praticien de Recherche Associé" award from APHP-CNRS. This study was supported by INSERM and grants from Ministère de la Recherche et de la Technologie (91.C.0055) and ARSEP to BZ and Association Française contre les Myopathies to Ch. B.

REFERENCES

1. Dubois-Dalcq, M., Behar, T., Hudson, L., and Lazzarini, R.A. Emergence of three myelin proteins in oligodendrocytes cultured without neurons. *J. Cell Biol.*, 102:384-392 (1986).
2. Monge, M., Kadiiski, D., Jacque, C., and Zalc, B. Oligodendroglial expression and deposition of four major myelin constituents in the myelin sheath during development : an in vivo study. *Dev. Neurosci.* 8:222-235 (1986).
3. Lees, M.B., and Brostoff, S.W. Proteins of myelin. In: Myelin, P. Morell (Ed.) Plenum Press, New York, pp. 197-224 (1984).
4. Takahashi, N., Roach, A., Teplow, D.B., Prusiner, S.B., and Hood, L. Cloning and characterization of the myelin basic protein gene from mouse: one gene can encode both 14-kD and 18.5-kD MBPs by alternate use of exons. *Cell* 42:139-148 (1985).
5. de Ferra, F., Engh, H., Hudson, L., Kamholz, J., Puckett, C., Molineaux, S., and Lazzarini, R.A. Alternative splicing accounts for the four forms of myelin basic protein. *Cell* 43: 721-727 (1985).
6. Aruga, J., Okano, H., and Mikoshiba, K. Identification of new isoforms of mouse myelin basic protein: the existence of exon 5a. *J. Neurochem.* 56:1222-1226 (1991).
7. Kitamura, K., Newman, S.L., Campagnoni, C.W., Verdi, J.M., Mohandas, T., Handley, V.W., and Campagnoni, A.T. Expression of a novel transcript of the myelin basic protein. *J. Neurochem.* 54: 2032-2041 (1990).
8. Miura, M., Tamura, T., Aoyama, A., and Mikoshiba, K. The promoter elements of the mouse myelin basic protein gene function efficiently in NG108-15 neuronal/glial cells. *Gene* 75:31-38 (1989).
9. Tamura, T., Sumita, K., Hirose, S., and Mikoshiba, K. Core promoter of the mouse myelin basic protein gene governs brain-specific transcription in vitro. *Embo J.* 9:3101-3108 (1990).
10. Devine-Beach, K., Haas, S., and Khalili, K. Analysis of the proximal transcriptional element of the myelin basic protein gene. *Nuc.Ac.Res.* 20:545-550 (1992).
11. Tamura, T., Aoyama, A., Inoue, T., Miura, M., Okano, H. and Mikoshiba, K.Tissue-specific in vitro transcription from the mouse myelin basic protein promoter. *Mol. Cell. Biol.* 9:3122-3126 (1989).
12. Ikenaka, K., Nakahira, K., Nakajima, K., Fujimoto, I., Kagawa, T., Ogawa, M., and Mikoshiba, K. Detection of brain-specific gene expression in brain cells in primary culture: a novel promoter assay based on the use of a retrovirus vector. *The New Biol.* 4: 53-60 (1991).
13. Goujet-Zalc, C., Ch. Babinet, M. Monge, S. Timsit, F. Cabon, A. Gansmüller, M. Miura, S. Sanchez, S. Pournin, K. Mikoshiba, and B. Zalc. The Proximal Region of the MBP Gene Promoter Is Sufficient to Induce Oligodendroglial Specific Expression in Transgenic Mice. *Eur. J. Neurosci.* 5:624-632 (1993).
14. Friedman, B., Hockfield, S., Black, J.A., Woodruff, K.A., and Waxman S.G. In situ demonstration of mature oligodendrocytes and their processes: an immunocytochemical study with a new monoclonal antibody, RIP. *Glia* 2: 380-390 (1989).
15. Allinquant, B., Staugaitis, S.M., D'Urso, D., and Colman, D. The ectopic expression of myelin basic protein isoforms in shiverer oligodendrocytes: Implications for myelinogenesis. *J. Cell. Biol.* 113: 393-403 (1990).
16. Colman, D., Kresbich, G., Frey, A.B., and Sabatini, D.D. Synthesis and incorporation of myelin polypeptides into CNS myelin. *J. Cell. Biol.* 95:598-608 (1982).
17. Sternberger, N.H., Itoyama, Y., Kies, M.W., and Webster H. DeF. Immunocytochemical method to identify basic protein in myelin-forming oligodendrocytes of newborn rat CNS. *J. Neurocytol.* 7:251-253 (1978).
18. Moody, S.A., Quigg, M.S., and Frankfurter, A. Development of the peripheral trigeminal system in the chick revealed by an isotype-specific anti-ß-tubulin monoclonal antibody. *J. Comp. Neur.* 279:567-580 (1989).
19. Lubetzki, C., C. Demerens, P. Anglade, H. Villaroya, A. Frankfurter, V. M.-Y. Lee and B. Zalc. Even in culture, oligodendrocytes myelinate solely axons. Proc. Natl. Acad.Sci. USA 90:6820-6824 (1993).
20. Foran, D.R., and Peterson, A.C. Myelin acquisition in the central nervous system of the mouse revealed by an MBP-lacZ transgene. *J. Neurosci.* 12: 4890-4897 (1992).
21. Gow, A., Friedrich Jr., V.L., and Lazzarini, R.A. Myelin basic protein gene contains separate enhancers for oligodendrocyte and Schwann cell expression. *J. Cell Biol.* 119:605-616 (1992).
22. Weis, J., Fine, S.M., David, C., Savarirayan, S., and Sanes, J. Integration site-dependent expression of a transgene reveals specialized features of cells associated with neuromuscular junctions. *J. Cell. Biol.* 113:1385-1397 (1991).
23. Lubetzki, C., Goujet-Zalc, C., Demerens, C., Danos, O., and Zalc B. Clonal segregation of oligodendrocytes and astrocytes during in vitro differentiation of progenitor cells. *Glia* 6:289-300 (1992).
24. De Waegh, S.M., Lee, V.M.-Y., and Brady, S.T. Local modulation of neurofilament phosphorylation, axonal caliber, and slow axonal transport by myelinating Schwann cells. *Cell* 68:451-464 (1992).

RETROVIRUS-MEDIATED GENE TRANSFER OF PMP22 IN SCHWANN CELLS: STUDIES ON CELL GROWTH

Georg Zoidl, Corinne Schmalenbach, and Hans Werner Müller

Molecular Neurobiology Laboratory
Department of Neurology
University of Düsseldorf
D-40225 Düsseldorf
Germany

INTRODUCTION

The recent discovery of the novel peripheral myelin protein PMP22 and the association of mutations of this gene with dominant inherited peripheral neuropathies in mice and man has raised particular interest in the biological function of this gene (for review: Lemke, 1993; Suter et al., 1993). Due to the separate occasions leading to the identification of this gene different names have been given for PMP22 cDNAs in the literature. The reported cDNA sequences are: gas3 (growth arrest-specific gene 3; Schneider et al., 1988), CD25 (crush denervated cDNA clone 25; Spreyer et al., 1991) SR13 (sciatic nerve regeneration clone 13; Welcher et al., 1991) or PMP22 (peripheral myelin protein 22kDa; Snipes et al., 1992). All sequences are variants of the murine growth arrest-specific gene gas3, which was originally cloned from NIH3T3 fibroblasts as one of a set of genes induced when these cells are growth-arrested (Schneider et al., 1988; Manfioletti et al., 1990). The demonstration that gas3 is also a prominent Schwann cell gene product came from the independent cloning of cDNAs that are either induced or repressed following rat sciatic nerve crush or transection (Spreyer et al., 1991; Welcher et al., 1991). It turned out that PMP22 expression is axonally regulated like the major myelin genes. The level of expression is high in fully differentiated Schwann cells, declines rapidly upon Schwann cell dedifferentiation and proliferation as a result of axonal degeneration after nerve injury, and is upregulated upon regeneration (Kuhn et al., 1993).

The open reading frame of the PMP22 cDNA encodes for a 17kDa core protein. The deduced primary structure together with *in vitro* translation experiments and more recent antibody surveys indicate that PMP22 is a glycosylated integral membrane protein with a molecular weight of 22kDa that is located within the compact myelin of the PNS (Snipes et al., 1992; Kuhn et al., 1992). Although no significant homology to any other protein sequence could be identified, similarities to the CNS myelin protein PLP exist at the putative tertiary structure level. Whether this may indicate functional homologies is unclear. Several groups have speculated previously that PMP22 may serve a dual function in myelination and regulation of cell growth (Spreyer et al., 1991; Snipes et al., 1992; Lemke, 1993; Suter et al. 1993).

The search for putative functions of PMP22 in myelination and/or cell growth has been stimulated by identification of mutations of the PMP22 gene. Recently, PMP22 could be linked to the dysmyelinating mouse mutants Trembler J(TrJ) and Trembler (Tr), where point mutations within the first and fourth putative transmembrane spanning regions could be identified (Suter et al., 1992 a, b). Moreover, PMP22 appears to be involved in the pathogenesis of the autosomal dominant inherited dysmyelinating peripheral neuropathies Charcot-Marie-Tooth Type 1 (CMT1A) and the Hereditary Neuropathy with Liability to Pressure Palsies (HNPP). It was shown that CMT1A patients carry a partial duplication (1.5Mb) of chromosome 17p11.2 in a region that contains the PMP22 gene (Matsunami et al., 1992; Patel et al., 1992; Timmermann et al., 1992; Valentijn et al., 1992). Conversely in HNPP patients the same 1.5Mb chromosomal region is deleted (Chance et al., 1993). It seems likely that both neurological diseases arise from unequal cross over events leading to abnormal PMP22 gene dosage, that is increased (2x to 3x) in CMT1A or decreased (2x to 1x) in HNPP, respectively. Pathological changes observed in Trembler mice and CMT1A patients include severe demyelination and continued Schwann cell proliferation ("onion bulb" formation), whereas biopsies from HNPP patients characteristically show tomaculous structures, i.e. sausage-like thickenings of the myelin sheath, but no significant enhancement of Schwann cell proliferation.

We have recently started to develop a strategy to modify experimentally the PMP22 gene dosage in recombinant Schwann cells in order to study the influence of altered PMP22 expression on cell growth and myelination.

RESULTS

Retrovirus-mediated gene transfer into a wide variety of cell types *in vivo* and *in vitro* has been proven as a powerful tool to stably alter the genome of mammalian cells in functional studies (for review: by MacDonald, 1990; Kaufman, 1990). We have exploited the advantages of this technique in order to express PMP22 in cultured Schwann cells from rat sciatic nerve. The retrovirus vectors pSLX-CMV/PMP22+/- were constructed by cloning a fragment of the CD25/PMP22 cDNA sequence (Spreyer et al., 1991), extending from nucleotide 1 to nucleotide 788, in sense (+) or antisense (-) orientation into the multicloning site of the vector pSLX-CMV (Fig. 1). The subcloned DNA fragment included the 5'-untranslated region and the entire coding sequence of PMP22. In both constructs PMP22 expression is under control of the strong internal human cytomegalovirus immediate-early promoter (Boshart et al., 1985). This promoter has been proven to direct high constitutive expression levels of foreign genes in a wide variety of mammalian cells (Foecking and Hofstetter, 1986).

Infectious ecotropic virus vector stocks of the plasmids pSLX-CMV/PMP22+/- and the vector pSLX-CMV were obtained from transfected Psi2 packaging cells (Mann et al., 1983). The productivity of several Psi2 packaging cell lines, tested by virus titration on established NIH3T3 cells as described by Cepko et al. (1984), demonstrated the production of high titer virus vector stocks of SLX-CMV (>10^5 CFU/ml) and SLX-CMV/PMP22+/-

(0,5-3x10^4 CFU/ml). Infectious cell culture supernatants from high titer Psi2 packaging cell lines were used to infect Schwann cell cultures prepared from sciatic nerve of neonatal Wistar rats (P1-P3). These Schwann cell cultures were depleted from endogenous fibroblasts by immunoselection as described by Brockes et al. (1979), expanded and infected during early passages. G418 resistent cell clones became visible within 8-10 days post infection and up to several hundred cell clones were pooled and expanded for molecular and functional analysis two weeks later. No single cell clones were expanded in order to avoid cell growth artifacts due to prolonged cell culture periods.

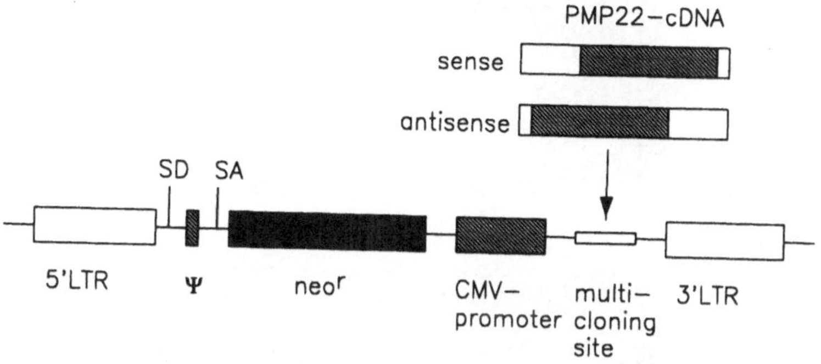

Fig. 1. Structure of the recombinant retrovirus vector pSLX-CMV and the subcloned PMP22-cDNA fragments (nt1-nt788). Abbreviations: LTR, Moloney murine leukemia virus long terminal repeat; CMV, cytomegalovirus immediate-early promoter; neo^r, neomycinphosphotransferase gene; SD/SA, splice donor/ splice acceptor site;ψ, packaging signal sequence

Gene transfer into infected Schwann cells was documented by analysing the integration of the virus vectors into the genome of infected cells by DNA-PCR as described by Higuchi et al. (1990). We were able to distinguish between the endogenous genomic PMP22 sequence and the exogenous PMP22 cDNA sequence by selecting a pair of primers that anneal to different exons of the PMP22 gene separated by large intron sequences (F. Bosse, personal communication). An amplification product of the expected size of 524bp obtained by DNA-PCR is shown in Fig. 2A, lanes 2 and 3. No amplification product could be derived from the endogenous PMP22 gene (Fig. 2A, lane 1). This result proves the integration of the transduced PMP22 cDNA sequences into the host cell genome. To determine the orientation of the integrated PMP22 sequences we have applied a different strategy using one primer located within the vector sequence and another within the cDNA sequence. Under these conditions DNA-PCR resulted in an amplification product of 598bp from those cells harbouring a vector construct with a PMP22 sequence in sense orientation (Fig. 2B, lane 2). As expected no amplification product of the antisense construct as well as from control cells was obtained (Fig. 2B, lane 1 and 3). A similar approach using different primers was successful in identifying the integration into the Schwann cell genome of vectors carrying the PMP22 sequence in antisense orientation (data not shown).

PMP22 cDNA sequences have been introduced into Schwann cells in order to establish cell cultures that either overexpress or underexpress PMP22. A crucial step in characterizing the infected Schwann cells was to demonstrate enhanced or reduced levels of PMP22 mRNA. For accurate quantification of the PMP22 mRNA expression we used a competitive RNA-PCR protocol based on a single tube reaction (Gilliland et al., 1990; Myers and Gelfand, 1991). In the experiment shown, mRNA preparations from cells grown to high

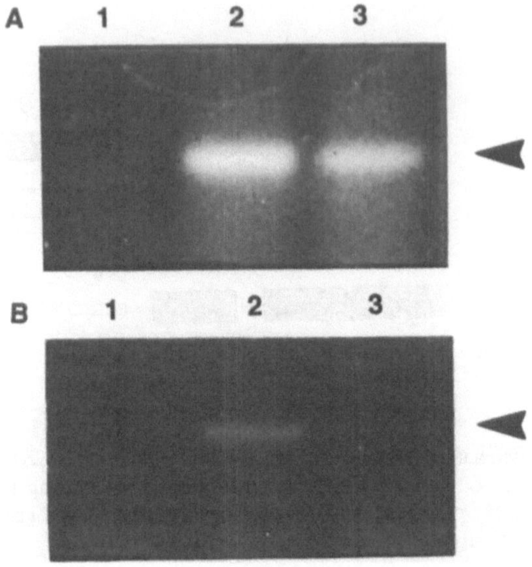

Fig. 2. Gel electrophoresis of PMP22-specific amplification products: Products were obtained from DNA-PCR of high molecular weight genomic DNA derived from infected Schwann cells. The PCR was performed using standard conditions recommended by the manufacturer (New England Biolabs). (A) Amplification of a 524bp product derived from the subcloned CD25/PMP22 fragment extending from nt 48 to nt 572. (B) Amplification of a 598bp product extending from the SLX-CMV multicloning site sequence 26bp upstream of the subcloned PMP22-cDNA fragment to nt 572 of the cDNA. Cells infected with the vector SLX-CMV (lane 1), SLX-CMV/PMP22+ (lane 2) and SLX-CMV- (lane 3).

density were used for competitive RNA-PCR. Schwann cells that express exogenous PMP22 transcripts in sense orientation show a more than 7 fold higher level of PMP22 mRNA than control cells harbouring the vector SLX-CMV (Fig. 3). This indicates that the constitutive expression of PMP22 mRNA significantly exceeds the endogenous PMP22 mRNA level even under growth conditions, when the endogenous level of PMP22 expression is high (Kuhn et al., 1993; Bosse et al., 1993). In contrast, cells that harbour the antisense PMP22 construct show a 18% reduction of the PMP22 mRNA expression level (Fig. 3).

In order to measure the effect of altered PMP22 expression on proliferation of recombinant Schwann cells we have either pulse labeled newly synthesized DNA with bromodeoxyuridine (BrdU) or counted cells at different time points after plating. The result of a cell count experiment is shown in Fig. 4. In this experiment cells show a typical growth

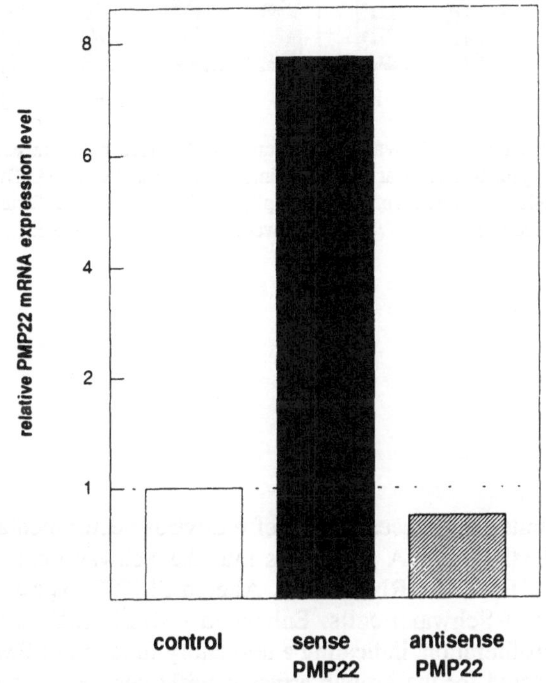

Fig. 3. Changes in PMP22 mRNA level associated with the expression of sense- and antisense PMP22 sequences revealed by competitive RNA-PCR. The amplification products were separated by gel electrophoresis and visualized by autoradiography. Quantification of the radiolabeled products was performed by densitometric scanning (Personal Densitometer, Molecular Dynamics). Relative PMP22 mRNA levels were calculated by normalizing optical densities for the control cell extinction.

behaviour starting with an initial phase of proliferation and reaching a saturation density around day 6. It is very interesting to note that the proliferation of Schwann cells overexpressing PMP22 is significantly reduced compared to control cells or cells underexpressing PMP22 mRNA. This observation strongly suggests interference of elevated PMP22 expression levels with normal Schwann cell growth behaviour. On the other hand, a reduced PMP22 mRNA level has only little effect upon Schwann cell growth.

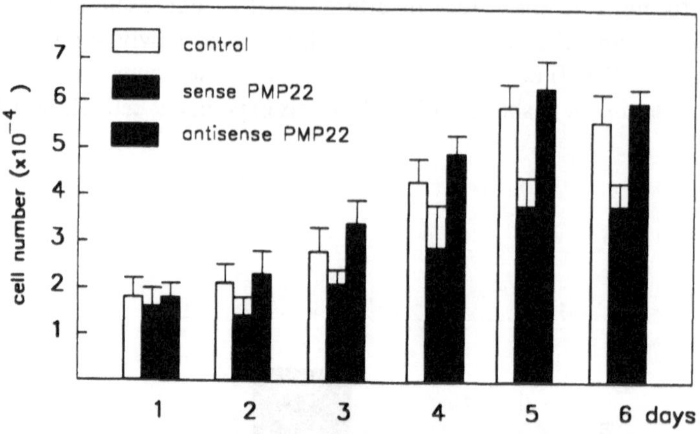

Fig. 4. Growth rates of recombinant Schwann cell cultures. Cells were plated at equal densities, harvested quantitatively after mild trypsin treatment at the timepoints indicated and counted. The graph is composed of values derived from a single experiment run in quadruplicate. Schwann cells infected with the virus vector SLX-CMV and the virus vectors SLX-CMV/PMP22+/- were compared (mean± SEM.).

CONCLUSIONS

We have demonstrated the successful use of retrovirus vector mediated gene transfer to integrate exogenous PMP22 cDNA sequences into the Schwann cell genome leading to abnormal expression of PMP22 mRNA levels. Altered PMP22 expression clearly changes the growth behaviour of Schwann cells. Enhanced PMP22 mRNA levels significantly reduce Schwann cell proliferation, indicating a regulatory function of PMP22 in cell growth, a role previously suspected for the growth arrest-specific gene gas3 in NIH3T3 fibroblasts (Schneider et al., 1988). In contrast, underexpression of PMP22 mRNA has only little effect on Schwann cell proliferation, suggesting that PMP22 may not be required to maintain cell division but to induce growth arrest.

The mechanism by which PMP22 exerts its function is unknown and remains rather obscure with respect to the pathology of the dysmyelinating neuropathy CMT1A. In CMT1A, a disease with an enhanced gene dosage (Timmerman et al., 1992), Schwann cell proliferation is abnormally high, in contrast to what we would have predicted from our present investigation. However, since the level of PMP22 mRNA and the protein expression in CMT1A has not yet been determined it is still unclear whether the PMP22 expression level is elevated or perhaps even reduced. On the other hand, the deletion of PMP22 among other genes in HNPP apparently does not significantly alter the Schwann cell proliferation. This *in situ* observation is consistent with the *in vitro* growth behaviour of recombiant Schwann cells underexpressing the PMP22 gene.

ACKNOWLEDGMENTS

The autors wish to thank Dr. Donatella D'Urso and Clemens Gillen for helpful comments during preparation of the manuscript. We also wish to thank Dr. Rainer Kuhn

(Ciba-Geigy) for kindly providing the retrovirus vector pSLX-CMV and Dr. Dieter Brockmann (University of Essen) for the cell lines Psi2 and NIH3T3. This work was supported by the BMFT and the F.Thyssen Stiftung.

REFERENCES

Boshart, M., Weber, F., Gerhard, J., Dorsch-Hasler, K., Fleckenstein, B., and Schaffner, W., 1985, A very strong enhancer located upstream of an immediate early gene human cytomegalovirus. *Cell* 41:521-530.

Bosse, F., Zoidl, G., Gillen, C., Wilms, S. and Müller, H.W., 1993, Differential expression of two PMP22(CD25/SR13) transcripts in vivo and in vitro, *EMBO J.* submitted.

Brockes, J.P., Fields, K.L., and Raff, M.C., 1979, Studies on cultured rat cells. I. Establishment of purified populations from cultures of peripheral nerve, *Brain Res.* 165:105-118.

Cepko, C.L., Roberts, B.E., and Mulligan, R.C., 1984, Construction and application of a highly transmissable murine retrovirus shuttle vector, *Mol. Cell. Biol.* 7:1053-1062.

Chance, P.F., Alderson, M.K., Leppig, K.A., Lensch, M.W., Matsunami, N., Smith, B., Swanson, P.D., Odelberg, S.J., Disteche, C.M., and Bird, T.D., 1993, DNA deletion associated with the Hereditary Neuropathy with Liability to Pressure Palsies, *Cell* .72:143-151.

Foecking, M.K., and Hofstetter, H., 1986, Powerful and versatile enhancer-promoter unit for mammailian expression vectors, *Gene* 45:101.

Gilliland, G., Perrin, S., Blanchard, K., and Bunn, H.F., 1990, Analysis of cytokine mRNA and DNA: detection and quantitation by competitive polymerase chain reaction, *Proc.Natl. Acad. Sci., USA*, 87:2725-2729.

Kaufmann, R.J., 1990, Strategies for obtaining high level expression in mammalian cells. *Technique* 2:221-236.

Kuhn, G., Lie, A., Zoidl, G. and Müller, H.W., 1992, Coexpression of a new 22kD Schwann cell protein with other peripheral myelin genes during regeneration and development, Abstract of the *Proceedings of the 19th Göttingen Neurobiology Conference*, ed. Elsner, N., and Penzlin, H., Georg Thieme Verlag, Stuttgart, New York.

Kuhn, G., Lie, A., Wilms, S., and Müller, H.W., 1993, Coexpression of the PMP22 gene with MBP and Po during de novo myelination and nerve repair, *Glia* : 8:256-264.

Higuchi, M., Wonh, C., Kochan, L., Olek, K., Aronis, S., Kaspar, C.K., Kazazian, H.H.Jr., and Antonarakis , S.E., 1990, Characterization of mutations in the factor VIII gene by direct sequencing from amplified genomic DNA, *Genomics* 6: 65-71.

Lemke, G., 1993, The molecular genetics of myelination: an update, *Glia* 7:263-271.

MacDonald, C., 1990, Development of new cell lines for animal cell biotechnology, *Biotechnology* 10:155-178.

Manfioletti, G., Ruaro, E., Del Sal, G., Philipson, L., and Schneider, C., 1990, A growth-arrest-specific gene codes for a membrane protein, *Mol. Cell. Biol.* 10:2924-2930.

Mann, R., Mulligan, R.C., and Baltimore, D., 1983, Construction of a retrovirus packaging mutant and its use to produce helper-free defective retrovirus, *Cell* 33:153-159.

Matsunami, N., Smith, B., Ballard, L., Lensch, M.W., Robertson, M., Albertsen, H., Hanemann, C.O., Müller, H.W., Bird, T.D., White, R., and Chance, P.F. ,1992, Peripheral myelin protein-22 gene maps to the duplication in chromosome 17p11.2 associated with Charcot-Marie-Tooth 1A, *Nature Genetics* 1:176-179.

Myers, T.W., and Gelfand, D.H., 1991, Reverse transcription and DNA amplification by the Thermus thermophilus DNA poymerase, *Biochemistry* 30: 7661-7666.

Patel, P.I., Benjamin, B.R., Welcher, A.A., Schoener-Scott, R., Trask, B.J., Pentao, L., Snipes, G.J., Garcia, C.A., Francke, U., Shooter, E.M., Lupski, J.R., and Suter, U., 1992, The gene for the peripheral myelin protein PMP-22 is a candidate for the Charcot-Marie-Tooth disease type 1A, *Nature Genetics* 1:159-165.

Schneider,C., King, R.M. and Philipson, L., 1988, Genes specifically expressed at growth arrest of mammalian cells, *Cell* 54:787-793.

Snipes, G.J., Suter, U., Welcher, A.A., and Shooter, E.M., 1992, Characterization of a novel peripheral nervous system myelin protein (PMP22/SR13), *J. Cell. Biol.* 117:225-238.

Spreyer, P., Kuhn, G., Hanemann, C.O., Gillen, C., Schaal, H., Kuhn, R., Lemke, G.E., and Müller, H.W. 1991, Axonal-regulated expression of a Schwann cell transcript that is homologous to a "growth-arrest-specific" gene, *EMBO J.* 10:3661-3668.

Suter, U., Moskow, J.J., Welcher, A.A., Snipes, G.J., Kosaras, B., Sidman, R.L., Buchberg, A.M., and

Shooter, E.M., 1992a, A leucine-to-proline mutation in the putative first transmembrane domain of the 22-kDa peripheral myelin protein in the Trembler-J mouse,. *Proc. Natl. Acad. Sci. USA*. 89:4382-4386.

Suter, U., Welcher, A.A., Özcelik, T., Snipes, G.J., Kosaras, B., Francke, U., Billings-Garliardi, S., Sidman, R.L., and Shooter, E.M., 1992b, Trembler mouse carries a point mutation in a myelin gene, *Nature* 356:241-244.

Suter, U., Welcher, A.A., and Snipes, G.J., 1993, Progress in the molecular understanding of hereditary peripheral neuropathies reveals insight into the biology of the peripheral nervous system, *TINS* 16:50-57.

Timmerman, V., Nelis, E., Van Hul, W., Nieuwenhuijsen, B.W., Chen, K.L., Wang, S., Othman, K.B., Cullen, B., Leach, R.J., Hanemann, C.O., De Jonghe, P., Raeymaekers, P., van Ommen, G.J.B., Martin, J.J., Müller, H.W., Vance, J.M., Fischbeck, K.H., and Van Broeckhoven, C., 1992, The peripheral myelin gene PMP22(GAS3) is duplicated in the Charcot-Marie-Tooth disease Type 1a duplication, *Nature Genetics* 1:171-175.

Valentijn, L.J., Bolhuis, P.A., Zorn, I., Hoogendijk, J.E., van den Bosch, N., Hensels, G.W., Stanton, V.P., Housman, D.E., Fischbeck, K.H., Ross, D.A., Nicholson, G.A., Meershoek, E.J., Dauwerse, H.G., van Ommen, G.J.B., and Baas, F., 1992, The peripheral myelin gene PMP22-22/GAS-3 is duplicated in Charcot-Marie-Tooth disease type 1A, *Nature Genetics* 1:166-170.

Welcher, A.A., Suter, U., De Leon, M., Snipes, G.J., and Shooter, E.M., 1991, A myelin protein is encoded by the homologue of a growth-arrest specific gene, *Proc. Natl. Acad. Sci. USA*. 88: 7195-7199.

PROTEIN SORTING AND TARGETING IN MYELIN-FORMING SCHWANN CELLS

Bruce D. Trapp,[1] Grahame Kidd,[1] and S. Brian Andrews[2]

[1]Department of Neurology
Johns Hopkins University School of Medicine
Baltimore, MD 21287-6965
[2]Laboratory of Neurobiology
National Institute of Neurological Diseases and Stroke
National Institutes of Health
Bethesda, MD 20892

INTRODUCTION

Structural and functional specialization of cell surface membranes is essential for the normal function and survival of all eurkaryotic cells. Understanding how cells establish and maintain specialized surface membrane domains is a major challenge to cell biologists. Moreover, the failure to polarize surface membranes into specialized domains can have drastic effects on cell function and survival. Even the simplest cells can organize their surface membranes into discrete domains; yeasts, for example, specialize their surface membranes into growing buds and mating projections (Drubin, 1991).

The best-characterized example of cell surface polarity in mammals is found in the most common type of tissue, the epithelium. Simple epithelial cells divide their surface membranes into two domains: an *apical domain,* which interacts with the external environment, and a *basolateral domain*, which contacts adjacent cells and connective tissue (Rodriguez-Boulan and Nelson, 1989; Simons and Wandinger-Ness, 1990; Mostov et al., 1992). In contrast to the simple organization of the epithelium, surface membrane specializations can be very complex in cells with long processes; neurons, for example, form two general surface domains, *axons* and *dendrites*, but both of these domains are divided into multiple subdomains. Specialized regions of the axonal plasma membrane include the *axon hillock*, *internodal axolemma*, *nodal axolemma*, and *synaptic terminal*. The latter can be further subdivided into even more specialized regions, e.g., active zones.

Polarization of cell surface membranes into discrete domains requires site-specific delivery and/or stabilization of membrane proteins (Rodriguez-Boulan and Nelson, 1989; Simons and Wandinger-Ness, 1990; Mostov et al., 1992; Sztul et al., 1992). How proteins are sorted, transported, integrated, and stabilized is just beginning to be

A Multidisciplinary Approach To Myelin Diseases II
Edited by S. Salvati, Plenum Press, New York, 1994

understood. Much of what is known has been obtained from *in vitro* experiments using simple polarized epithelial cells. As studies explore how surface membranes of more complex cells achieve polarization, additional mechanisms will be identified. The purpose of this review is to summarize current knowledge about protein sorting and targeting in myelinating Schwann cells and to highlight some of the advantages that myelin-forming cells offer for investigating mechanisms of protein targeting.

SORTING AND TARGETING OF PROTEINS IN EPITHELIAL CELLS

Two pathways have been identified for delivering protein from Golgi membranes to the appropriate surface domains in polarized epithelial cells (Figure 1). In one pathway, basolateral and apical proteins are sorted into separate carrier vesicles as they exit the trans-Golgi network (Rindler et al., 1984; Griffiths and Simons, 1986; Rodriguez-Boulan and Nelson, 1989) and are then targeted to the correct surface membrane. In the second pathway, apical and basolateral proteins are transported together to the basolateral surface; apical proteins are then removed from the basolateral membrane by endocytosis and relocated to the apical surface (Hubbard and Stieger, 1989). Different epithelial cells use these two transport pathways to different degrees. Mardin-Darby canine kidney (MDCK) cells, for instance, sort most proteins by the first pathway, i.e., trans-Golgi sorting (Rodriguez-Boulan and Nelson, 1989; Wandinger-Ness et al., 1990), whereas hepatocytes sort proteins by the endocytic/transcytotic route (Hubbard and Stieger, 1989). Other cells, such as the intestinal cell line CaCo-2, appear to use both pathways equally (LeBivic et al., 1990; Matter et al., 1990).

Most earlier studies of protein sorting investigated the sorting of viral glycoproteins in polarized epithelial cells (Rodriguez-Boulan and Sabatini, 1978; ; Rindler et al., 1984; Griffiths and Simons, 1986), but did not delineate the nature of sorting signals. When nonviral proteins were studied, it became clear that both the cell type and the protein influenced cell sorting and targeting pathways (Wessels et al., 1990;Low et al., 1991a,b). Although conflicting hypotheses have arisen from these studies, it is clear that both the extracellular domain and the cytoplasmic domain of transmembrane proteins can contain sorting and/or targeting information (Mostov et al., 1986; Hopkins, 1991), that glycerophosphatidylinositol (GPI)-linked proteins are sorted via interaction with lipids (Lisanti and Rodriguez-Boulan, 1990; Lisanti et al., 1990), and that microtubules are required for the targeting of some proteins (Hugon et al., 1987; Achler et al., 1989; Gilbert et al., 1991). Some proteins contain a hierarchy of signals that exist in both the extracellular and cytoplasmic domains (LeBivic et al., 1991). Confusion as to how proteins are sorted and targeted is due in part to the focus on proteins that are foreign to cells and to the use of cells that are not in their natural setting.

The natural, *in vivo* environment is crucial to the normal structure and function of polarized epithelial cells (Rodriguez-Boulan and Nelson, 1989). Polarization of all epithelial cells depends on cell-cell and/or cell-substrate contact. Such interactions induce the differentiation of a non-polarized precursor cell into a differentiated, polarized cell and includes the expression of a differentiation-specific genes, the redistribution of constitutively expressed proteins, the polarization of cytoskeletal networks, and the development of specialized transport pathways.

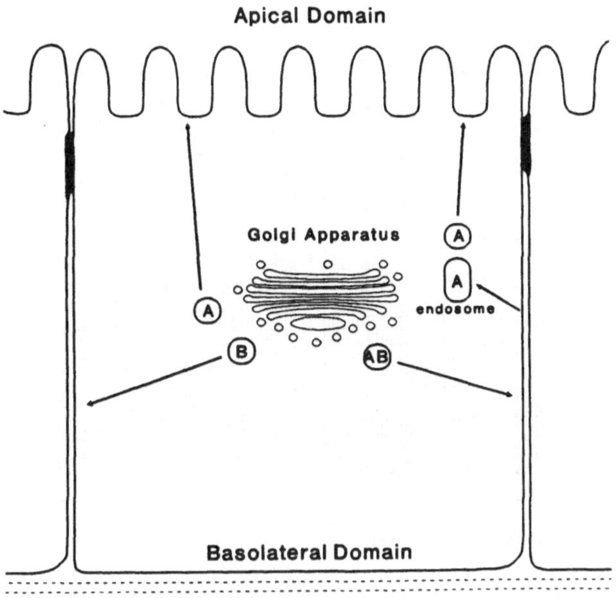

Figure 1. Schematic representation of protein targeting pathways in epithelial cells. In some cells, proteins destined for apical and basolateral surfaces are sorted into separate carrier vesicles in the trans-Golgi network and then transported directly to their respective target membranes. In other cells, both apical and basolateral-specific proteins are initially delivered to the basolateral surface. Apical proteins are subsequently recovered into endosomes and redirected to the apical domain.

MYELINATING SCHWANN CELLS AS A MODEL FOR INVESTIGATING PROTEIN SORTING AND TARGETING

Schwann cells, the myelin-forming cells of the peripheral nervous system, are well-suited for studying mechanisms of cell surface polarization. Because peripheral nerves are accessible, myelinating Schwann cells can be examined and manipulated within their normal *in vivo* environment. One of the major advantages Schwann cells offer in investigating surface membrane polarity is the large amount of surface membrane they synthesize. The myelin internode can be as long as 2 mm and can consist of more than 75 spiral wraps of myelin. Formation of the myelin sheath occurs in an orderly and predictable manner over a relatively short period of time. The myelin internode is not a simple extension of the Schwann cell plasma membrane but consists of biochemically and ultrastructurally distinct membrane domains that include compact myelin, Schmidt-Lanterman incisures, paranodal loops, periaxonal membranes, and the Schwann cell plasma membrane (Figure 2). Several molecules enriched in or specific for myelin membranes are well-characterized with respect to their amino acid sequence, their site of synthesis, and their ultimate location within the myelin internode. This information represents the starting- and ending-points of targeting pathways and provides a foundation for elucidating cellular and molecular mechanisms involved in polarization of Schwann cell surface membranes. The cytoskeletal transport pathways that mediate myelin protein transport in myelinating Schwann cells have also been characterized (Kidd et al., 1991; Kidd et al., 1992). In

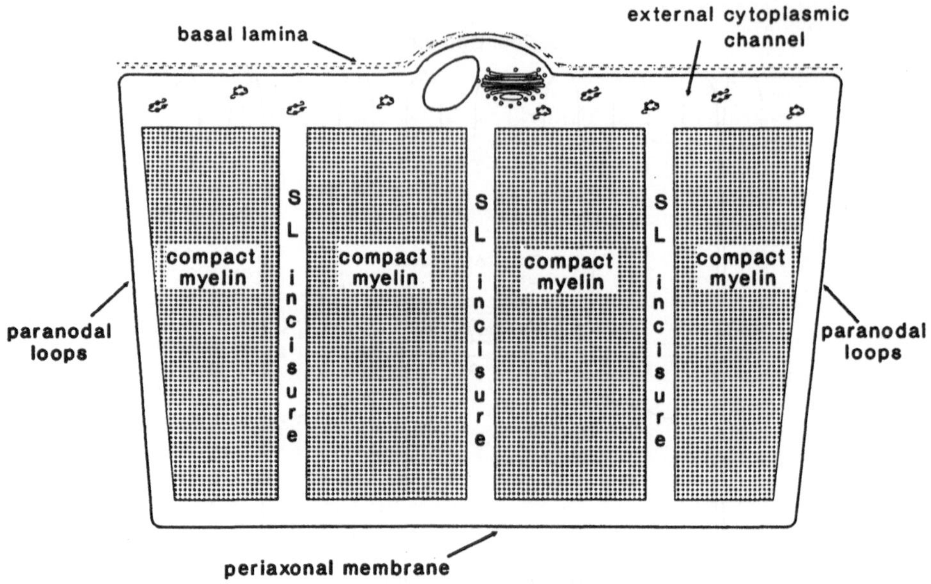

Figure 2. Schematic representation of a Schwann cell and its "unrolled" myelin internode. The clear areas represent Schwann cell cytoplasm; the stippled areas represent compact myelin. Most of the myelin internode consists of compact myelin. Schmidt-Lanterman incisures are channels of Schwann cell cytoplasm that traverse the compact myelin. The myelin internode terminates at the paranodal loops. The periaxonal membrane directly apposes the axolemma. The Schwann cell plasma membrane is covered by a basal lamina.

particular, microtubules appear to be essential for translocating proteins and possibly mRNAs and ribosomes to various regions of the myelin internode (Trapp et al., 1991) and for organizing the distribution of endoplasmic reticulum and intermediate filaments (Hansson and Sjostrand, 1971; Jacobs et al., 1972; Roytta et al., 1984).

LOCALIZATION OF MYELIN PROTEINS

Figure 3 schematically illustrates the distribution of several myelin proteins. Compact myelin contains three proteins, P_0, *myelin basic protein* (MBP), and *PMP22* (Trapp et al., 1981; Omlin et al., 1982; Heath et al., 1991; Snipes et al., 1992). P_0 protein, the major structural protein of compact myelin, accounts for more than 50% of the total PNS myelin protein, has an apparent molecular weight of 28 kD, and contains a single glycosylation site that is occupied by a oligosaccharide. Myelin basic protein is a family of extrinsic membrane proteins produced from a single gene by alternate splicing (de Ferra et al., 1985; Campagnoni, 1988). Four major forms, with molecular weights of 14 kD, 17 kD, 18.5 kD, and 21 kD, are present in rodent myelin (Barbarese et al., 1978) and they account for about 5% of the total myelin protein in the peripheral nervous system. Like P_0, PMP22 is an integral membrane glycoprotein (Snipes et al., 1992); it has an apparent molecular weight of 22 kD and accounts for approximately 5% of the total protein in rodent myelin. P_0 and PMP22 are specific for PNS myelin, whereas MBP is also found in myelin isolated from the central nervous system.

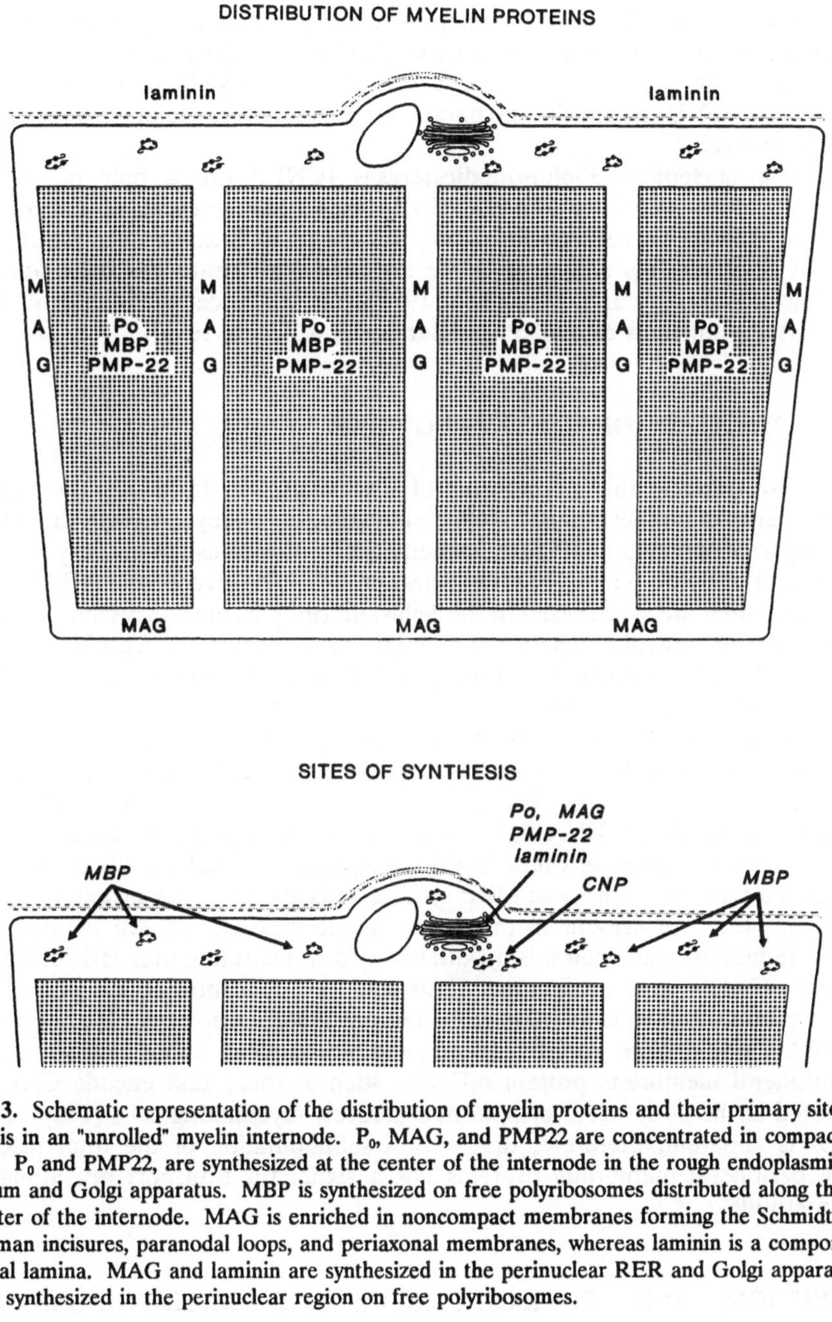

Figure 3. Schematic representation of the distribution of myelin proteins and their primary sites of synthesis in an "unrolled" myelin internode. P_0, MAG, and PMP22 are concentrated in compact myelin. P_0 and PMP22, are synthesized at the center of the internode in the rough endoplasmic reticulum and Golgi apparatus. MBP is synthesized on free polyribosomes distributed along the outer perimeter of the internode. MAG is enriched in noncompact membranes forming the Schmidt-Lanterman incisures, paranodal loops, and periaxonal membranes, whereas laminin is a component of the basal lamina. MAG and laminin are synthesized in the perinuclear RER and Golgi apparatus. CNP is synthesized in the perinuclear region on free polyribosomes.

Two other proteins in peripheral myelin internodes are the *myelin-associated glycoprotein* (MAG) and *laminin*. MAG accounts for less than 0.1% of the total protein in purified PNS myelin fractions (Figlewicz et al., 1981) and is located in the non-compact membranes of the myelin internode (Trapp and Quarles, 1982; Trapp, 1990). These membranes include the periaxonal membrane, the Schmidt-Lanterman incisures, the paranodal loops, and the inner and outer mesaxons. MAG has an apparent molecular weight of approximately 100 kD; it is heavily glycosylated, with sugar moieties accounting for approximately 30 kD of its molecular mass (Quarles et al., 1992). Laminin, another glycoprotein synthesized by myelinating Schwann cells (Bunge et al., 1986), is enriched in the basal lamina that covers the plasma membrane of the Schwann cell.

2',3'-cyclic nucleotide 3'-phosphodiesterases (CNPs) are a pair of extrinsic membrane proteins (CNP_I = 46 kD; CNP_{II} = 48 kD) present in noncompact membranes of Schwann cell myelin (Trapp et al., 1988; Braun et al., 1988). The function of CNP remains enigmatic, since no natural substrate has been found in myelinating cells (Vogel and Thompson, 1988), but sequence homologies with ras suggest possible functions as a GTPase (Bernier and Braun, 1991).

SITES OF SYNTHESIS OF MYELIN PROTEINS

When investigating the mechanisms of myelin protein targeting, the primary intracellular sites of synthesis must first be delineated, since they represent the starting point of any transport system. Figure 3 schematically illustrates the primary synthetic sites of several proteins in a myelinating Schwann cell. The glycoproteins P_0, PMP22, MAG, and laminin are synthesized in the Schwann cell perinuclear region where the rough endoplasmic reticulum (RER) and Golgi are concentrated. This is confirmed by *in situ* hybridization studies that localized mRNAs encoding these proteins (Trapp et al., 1987; Griffiths et al., 1989).

In contrast, a number of studies have established that myelin basic protein is transported to myelin by the translocation of its mRNA. Evidence for this mode of transport was first presented by Colman and collaborators (1982), who showed a 20-fold enrichment of MBP mRNA in RNA extracts of purified myelin fractions when compared to RNA extracts of whole brain homogenate. Based on this observation, they proposed that the majority of MBP synthesis occurs along the myelin internode on polysomes that are present in the outer tongue process. *In situ* hybridization studies subsequently confirmed this hypothesis by demonstrating that MBP mRNA is distributed diffusely over myelinated fibers and not concentrated on perinuclear regions of myelin-forming cells (Kristensson et al., 1986; Trapp et al., 1987; Griffiths et al., 1989). Additional *in situ* hybridization and biochemical studies have shown that other peripheral membrane protein mRNAs, such as those that encode CNP, are concentrated in the perinuclear cytoplasm of actively myelinating cells (Trapp et al., 1988). It is apparent, therefore, that only a subpopulation of the total mRNAs encoding peripheral membrane proteins has a signal for transport out along the myelin internode.

DISTRIBUTION AND CHARACTERISTICS OF MICROTUBULES IN MYELINATING SCHWANN CELLS

Epithelial cells have specialized transport pathways that establish and maintain their polarized surface membranes. Microtubules are a major component of these transport pathways and are organized in diverse and specialized ways, in different cell types. Two features of microtubules that are important in protein transport are their *polarity* and their *organization*. Microtubules have an intrinsic polarity, so that each has a (+) end and a (-) end (Dustin, 1984). Plus- and (-)-end-specific motor proteins, e.g., kinesin and dynein, respectively, have been identified, signifying that microtubule polarity can vectorially direct protein targeting (Vale et al., 1985; Paschal et al., 1987). In neurons, for example, axonal microtubules have a uniform polarity, with (+) ends distal to the cell body (Heidemann et al., 1981; Baas et al., 1988), whereas dendritic microtubules have a mixed polarity, with equal numbers of each orientation (Baas et al., 1988; Burton, 1988). Microtubule polarity is clearly essential in directing specific organelles and/or membrane traffic in axons, and is likely to be important in dendrites as well.

Microtubule organization can also influence protein transport pathways. In some cells, microtubules originate from a single organizing center (MTOC) located close to the nucleus and associated with the centrosome (Dustin, 1984). In other cells, microtubules originate from multiple sites (Spiegelman et al., 1979a; Spiegelman et al., 1979b; Burton, 1988; Achler et al., 1989; Baas et al., 1988; Bacallao et al., 1989). The location of these originating sites can play a major role in determining the overall distribution and orientation of microtubules (Spiegelman et al., 1979b; Bacallao et al., 1989; Bre et al., 1990); however, as described below, other factors besides the site of microtubule nucleation can play a role in determining microtubule organization.

In myelinating Schwann cells, microtubules are abundant in the major cytoplasmic compartments, and thus they have the potential to transport and target myelin components to all membrane domains of the myelin internode. The sites of microtubule nucleation and the polarity of microtubules in myelinating Schwann cells have been investigated and provide insight into how microtubules might help target myelin proteins. Schwann cell microtubules originate from multiple sites concentrated in the perinuclear cytoplasm (Kidd et al., 1992) (Figure 4). The nature of this perinuclear microtubule nucleating material is unknown; the centrosome, however, is not a major MTOC. Although studies indicate that all Schwann cell microtubules are nucleated in Schwann cell perinuclear cytoplasm, microtubules located in the transport channels along the outer margin of the myelin sheath have a mixed polarity (Kidd et al., 1991) (Figure 4). The tubulin hook labeling method has shown that 75% of the microtubules have (–) ends pointing toward the perinuclear region, whereas 25% have (–) ends pointing away from the perinuclear region. This 3:1 ratio was observed uniformly along the internode. These observations indicate that myelin proteins could be transported toward either the (+) or (–) ends of microtubules. The lack of MTOCs or major nucleating sites within the external cytoplasmic channels makes it unlikely that (–)-end-directed transport targeted to MTOCs plays a role in guiding proteins distally to discrete sites. The functions of microtubules in protein transport and trafficking are discussed in greater detail below.

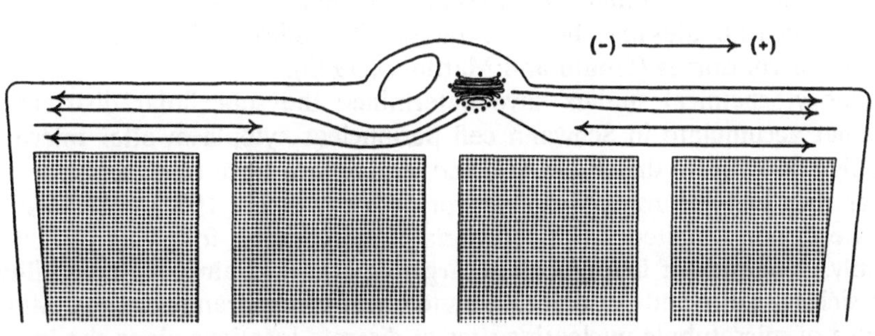

Figure 4. Schematic representation of microtubule polarity in the external cytoplasmic channels of a myelinating Schwann cell. Seventy-five percent of microtubules have their (+) ends directed away from the perinuclear region, whereas 25% have (+) ends directed toward the nucleus.

Another important and related issue regarding microtubule polarity, however, occurs in the context of microtubule-microtubule interactions. It has been recognized for some time that some cellular motor proteins, e.g., ciliary dynein, could mediate force-producing microtubule-microtubule interactions. Recent studies have shown that kinesins are able to crossbridge microtubules (Navone et al., 1992; Andrews et al., 1993), and this bridging should lead to the extension of antiparallel microtubule-microtubule arrays (by microtubule sliding) if kinesin acts as a (+)-end-directed motor. Such a mechanism could readily explain the essential role of kinesin-like proteins in the elongation of mitotic spindles (Goldstein, 1993). Since myelinating Schwann cells have a significant complement of MTs with antiparallel orientations, the machinery exists for kinesin-driven microtubule extension, either accompanying or even promoting expansion of the myelin sheath. Thus, kinesin-like motors may have a role beyond those of mediating vesicle transport and organizing networks of cellular organelles.

MECHANISMS FOR PROTEIN SORTING AND TARGETING IN SCHWANN CELLS

Recent studies have investigated how P_0, MAG, and laminin are sorted and targeted to different surface membranes (Trapp et al., 1991). All three proteins are synthesized in rough endoplasmic reticulum and Golgi membranes, but are targeted to different surface membranes (Figure 3). P_0 and MAG are membrane-bound in their surface membranes and during synthesis and transport. Therefore, they are presumably transported as vesicle-packaged proteins. Laminin, an extracellular matrix component that is secreted from Schwann cells, also undergoes vesicular transport from Golgi membranes to the cell surface.

Microtubule disassembly studies carried out in a variety of cell types have helped to elucidate mechanisms of protein sorting and targeting. Microtubule disassembly has provided evidence that microtubules organize the distribution of Golgi membranes, endoplasmic reticulum, and intermediate filaments (Wehland et al., 1983; Rogalski and Singer, 1984; Terasaki et al., 1986), and that microtubules are required for site-specific transport of many but not all proteins (Rindler et al., 1987; Achler et al., 1989; Eilers et al., 1989). Following microtubule disassembly, P_0, MAG, and laminin accumulate exclusively in Schwann cell perinuclear cytoplasm (Trapp et al., 1991), indicating that the translocation of these molecules from Golgi membranes out along the myelin internode requires microtubules. In contrast, significant accumulations of myelin basic protein were not detected after microtubule disassembly. This was not surprising, since the targeting of MBP to myelin is achieved by mRNA translocation (Colman et al., 1982; Trapp et al., 1987; Griffiths et al., 1989). Schwann cell microtubules may, however, target MBP indirectly by translocating MBP mRNA and/or ribosomes (Gould and Mattingly, 1990).

Immunocytochemical studies have determined the subcellular distribution of proteins that accumulate in Schwann cell perinuclear cytoplasm after microtubule disassembly and indicate that P_0, MAG, and laminin are sorted into separate carrier vesicles as they exit the trans-Golgi network (Trapp et al., 1991). P_0, MAG, and laminin were not targeted to inappropriate membranes following microtubule disassembly. The lack of inappropriate targeting argues against the possibility that carrier vesicles are directed by microtubules to specific membrane domains. Similarly, the absence of microtubule nucleating sites at discrete locations along the internode (i.e., Schmidt-Lanterman incisures or paranodal loops) (Kidd et al., 1992), argues against microtubule (-)-end-directed vesicular transport as a mechanism for site-specific protein targeting. These negative results suggest that outward vesicular transport of P_0-rich, MAG-rich, and laminin-rich carrier vesicles from the Schwann cell perinuclear cytoplasm occurs by (+)-end-directed transport. We consider this most likely because microtubules having (+) ends away from the perinuclear region outnumber the other microtubules by 3 to 1, and because transport demands are substantial during peak periods of myelination. Microtubules with (-) ends directed away from the perinuclear regions could provide a specialized transport pathway, could modulate (+)-end-directed transport, or could organize the distribution of cytoplasmic organelles within Schwann cell transport channels. One specialized transport function attributed to (-)-end-directed microtubule movement and relevant to Schwann cell myelination and myelin protein targeting is mRNA and ribosomal translocation. Ribosomal RNA is transported from the Schwann cell nucleus to points along the myelin internode at a rate of 0.1-0.3 mm/day (Gould and Mattingly, 1990). It is not known, however, if MBP mRNA is transported with ribosomes.

Another observation from studies of microtubule disassembly provides further insight into the mechanisms of targeting P_0, MAG, and laminin to different membranes. In Schwann cell perinuclear cytoplasm, P_0-rich carrier vesicles fuse with one another to form compact myelin-like membrane whorls (Trapp et al., 1991), whereas MAG-rich carrier vesicles fuse with each other to form mesaxon-like membrane whorls. These observations, and the observation that these proteins are not mistargeted after microtubule disassembly, are most consistent with the direct insertion of P_0-rich carrier vesicles into compact myelin, of MAG-rich carrier vesicles into non-compact myelin, and of laminin-rich carrier vesicles into the plasma membrane. Implicit in these conclusions is the hypothesis that each class of carrier vesicle contains not only site-specific signals that enhance fusion with the appropriate membrane but also other signals that prevent fusion with inappropriate membranes. The molecules that govern membrane fusion are just beginning to be defined. Fusogenic proteins, receptors for fusogenic proteins, the rab family of GTPases, and the submembranous cytoskeleton are but a few of the molecules thought to be involved in site-specific membrane fusion events (Sztul et al., 1992).

ACKNOWLEDGEMENTS

Grahame Kidd is a postdoctoral fellow of the National Multiple Sclerosis Society. This work was supported by grants NS22849 and NS29818 from the National Institutes of Health. The authors thank Rod Graham for editorial assistance and manuscript preparation.

REFERENCES

Achler, C., D. Filmer, C. Merte, and D. Drenckhahn, 1989, Role of microtubules in polarized delivery of apical membrane proteins to the brush border of the intestinal epithelium, *J. Cell Biol.* 109:179.

Andrews, S.B., R.D. Leapman, P.E. Gallant, B.J. Schnapp, and T.S. Reese, 1993, Single kinesin molecules crossbridge microtubules in vitro, *Proc. Natl. Acad. Sci. USA* in press.

Baas, P.W., J.S. Deitch, M.M. Black, and G.A. Banker, 1988, Polarity orientation of microtubules in hippocampal neurons: Uniformity in the axon and nonuniformity in the dendrite, *Proc. Natl. Acad. Sci. USA* 85:8335.

Bacallao, R., C. Antony, C. Dotti, E. Karsenti, E.H.K. Stelzer, and K. Simons, 1989, The subcellular organization of Madin-Darby canine kidney cells during the formation of a polarized epithelium, *J. Cell Biol.* 109:2817.

Barbarese, E., J.H. Carson, and P.E. Braun, 1978, Accumulation of the four myelin basic proteins in mouse brain during development, *J. Neurochem.* 31:779.
Bernier, L. and P. Braun, 1991, Relationship of CNP to regulated events in myelinogenesis, *Trans. Amer. Soc. Neurochem.* 22:265 (Abstract).
Braun, P.E., F. Sandillon, A. Edwards, J.-M. Matthieu, and A. Privat, 1988, Immunocytochemical localization by electron microscopy of 2',3'-cyclic nucleotide 3'-phosphodiesterase in developing oligodendrocytes of normal and mutant brain, *J. Neurosci.* 8:3057.
Bre, M.-H., R. Pepperkok, A.M. Hill, N. Levilliers, W. Ansorge, E.H.K. Stelzer, and E. Karsenti, 1990, Regulation of microtubule dynamics and nucleation during polarization in MDCK II cells, *J. Cell Biol.* 111:3013.
Bunge, R.P., M.B. Bunge, and C.F. Eldridge, 1986, Linkage between axonal ensheathment and basal lamina production by Schwann cells, *Ann. Rev. Neurosci.* 9:305.
Burton, P.R., 1988, Dendrites of mitral cell neurons contain microtubules of opposite polarity, *Brain Res.* 473:107.
Campagnoni, A.T., 1988, Molecular biology of myelin proteins from the central nervous system, *J. Neurochem.* 51:1.
Colman, D.R., G. Kreibich, A.B. Frey, and D.D. Sabatini, 1982, Synthesis and incorporation of myelin polypeptide into CNS myelin, *J. Cell Biol.* 95:598.
de Ferra, F., H. Engh, L. Hudson, J. Kamholz, C. Puckett, S. Molineaux, and R. Lazzarini, 1985, Alternative splicing accounts for the four forms of myelin basic protein, *Cell* 43:721.
Drubin, D.G., 1991, Development of cell polarity in budding yeast, *Cell* 65:1093.
Dustin, P., 1984, "Microtubules," Springer-Verlag, Berlin.
Eilers, U., J. Klumperman, and H.-P. Hauri, 1989, Nocodazole, a microtubule-active drug, interferes with apical protein delivery in cultured intestinal epithelial cells (Caco-2), *J. Cell Biol.* 108:13.
Figlewicz, D.A., R.H. Quarles, D. Johnson, G.R. Barbarash, and N.H. Sternberger, 1981, Biochemical demonstration of the myelin-associated glycoprotein in the peripheral nervous system, *J. Neurochem.* 37:749.
Gilbert, T., A. Le Bivic, A. Quaroni, and E. Rodriguez-Boulan, 1991, Microtubular organization and its involvement in the biogenetic pathways of plasma membrane proteins in Caco-2 intestinal epithelial cells, *J. Cell Biol.* 113:275.
Goldstein, L.S.B., 1993, Functional redundancy in mitotic force generation, *J. Cell Biol.* 120:1.
Gould, R.M. and G. Mattingly, 1990, Regional localization of RNA and protein metabolism in Schwann cells in vivo, *J. Neurocytol.* 19:285.
Griffiths, G. and K. Simons, 1986, The trans Golgi network: sorting at the exit site of the Golgi complex, *Science* 234:438.
Griffiths, I.R., L.S. Mitchell, K. McPhilemy, S. Morrison, E. Kyriakides, and J.A. Barrie, 1989, Expression of myelin protein genes in Schwann cells, *J. Neurocytol.* 18:345.
Hansson, H.-A. and J. Sjostrand, 1971, Ultrastructural effects of colchicine on the hypoglossal and dorsal vagal neurons of the rabbit, *Brain Res.* 35:379.
Heath, J.W., T. Inuzuka, R.H. Quarles, and B.D. Trapp, 1991, Distribution of P_0 protein and the myelin-associated glycoprotein in peripheral nerves from Trembler mice, *J. Neurocytol.* 20:439.
Heidemann, S.R., J.M. Landers, and M.A. Hamborg, 1981, Polarity orientation of axonal microtubules, *J. Cell Biol.* 91:661.
Hopkins, C.R., 1991, Polarity signals, *Cell* 66:827.
Hubbard, A.L. and B. Stieger, 1989, Biogenesis of endogenous plasma membrane proteins in epithelial cells, *Annu. Rev. Physiol.* 51:755.
Hugon, J.S., G. Bennett, P. Pothier, and Z. Ngoma, 1987, Loss of microtubules and alteration of glycoprotein migration in organ cultures of mouse intestine exposed to nocodazole or colchicine, *Cell Tissue Res.* 248:653.
Jacobs, J.M., J.B. Cavanagh, and F.C.-K. Chen, 1972, Spinal subarachnoid injection of colchicine in rats, *J. Neurol. Sci.* 17:461.
Kidd, G.J., S.B. Andrews, and B.D. Trapp, 1991, Microtubule organization in myelinating Schwann cells, *Trans. Amer. Soc. Neurochem.* 22:262 (Abstract).
Kidd, G.J., S.B. Andrews, D.L. Curley, and B.D. Trapp, 1992, Microtubule organizing centers in myelinating Schwann cells, *Trans. Amer. Soc. Neurochem.* 23:210 (Abstract).
Kristensson, K., N.K. Zeller, M.E. Dubois-Dalcq, and R.A. Lazzarini, 1986, Expression of myelin basic protein gene in the developing rat brain as revealed by in situ hybridization, *J. Histochem. Cytochem.* 34:467.
LeBivic, A., A. Quaroni, B. Nichols, and E. Rodriguez-Boulan, 1990, Biogenic pathways of plasma membrane proteins in Caco-2, a human intestinal epithelial cell line, *J. Cell Biol.* 111:1351.

LeBivic, A., Y. Sambuy, A. Patzak, N. Patil, M. Chao, and E. Rodriguez-Boulan, 1991, An internal deletion in the cytoplasmic tail reverses the apical localization of human NGF receptor in transfected MDCK cells, *J. Cell Biol.* 115:607.

Lisanti, M.P., A. LeBivic, A.R. Saltiel, and E. Rodriguez-Boulan, 1990, Preferred apical distribution of glycosyl-phosphatidylinositol (GPI) anchored proteins: A highly conserved feature of the polarized epithelial cell phenotype, *J. Memb. Biol.* 113:155.

Lisanti, M.P. and E. Rodriguez-Boulan, 1990, Glycophospholipid membrane anchoring provides clues to the mechanism of protein sorting in polarized epithelial cells, *TIBS* 15:113.

Low, H.L., S.H. Wong, B.L. Tang, V.N. Subramaniam, and W. Hong, 1991a, Apical cell surface expression of rat dipeptidyl peptidase IV in transfected Madin-Darby canine kidney cells, *J. Biol. Chem.* 266:13391.

Low, S.H., S.H. Wong, B.L. Tang, and W. Hong, 1991b, Involvement of both vectorial and transcytotic pathways in the preferential apical cell surface localization of rat dipeptidyl peptidase IV in transfected LLC-PK1 cells, *J. Biol. Chem.* 266:19710.

Matter, K., M. Brauchbar, K. Bucher, and H.-P. Hauri, 1990, Sorting of endogenous plasma membrane proteins occurs from two sites in cultured human intestinal epithelial cells (Caco-2), *Cell* 60:429.

Mostov, K., G. Apodaca, B. Aroeti, and C. Okamoto, 1992, Plasma membrane protein sorting in polarized epithelial cells, *J. Cell Biol.* 116:577.

Mostov, K.E., A. de Bruyn Kops, and D.L. Deitcher, 1986, Deletion of the cytoplasmic domain of the polymeric immunoglobulin receptor prevents basolateral localization and endocytosis, *Cell* 47:359.

Navone, F., J. Niclas, N. Hom-Booher, L. Sparks, H.D. Bernstein, G. McCaffrey, and R.D. Vale, 1992, Cloning and expression of a human kinesin heavy chain gene: Interaction of the C-terminal domain with cytoplasmic microtubules in transfected CV-2 cells, *J. Cell Biol.* 117:1263.

Omlin, F.X., H.deF. Webster, C.G. Palkovits, and S.R. Cohen, 1982, Immunocytochemical localization of basic protein in major dense line regions of central and peripheral myelin, *J. Cell Biol.* 95:242.

Paschal, B.M., H.S. Shpetner, and R.B. Vallee, 1987, MAP 1C is a microtubule-activated ATPase which translocates microtubules in vitro and has dynein-like properties, *J. Cell Biol.* 105:1273.

Quarles, R.H., D.R. Colman, J.L. Salzer, and B.D. Trapp, 1992, Myelin-associated glycoprotein: Structure-function relationships and involvement in neurological diseases. *in:* "Myelin: Biology and Chemistry," R.E. Martenson, ed., CRC Press, Boca Raton.

Rindler, M.J., I.E. Ivanov, H. Plesken, E. Rodriguez-Boulan, and D.D. Sabatini, 1984, Viral glycoproteins destined for apical or basolateral plasma membrane domains transverse the same Golgi apparatus during their intracellular transport in Madin-Darby Canine Kidney cells, *J. Cell Biol.* 98:1304.

Rindler, M.J., I.E. Ivanov, and D.D. Sabatini, 1987, Microtubule-acting drugs lead to the nonpolarized delivery of the influenza hemagglutinin to the cell surface of polarized Madin-Darby canine kidney cells, *J. Cell Biol.* 104:231.

Rodriguez-Boulan, E. and W.J. Nelson, 1989, Morphogenesis of the polarized epithelial cell phenotype, *Science* 245:718.

Rodriguez-Boulan, E. and D. Sabatini, 1978, Asymmetric budding of viruses in epithelial monolayers: a model system for study of epithelial polarity, *Proc. Natl. Acad. Sci. USA* 75:5071.

Rogalski, A.A. and S.J. Singer, 1984, Associations of elements of the Golgi apparatus with microtubules, *J. Cell Biol.* 99:1092.

Roytta, M., S.B. Horwitz, and C.S. Raine, 1984, Taxol-induced neuropathy: short-term effects of local injection, *J. Neurocytol.* 13:685.

Simons, K. and A. Wandinger-Ness, 1990, Polarized sorting in epithelia, *Cell* 62:207.

Snipes, G.J., U. Suter, A.A. Welcher, and E.M. Shooter, 1992, Characterization of a novel peripheral nervous system myelin protein (PMP-22/SR13), *J. Cell Biol.* 117:225.

Spiegelman, B.M., M.A. Lopata, and M.W. Kirschner, 1979a, Multiple sites for the initiation of microtubule assembly in mammalian cells, *Cell* 16:239.

Spiegelman, B.M., M.A. Lopata, and M.W. Kirschner, 1979b, Aggregation of microtubule initiation sites preceding neurite outgrowth in mouse neuroblastoma cells, *Cell* 16:253.

Sztul, E.S., P. Melancon, and K.E. Howell, 1992, Targeting and fusion in vesicular transport, *Trend. Cell Biol.* 2:381.

Terasaki, M., L.B. Chen, and K. Fujiwara, 1986, Microtubules and the endoplasmic reticulum are highly interdependent structures, *J. Cell Biol.* 103:1557.

Trapp, B.D., Y. Itoyama, N.H. Sternberger, R.H. Quarles, and H.deF. Webster, 1981, Immunocytochemical localization of Po protein in Golgi complex membranes and myelin of developing rat Schwann cells, *J. Cell Biol.* 90:1.

Trapp, B.D., T. Moench, M. Pulley, E. Barbosa, G. Tennekoon, and J.W. Griffin, 1987, Spatial segregation of mRNA encoding myelin-specific proteins, *Proc. Natl. Acad. Sci. USA* 84:7773.

Trapp, B.D., L. Bernier, S.B. Andrews, and D.R. Colman, 1988, Cellular and subcellular distribution of 2',3' cyclic nucleotide 3' phosphodiesterase and its mRNA in the rat nervous system, *J. Neurochem.* 51:859.

Trapp, B.D., 1990, The myelin-associated glycoprotein: Location and potential functions. In *Myelination and Dysmyelination*, I.D. Duncan, R.P. Skoff and D.R. Colman, eds., pp. 29-43, New York Academy of Science, New York.

Trapp, B.D., G.J. Kidd, P.E. Hauer, and E. Mulrenin, 1991, Disassembly of Schwann cell microtubules alters myelin protein transport, *Trans. Amer. Soc. Neurochem.* 22:263 (Abstract).

Trapp, B.D. and R.H. Quarles, 1982, Presence of the myelin-associated glycoprotein correlates with alterations in the periodicity of peripheral myelin, *J. Cell Biol.* 92:877.

Vale, R.D., T.S. Reese, and M.P. Sheetz, 1985, Identification of a novel force-generating protein, kinesin, involved in microtubule-based motility, *Cell* 42:39.

Vogel, U.S. and R.J. Thompson, 1988, Molecular structure, localization, and possible functions of the myelin-associated enzyme 2',3'-cyclic nucleotide 3'-phosphodiesterase, *J. Neurochem.* 50:1667.

Wandinger-Ness, A., M.K. Bennett, C. Antony, and K. Simons, 1990, Distinct transport vesicles mediate the delivery of plasma membrane proteins to the apical and basolateral domains of MDCK cells, *J. Cell Biol.* 111:987.

Wehland, J., M. Henkart, R. Klausner, and I.V. Sandoval, 1983, Role of microtubules in the distribution of the Golgi apparatus: Effect of taxol and microinjected anti-alpha-tubulin antibodies, *Proc. Natl. Acad. Sci. USA* 80:4286.

Wessels, H.P., G.H. Hansen, C. Fuhrer, A.T. Look, H. Sjostrom, O. Noren, and M. Spiess, 1990, Aminopeptidase N is directly sorted to the apical domain in MDCK cells, *J. Cell Biol.* 111:2930.

CELLULAR AND MOLECULAR CHARACTERISTICS OF CNP SUGGEST REGULATORY MECHANISMS IN MYELINOGENESIS

Dino De Angelis, Martha Cox, Enoch Gao, and Peter E. Braun

Department of Biochemistry
McGill University
Montreal, Quebec, Canada H3G 1Y6

INTRODUCTION

A variety of observations call attention to the potential significance of CNP in CNS myelination[1-6]. Although this protein is not unique to oligodendrocytes, these cells are the only ones to express it in the CNS. CNP is the earliest of all known myelination-related proteins to appear developmentally in oligodendrocytes; it accumulates in the cytoplasm-containing compartments of the myelin sheath, and is absent from compact lamellae[4,7]. Despite several interesting primary structural domains that are shared by other proteins, including several that are involved in signal transduction[2,3,5] and an isoprenylated domain[6], CNP is not assignable to any family of known proteins by virtue of overall sequence homology. This and the catalytic potential for hydrolysis of 2′-3′-cyclic nucleotides to yield 2′-nucleotides (a reaction of no known physiological significance for events in the cytosol) have left the problem of a functional role unresolved.

We are currently pursuing this question in several ways: (1) by production of transgenic mice that manifest altered expression of CNP; (2) by using homologous recombination to knock out the CNP gene in mice; and (3) by studying the consequences of CNP expression in transfected cells that normally don't express this gene. In our presentation here we focus on this last approach and show that: (1) wild-type CNP1 and CNP2, when expressed in non-glial cells, induce filopodia and process extension in these cells; (2) a cys→ ser mutation in the C-terminal domain that renders CNP incapable of being isoprenylated or carboxylmethylated does not alter cell morphology to the same extent, and influences the intracellular distribution of the expressed CNP; (3) wild-type, but not mutant (cys→ ser) CNP may interact with and alter the actin-based cytoskeleton; (4) CNP, like isoprenylated G-protein, *ras*, is carboxylmethylated on the C-terminal cysteine, provided the protein is first isoprenylated; (5) palmitoylation of CNP, presumably on cysteine, is not dependent on prior isoprenylation.

A Multidisciplinary Approach To Myelin Diseases II
Edited by S. Salvati, Plenum Press, New York, 1994

MATERIALS AND METHODS

Preparation of cDNAs for Transfection

The cDNA inserts encoding rat CNP1 and the cys→ ser mutation at position 397 of rat CNP1 were subcloned into the Rc/RSV plasmid from Invitrogen Corp. This vector contains the Rous sarcoma virus promoter, recommended for high levels of expression in eucaryotic cells. All plasmid preparations were extensively purified.

Protocol for Transient Transfection

Fibroblasts (L cells) and hepatoma (HepG2) cells were seeded at a density of 15,000 cells per 9 mm plastic coverslip in Dulbecco's Modified Eagle's medium containing 10% fetal bovine serum. Calcium phosphate mediated DNA transfers were performed according to standard procedures. Briefly, a precipitate containing 5 μg of the plasmid was added to coverslips, followed 24 hr later by a medium change. Coverslips were then examined by immunofluorescence every 24 hr for 7 days.

Immunofluorescence and Confocal Microscopy

After 1 to 7 days post-transfection, cells were fixed in 3.7% paraformaldehyde, permeabilized with 0.3% Triton X-100, and incubated for 90 min with affinity-purified rabbit antisera against CNP or mouse antiß-tubulin, in the presence of 0.15% Triton X-100 and 10% normal goat serum as blocking agents. They were then treated with goat anti-rabbit (FITC-) or (mouse TRITC-conjugated) antisera. For visualization of filamentous actin, cells were treated with Texas Red-conjugated phallacidin. Coverslips were mounted on slides for routine epifluorescence or for confocal microscopy.

Confocal imaging of cells were carried out with Leica Confocal Laser Scanning Microscope (CLSM).

Diffusion Assay for Carboxylmethylation

Cultured Sf9 cells were infected with recombinant baculovirus containing the cDNA for either CNP1 or the mutant CNP1 in which the codon for cysteine at position 397 is altered to produce serine. After 40-60 hr post-infection, at the peak of active recombinant CNP expression, the culture medium was supplemented with ^{3}H-methyl-methionine, the substrate for biological methyl transfer from S-adenosyl methionine. Following a 2 hr labelling period the cells were harvested and their proteins separated and detected by SDS-PAGE and autoradiography. Gels were sliced into 3 mm pieces and treated with NaOH to hydrolyze the labile carboxylmethyl groups; these were trapped as methanol in scintillation fluid[8]. Radioactivity was determined by scintillation counting, and results were expressed as a ratio of volatile methyl groups released to ^{3}H-methionine incorporated into the polypeptide.

Palmitoylation of CNP *In Vivo*

Cultured Sf9 cells were infected with recombinant baculovirus containing the cDNA for either CNP1 or the mutant CNP1 in which the codon for cysteine at amino acid 397 was altered to encode serine. After 24 hr post-infection, the culture medium was supplemented with ^{14}C-labelled palmitic acid (50 μCi per dish; 5×10^{6} cells). Following a

16 hr labelling period, cells were harvested and their proteins separated and detected by SDS-PAGE and autoradiography. Aliquots of cell homogenates in each instance were also subjected to immunoprecipitation with affinity purified anti-CNP.

RESULTS AND DISCUSSION

Expression of CNP1 and CNP1$^{cys \rightarrow ser}$ in Non-Glial Cells

Figure 1a shows that CNP1, when ectopically expressed in hepatoma cells, distributes intracellularly similar to the normal localization of this protein in oligodendrocytes[5,7]. It is found unevenly distributed throughout the cytoplasm, but with pronounced concentration in some regions, especially near the margins of the cell. These transfected cells exhibit markedly

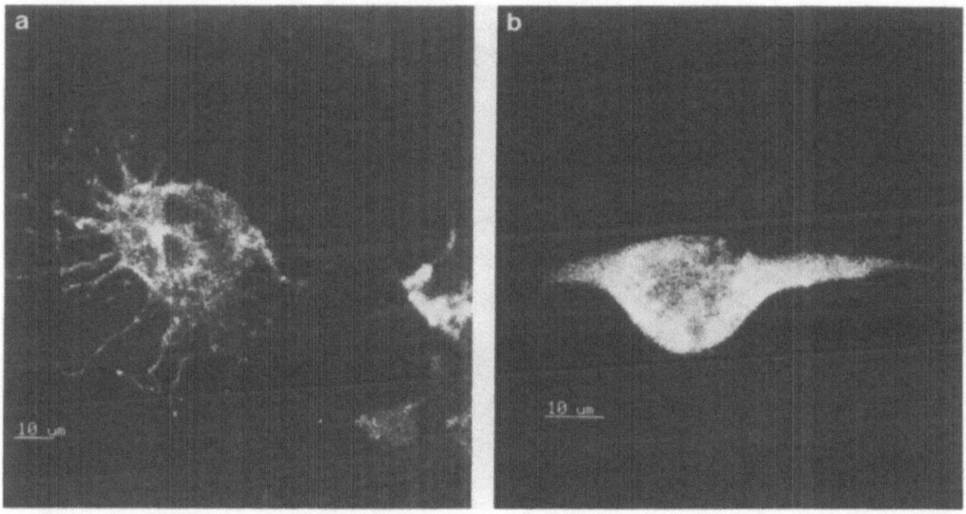

Figure 1. Transfected hepatoma (Hep G2) cells that express CNP1 (panel a) or the cys→ ser mutant CNP1 (panel b). Original renditions of these CLSM images were in color, in which a spectrum of pseudocolor represented varying concentrations of the fluorophore (CNP).

altered morphological features at their surfaces; they extend numerous filopodia that are filled with CNP. By comparison (Figure 1b), when transfected cells express the cys→ ser mutant form of CNP, that is incapable of being isoprenylated or carboxylmethylated, the cells retain a more normal appearance, with little alteration at the cell surface. Moreover, the mutant CNP is generally localized more uniformly throughout the cytoplasm although some "pockets" of higher concentration are evident. It should be noted that variations in protein localization are not readily detected by conventional epifluorescence, but are readily apparent when analyzed by "optical sectioning" using the CLSM system, and presenting the composite of digitized information in a pseudocolor format. Here, the lighter regions of the cell are the sites of highest CNP concentration.

Transfection of another cell type, namely fibroblasts, (Figures 2 and 3, panels a and c), confirmed the intracellular localizational differences between CNP1 and the mutant CNP1 and again demonstrated that expression of CNP1 but not the mutant protein is accompanied by the formation of CNP-containing filopodia and even lamellipodia (in some cells).

We then asked what effect the expression of CNP might have on the cytoskeleton. In Figure 2 (panel b) we show that staining of actin filaments is greatly reduced after CNP1 expression, presumably as a consequence of depolymerization. In contrast, the C-terminally unmodified CNP (panel d) has no effect on the actin-based cytoskeleton. This phenomenon

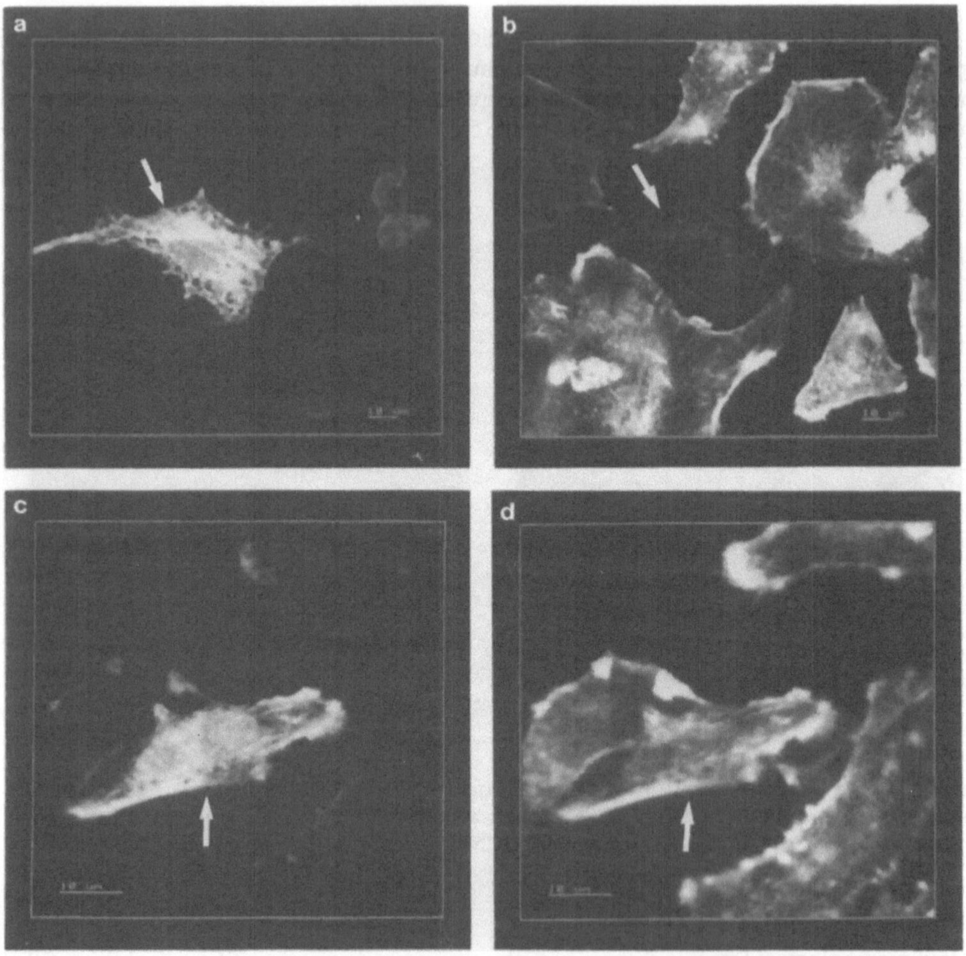

Figure 2. Transfected fibroblasts (NIH 3T3 cells) that express CNP1 (panel a and b) or the cys→ ser mutant CNP1 (panel c and d), and the effect of these proteins on cellular actin filaments (panels b and d) stained for F-actin with Texas Red-conjugated phallacidin).

was more pronounced in fibroblasts than in hepatoma cells (not shown). Apparently, then, isoprenylation and/or carboxylmethylation confers on CNP the ability to associate with microfilaments and to depolymerize them, by either a direct or indirect mechanism.

When we examined the effect of CNP expression on the microtubular cytoskeleton, we detected no differences in tubulin staining that could be attributed to either CNP1 or the mutant form of CNP1 (Figure 3, panels b and d).

Given that signal transducing proteins possessing an isoprenylated C-terminal cysteine, like the oncoprotein *ras*, are also methylated on the carboxyl group of this

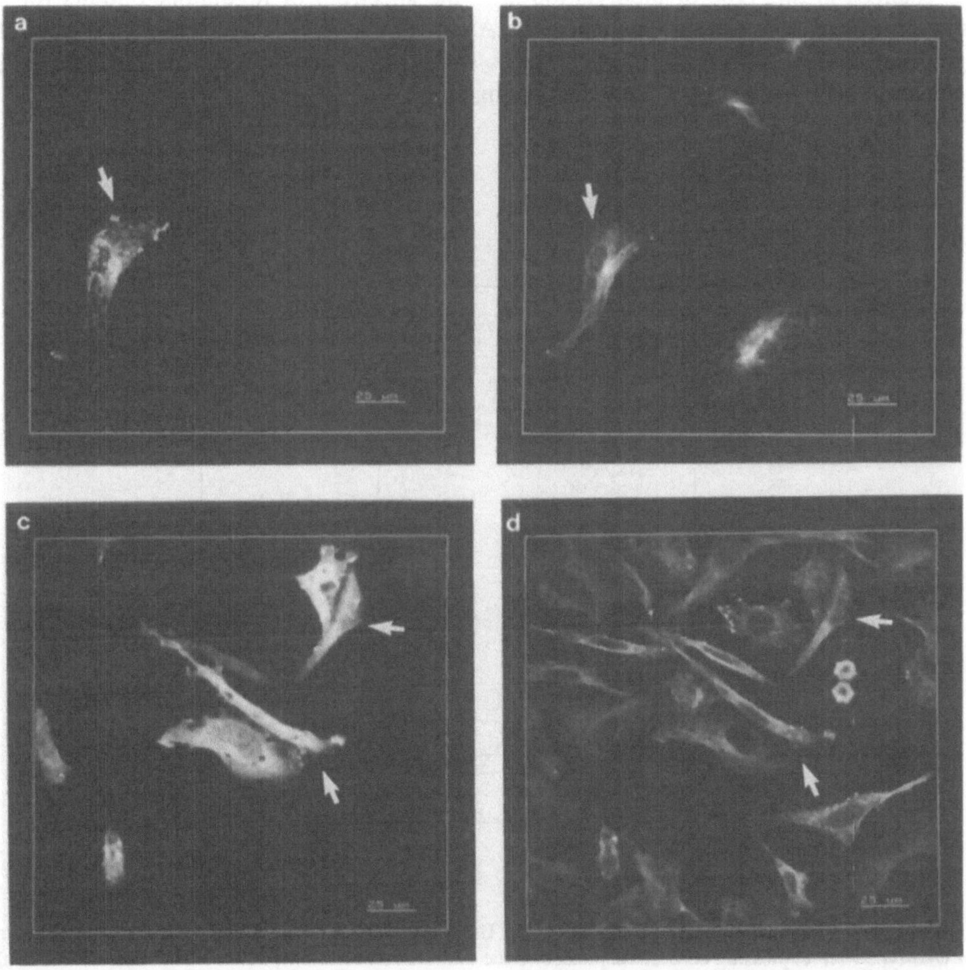

Figure 3. Transfected fibroblasts (L cells) that express CNP1 (panel a and b) or the cys→ ser mutant CNP1 (panels c and d) and the effect of these proteins on microtubules (panels b and d; immunostained for tubulin).

Carboxylmethylation of CNP

modified residue, it is important to determine if CNP is also modified in this way. In preliminary *in vitro* experiments we have been able to demonstrate that CNP can be carboxylmethylated (data not shown). We now needed to establish that this reaction also takes place in an intracellular environment. Figure 4 shows that the synthesis of CNP1 in cultured eukaryotic cells (Sf9) is accompanied by the incorporation of alkali-labile, tritiated methyl groups derived from the biological methyl donor S-adenosylmethionine. In contrast, cells that were infected with the cys→ ser mutant CNP-containing recombinant virus expressed abundant mutant CNP that failed to be carboxymethylated, as expected. Thus the methyl esterification that can occur in CNP shows that this protein is processed at the C-terminal (isoprenylation→ proteolysis→ carboxylmethylation) in a manner completely analogous to the modifications of signal-transducing proteins like *ras*. It has been suggested that changes in this type of methylation alter the spectrum of regulatory outputs for a particular signal transduction component[9,10]. Although we don't know yet in what kind of signalling cascades CNP might participate, we surmise that it could be an important regulator of cellular events that underlie myelination.

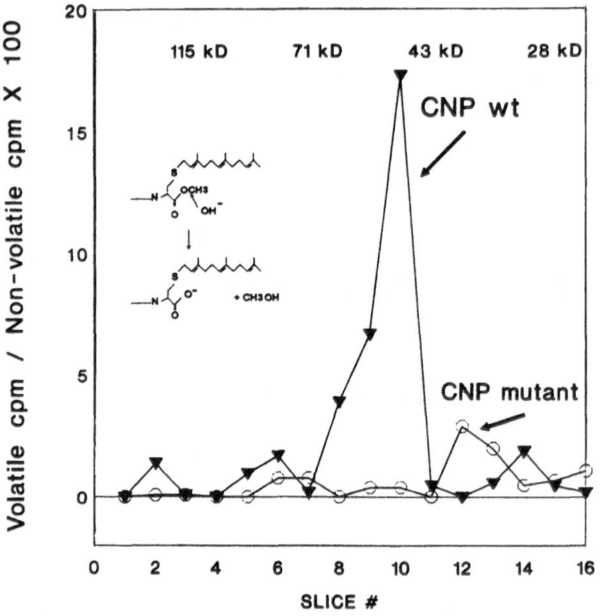

Figure 4. Carboxylmethylation of CNP1 in Sf9 cells. The major peak of alkali-labile, volatile radioactivity in the wild-type CNP expressing cells corresponded to the major Coomassie blue stained band of CNP (46kDa) in the gel. An equivalent visible band of cys→ ser mutant CNP was devoid of volatile methyl-derived tritium.

Palmitoylation of CNP

Fatty acylation of many proteins is known to occur, but in few cases is it known what effect this post-translational modification has on protein function, or what its relationship is to other modifications. In some isoforms of the signal-transducing oncoprotein *ras*, palmitoylation of specific cysteine residues is contingent upon prior isoprenylation at the C-terminus, and this influences targeting of the protein to membrane domains[10,12]. In Figure 5, we show that palmitoylation of CNP1 occurs independently of isoprenylation, since the cys→ ser mutation near the C-terminus that obviates isoprenylation, does not prevent palmitoylation elsewhere in the polypeptide. Experiments in progress are designed to determine the exact site(s) of this acylation, and to delineate its functional consequences.

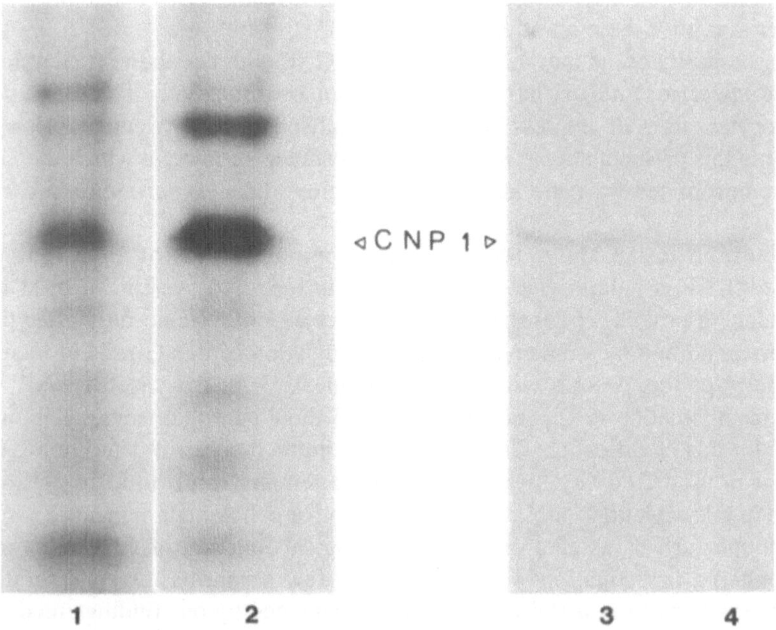

Figure 5. Autoradiograph showing fatty acylation with ^{14}C-palmitate of CNP1 and the cys→ ser mutant CNP1 in Sf9 cells. Lane 1, palmitoylated proteins of the Sf9 cell homogenate, with dominant labelling of CNP1; lane 2 same as lane 1, except that cells were infected with recombinant baculovirus to express the cys→ ser mutant CNP1; lane 3, immunoprecipitate of palmitoylated CNP1; lane 4, immunoprecipitate of palmitoylated cys→ ser mutant CNP1.

CONCLUSIONS

Earlier studies reported strong indications that CNP is associated with cytoskeletal elements of oligodendrocytes[13-16]. The capacity of CNP expression in transfected cells to promote major alterations of cellular morphology, with the attendant appearance of myriad filopodia and processes reminiscent of those present on oligodendrocytes is further evidence for the interaction of CNP with cellular proteins that are normally responsible for such surface features. The fact that a CNP mutant, with a single amino acid substitution (cys→ ser) fails to generate these major changes in cell surface morphology indicates that this interaction with the cytoskeleton is dependent on specific post-translational modifications at the C-terminus. Since we have demonstrated both isoprenylation and carboxylmethylation at this site, we can only suggest that either event, or both, are critical to this cytoskeletal association. Recently, it has been shown that isoprenylated proteins (identities not known) participate in the regulation of cellular actin polymerization in cultured K1 cells[17], and that carboxylmethylation may play a role in the regulation of some signal transduction events in fibroblasts[9]. Our observations that transfection of L cells with the wild-type CNP, results in the disappearance of much of the filamentous actin with sparing of microtubules in CNP-expressing cells is further support for our contention that CNP associates with some elements of the cytoskeleton. Again, the cys→ ser mutant CNP, which is not isoprenylated or carboxylmethylated, behaves differently, without an apparent disruption of actin filaments. We are unable at this point to delineate the exact nature of this interaction, whether it occurs directly with actin cables, or is mediated by other actin-binding proteins. We have now begun a biochemical investigation of this phenomenon *in vitro* in order to resolve these questions.

The possibility exists that the association of CNP with the actin-based cytoskeleton is cell-type specific, and/or that it is dependent on the level of CNP expression. Oligodendrocytes, after all, express both CNP and abundant actin filaments. New insights with respect to this problem should be derived from studies of mice in which the CNP gene has been compromised by gene knockout strategies; these experiments are currently underway.

In addition to the above two C-terminal modifications of CNP that we have demonstrated, CNP is palmitoylated on an as yet unspecified cysteine[18]. In at least one signal transducing protein, *H-ras*, palmitoylation occurs on cysteine only after the *H-ras* protein is first modified by isoprenylation[11]. Palmitoylation of this G-protein increases the binding avidity of this protein with cell membranes[11,12]. In the case of CNP, we have verified its palmitoylation in C_6 and SF9 cells, and show here that isoprenylation is not a prerequisite for this modification. Currently, we are in the process of determining the exact residue that is modified (by site-directed mutagenesis) and hope to establish the significance of this acylation for binding of CNP to other cellular components.

Although further insights into the physiological function of CNP must await the results of ongoing investigation we can suggest a few scenarios for consideration. The available observations from a variety of studies, augmented by our findings recorded here, point to a role for CNP in some dynamic cellular process that requires its association with the actin-based microfilaments. Cell surface plasticity is normally associated with dynamic actin networks, and the extension of filopodia and lamellipodia by oligodendrocytes is no exception[19]. Since these projections are elaborated by oligodendrocytes in the CNS well before myelination[4], concordant with the early appearance of CNP we surmised that CNP expression may be a functionally related event. Moreover, the emergent views of carboxylmethylation as a modulator of regulated cellular processes[9,10], when taken in the context of our observations on a similar modification of CNP, further compel us to invoke an important, regulatory role for CNP in myelinogenesis. Actin-binding proteins (e.g. gelsolin) are sure to be involved in regulating cytoskeletal architecture in oligodendrocytes[20], but it is premature to suggest just what their interplay with actin filaments and CNP might be.

Our earlier observations showing that isoprenylation is a prerequisite for binding CNP to membranes[6], was not as unexpected as our present findings that isoprenylation influences the interaction of CNP with actin filaments. Taken together, we can suggest the possibility that the C-terminus of CNP binds to some element of the actin-derived cytoskeleton near the surface of membranes (of the plasma membrane or endoplasmic reticulum) in a manner that is contingent upon binding of the hydrophobic isoprene moiety with the bilayer structure. Alternatively, the membrane-associated cytoskeleton (membraneskeleton) itself possesses sites for the binding of isoprenylated CNP. In this regard, it is worth recalling the recent report of a role for prenylated proteins in the regulation of actin polymerization[17].

In an alternative scenario, we envision that the association of CNP with microfilaments may subserve either a participatory or regulatory role in the protein synthesizing, cellular machinery that involves the interaction of mRNA with the cytoskeleton[20], or in the targeted transport of mRNA within the oligodendrocyte-myelin complex. Such a role is based not only on the CNP-microfilament association but also on the nucleotide binding domains present in this protein. In this regard, rudimentary evidence has been reported for the association of MBP mRNA with the cytoskeleton[21].

These are by no means the only possible functions of CNP; others are being considered, but ultimately this question will be resolved by contributions from many laboratories exploring the multi-faceted problem of myelinogenesis.

ACKNOWLEDGEMENTS

We thank Elsa Horvath for technical assistance and Marlene Gilhooly for superb manuscript typing and assembly. This work was supported by grants from the Medical Research Council and Multiple Sclerosis Society of Canada, and the Network of Centres of Excellence in Neural Regeneration.

REFERENCES

1. U.S. Vogel and R.J. Thompson, Molecular structure, localization and possible functions of the myelin associated enzyme 2′,3′-cyclic nucleotide 3′-phosphodiesterase, *J. Neurochem.* 50:1667-1677 (1988).
2. R.J. Thompson, 2′,3′-cyclic nucleotide 3′-phosphodiesterase and signal transduction in CNS myelin, *Biochem. Soc. Trans.* 20:621-626 (1992).
3. J.J. Sprinkle, 2′,3′-cyclic nucleotide 3′-phosphodiesterase, an oligodendrocyte-Schwann cell and myelin-associated enzyme of the nervous system, *CRC Crit. Rev. Neurobiol.* 4:235-301 (1989).
4. P.E. Braun, D. Sandillon, A. Edwards, J.M. Matthieu, and A. Privat, Immunocytochemical localization by electron microscopy of 2′,3′-cyclic nucleotide 3′-phosphodiesterase in developing oligodendrocytes of normal and mutant brain, *J. Neurosci.* 8:3057-3066 (1988).
5. P.E. Braun, L.L. Bambrick, A.M. Edwards, and L. Bernier, 2′,3′-cyclic nucleotide 3′-phosphodiesterase has characteristics of cytoskeletal proteins; a hypothesis for its function, *Annals N.Y. Acad. Sci.* 605: 55-65 (1990).
6. P.E. Braun, D. De Angelis, W.W. Shtybel, and L. Bernier, Isoprenoid modification permits 2′,3′-cyclic nucleotide 3′-phosphodiesterase to bind to membranes, *J. Neurosci Res.* 30:540-544 (1991).
7. B.D. Trapp, L. Bernier, S.B. Andrews, and D. Colman, Cellular and subcellular distribution of CNP and its mRNA in the rat CNS, *J. Neurochem.* 51:859-868 (1988).
8. D. Chelsky, N.I. Gutterson, and D.E. Koshland, Jr., A diffusion assay for detection and quantitation of methyl-esterified proteins on polyacrylamide gels, *Anal. Biochem.* 141:143-148 (1984).
9. C. Volker, R.A. Miller, W.R. McCleary, A. Rao, M. Poenie, J.M. Backer, and J.B. Stock, Effect of farnesylcysteine analogs on protein carboxyl methylation and signal transduction, *J. Biol. Chem.* 266:21515-21522 (1991).
10. J. Stock, Balancing effector outputs, *Current Biology* 1:154-156 (1991).
11. J.F. Hancock, A.I. Magee, J.E. Childs, and C.J. Marshall, All *ras* proteins are polyisoprenylated but only some are palmitoylated, *Cell* 57:1167-1177 (1989).
12. Y. Kuroda, N. Suzuki, and T. Kataoka, The effect of post-translational modifications on the interaction of Ras2 with adenylyl cyclase, *Science* 259:683-686 (1993).
13. P.M. Pereyra, E. Horvath, and P.E. Braun, Triton X-100 extractions of CNS myelin indicate a possible role for the minor myelin proteins in the stability of lamellae, *Neurochem. Res.* 13:583-595 (1988).
14. C.S. Gillespie, R. Wilson, A. Davidson, and P.J. Brophy, Characterization of a cytoskeletal matrix associated with myelin from rat brain, *Biochem. J.* 260:689-696 (1989).
15. R. Wilson and P.J. Brophy, Role for the oligodendrocyte cytoskeleton in myelination, *J. Neurosci. Res.* 22:439-448 (1989).
16. C.A. Dyer and J.A. Benjamins, Organization of oligodendroglial membrane sheets. I. Association of myelin basic protein and CNP with cytoskeleton, *J. Neurosci. Res.* 24:201-211 (1989).

17. R.G. Fenton, H-f. Kung, D.L. Longo, and M.R. Smith, Regulation of intracellular actin polymerization by prenylated cellular proteins, *J. Cell Biol.* 117:347-356 (1992).
18. H.C. Agrawal, T.J., Sprinkle, and D. Agrawal, 2′-3′-cyclic nucleotide 3′-phosphodiesterase in the CNS is fatty acylated by thioester linkage, *J. Biol. Chem.* 265:11849-11853 (1990).
19. B. Kacher, T. Behar, and M. Dubois-Dalcq, The oligodendrocyte cultured without neurons. Cell shape and motility, *Cell Tissue Res.* 244:27-38 (1986).
20. J.F. Hesketh and I.F. Pryme, Interaction between mRNA, ribosomes and the cytoskeleton, *Biochem. J.* 277:1-10 (1991).
21. S.J. Hill and E. Barbarese, Myelin basic protein mRNA is associated with the oligodendrocyte cytoskeleton, *Soc. Neurosci. Abs.* 18:272.9 (1992).

MYELIN BASIC PROTEIN mRNA TRANSLOCATION IN OLIGODENDROCYTES INVOLVES MICROTUBULES AND IS INHIBITED BY ASTROCYTES *IN VITRO*

Shashi Amur-Umarjee and Anthony T. Campagnoni

Mental Retardation Research Center
Neuropsychiatric Institute
U.C.L.A. School of Medicine
760 Westwood Plaza
Los Angeles, CA 90024 U.S.A.

INTRODUCTION

Cells are capable of targeting newly synthesized proteins to their appropriate subcellular locations by macromolecular sorting mechanisms. Recent findings indicate that some protein targeting is achieved through differential sorting at the mRNA level (Lawrence and Singer, 1986; see reviews by Hesketh and Pryme, 1991) and that this process can reduce the possibility of incorrect targeting of proteins to the desired subcellular compartments. In the central nervous system (CNS), oligodendrocytes synthesize several proteins that are incorporated into the myelin membrane that ensheathes axons. Myelin basic proteins (MBPs) constitute about 30% of the total myelin proteins. They are highly cationic and interact with virtually any negatively-charged molecule. Thus, targeting the MBP mRNAs to the oligodendrocyte processes prior to translation would help prevent inappropriate interactions of MBP polypeptides with other cellular components. MBP mRNA has been shown to be translocated into the processes of oligodendrocytes both *in vivo* (Trapp et al., 1987; Verity and Campagnoni, 1988) and *in* vitro (Amur-Umarjee et al., 1990a). The mechanism by which MBP mRNAs are translocated within the oligodendrocyte and an understanding of factors that regulate the process are yet to be elucidated.

MBP polypeptides are synthesized on free polysomes and several reports indicate that a fraction of the 'free' polysomes are, in fact, attached to cytoskeletal elements (Lenk et al.,1977; Van Venrooij et al., 1981; Hesketh and Pryme, 1991). Both microfilaments and microtubules have been reported to be involved in the translocation of various mRNAs (Sundell and Singer, 1991; Pokrywka and Stephenson, 1991; Raff et al., 1990; see reviews by Hesketh and Pryme, 1991 and Steward and Banker, 1992). It seemed possible to us that MBP mRNA also might be associated with cytoskeletal elements. In support of this notion, MBP polypeptides recently have been reported to be associated with microtubules (Dyer and Benjamins, 1989; Wilson and Brophy, 1989). We have utilized pharmacological agents to specifically disrupt microfilaments and microtubules and have studied the effect of these disruptive agents on the distribution of MBP mRNA in primary glial cultures. Our results suggest that the integrity of microtubules is important for the translocation of MBP mRNA.

Myelination and translocation of MBP mRNAs are developmentally-regulated events *in* vivo and *in* vitro (Verity and Campagnoni, 1988; Amur-Umarjee et al., 1990a), and it has recently been reported that astrocytes are capable of inhibiting myelination *in vitro* (Rosen et al., 1989). These findings suggested to us that other neural cells, such as astrocytes, might possibly influence molecular events involved in myelination, such as MBP mRNA translocation. Consistent with this notion, in this and in previous studies (Amur-Umarjee et al., 1990a) we observed that MBP mRNA is translocated into the processes of only about 25% of the oligodendrocytes that express the gene in mixed glial cultures. Yet the MBP mRNAs were translocated into the processes of almost all oligodendrocytes in cultures of enriched oligodendrocytes containing less than 5% other cell types. In this report, we present evidence that physical contact of oligodendrocytes with astrocytes inhibits MBP mRNA movement into oligodendrocyte processes.

MATERIALS AND METHODS

Preparation of primary mixed glial cultures

The method used for preparing primary mixed glial cultures of neonatal BALB/cByJ mouse brains has been described previously (Amur-Umarjee et al., 1990b). Briefly, cerebral hemispheres were dissected from 1-3 day-old mice and were mechanically dissociated into medium (1:1 mixture of Dulbecco's modified Eagle's medium: Ham's F-12 medium, pH 7.5, containing 6 g/l glucose and 10% fetal calf serum). The cells were seeded into poly L-lysine coated 75 cm^2 flasks at a density of about 15 million cells/flask.

Preparation of enriched oligodendrocytes from primary cultures

Enriched oligodendrocytes were prepared essentially by the method of Suzumura et al. (1984). Flasks containing primary cultures at 14 DIV were shaken at 200 rpm for 1 hour to remove microglia and the medium was replaced with fresh medium. The cells were then shaken at 280 rpm for 18-20 hours and the detached oligodendrocytes were purified by plating on a petri dish for 30 minutes at 37°C. The non-detached cells, most of which were oligodendrocytes, were collected, counted and plated at a density of about 30,000-40,000 per chamber on 8-chamber slides.

Fluorescence immunohistochemistry

Rabbit anti-MBP polyclonal antibody was prepared in our laboratory (Reidl et al., 1981) and was further purified by absorption on fixed astrocytes. Fluorescence immunocytochemistry was then carried out exactly as we have described previously (Amur-Umarjee et al., 1991b).

Preparation of oligodendrocytes for *in situ* hybridization histochemistry

Cells at various days *in vitro* (DIV) were washed with phosphate-buffered saline (PBS) and fixed in freshly prepared 4% paraformaldehyde solution for 1 hour. The cells were then washed with PBS and stored in PBS at 4°C until used. The cells can be stored for about 1-2 months under these conditions.

cDNA Probes

The mouse MBP cDNA for the MBP probe has been described previously (Roth et al., 1986) and the rat CNPase cDNA was a gift of Dr. David Colman (Bernier et al., 1987).

***In situ* hybridization histochemistry**

A non-radioactive *in situ* hybridization method that utilizes digoxigenin-labeled cDNA probes was employed. Chambers were removed from the slides and the cells were exposed to 70% ethanol at 4°C for 1 hour followed by treatment with 0.2 N HCl for 15 min. at room temperature and for 30 min. at 65°C. The cells were then washed in 2X SSC for 5 min and incubated in a solution containing 40% formamide, 6X SSC, 5X Denhardt's solution and 100 μg/ml denatured salmon sperm DNA for 1 hour at room temperature. Twenty microliters of hybridization mixture containing 45% formamide, 6X SSC, 5X Denhardt's solution, 10% dextran sulfate and 20 ng of digoxigenin-labeled DNA probe were then placed on the cells and incubated overnight at 42°C.

The cells were washed twice with 45% formamide, 6X SSC at 42°C for 15 min each, twice with 2X SSC at room temperature for 5 min each and twice with 0.2X SSC at 50°C for 15 min each. The cells were then incubated with 0.5% blocking buffer (supplied with the Genius kit) to prevent non-specific binding of the probes. After this step, antibody conjugated to alkaline phosphatase was added to the cells. After washes to remove unbound antibody, the alkaline phosphatase substrate was added along with dye (nitroblue, tetrazolium salt) and incubated overnight at room temperature in a dark chamber. The reaction was then stopped by the addition of TE Buffer (10mM Tris, pH 7.6, 1 mM EDTA) and the cells were mounted in vinol mounting medium.

Treatment of oligodendrocytes with colchicine or cytochalasin B

Colchicine and cytochalasin B were purchased from Sigma Chemical Company. Oligodendrocytes were grown for 6 DIV and either colchicine or cytochalasin B was added at a concentration of 10 μg/ml. The cells were fixed after 4 hours and processed for *in situ* hybridization.

Astrocyte-reconstitution experiments

Cells were shaken at 14 DIV at 280 rpm overnight to remove oligodendrocytes and the remaining astrocytes were trypsinized, centrifuged and resuspended in medium. The cells were plated on chamber slides and allowed to attach and divide for one day. Primary cultures from the same batch of cells were shaken at 15 DIV to obtain oligodendrocytes and the cells thus obtained were plated either on poly L-lysine-coated chamber slides or on astrocytes in the same proportion as in primary cultures (75% astrocytes and 25% oligodendrocytes). The cells were then allowed to grow for 3-4 DIV after which they were fixed and processed for *in situ* hybridization or fluorescence immunohistochemistry.

Astrocyte conditioned medium was collected from astrocytes that were replated after the shake-off procedure and grown for 2 DIV. The conditioned medium was added to oligodendrocytes after they settle down (after 1 DIV). The cells were fixed and processed for *in situ* immunocytochemistry,

Astrocyte matrix ('astromatrix') was made by the method of Rome et al., 1986, by lysing astrocytes with water. Enriched oligodendrocytes were plated on the matrix for 3-4 DIV and then, fixed and processed for *in situ* hybridization.

RESULTS

Distribution of MBP mRNA in enriched oligodendrocyte cultures and involvement of microtubules in MBP mRNA movement

Enriched oligodendrocytes were prepared by a standard shaking procedure to eliminate astrocytes and microglia. Figure 1A illustrates the distribution of MBP mRNA in enriched oligodendrocyte cultures. In almost all the cells in these cultures, MBP mRNA was widely distributed throughout the somatas and the elaborate processes. Many mRNAs are associated with cytoskeletal elements in the cell (see reviews by Hesketh and Pryme, 1991; Steward and Banker, 1992), possibly as polyribosomal complexes. There have been reports that myelin basic proteins can bind to tubulin in tissue sections (Omlin et al., 1982;

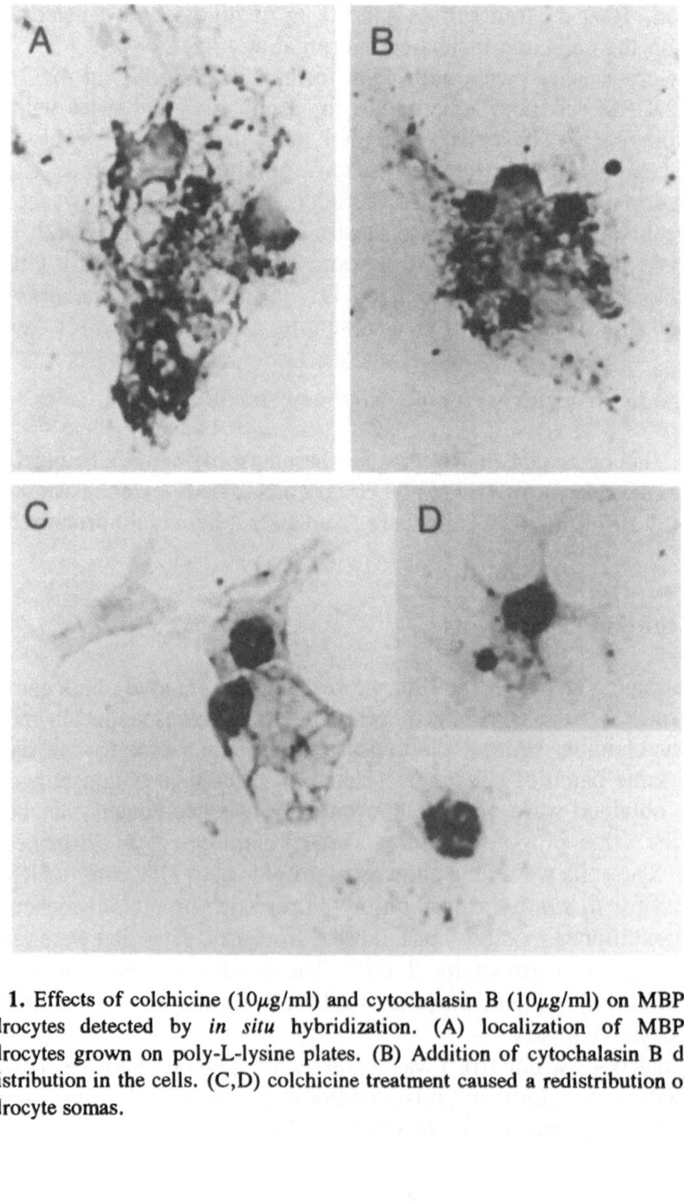

FIGURE 1. Effects of colchicine (10μg/ml) and cytochalasin B (10μg/ml) on MBP mRNA distribution in oligodendrocytes detected by *in situ* hybridization. (A) localization of MBP mRNA in enriched oligodendrocytes grown on poly-L-lysine plates. (B) Addition of cytochalasin B did not affect the MBP mRNA distribution in the cells. (C,D) colchicine treatment caused a redistribution of MBP mRNA into the oligodendrocyte somas.

Miyamoto et al., 1984) and in cultured oligodendrocytes (Dyer and Benjamins, 1989; Wilson and Brophy, 1989). These findings seemed to be consistent with the possibility that MBP mRNA and/or MBP polypeptide nascent chains on polyribosomes might be associated with cytoskeletal elements within oligodedendrocytes.

We investigated the possibility that MBP mRNA was associated with cytoskeletal elements by using pharmacological agents that disrupt specific components of the cytoskeleton and examining the subsequent distribution of MBP mRNA in enriched oligodendrocytes. We used colchicine and cytochalasin B to disrupt microtubules and microfilaments, respectively. The drugs were used at relatively low concentrations (10 μg/ml) for 4 hours according to the protocol of Dyer and Benjamins (1989) developed for oligodendrocytes. These investigators have shown the disruption of microtubules and microfilaments in enriched oligodendrocytes under these conditions. Figure 1B shows that cytochalasin B has no effect on the distribution of MBP mRNA in enriched oligodendrocytes. In cells treated with this drug MBP mRNA was distributed within the cell bodies and processes in a fashion indistinguishable from untreated oligodendrocytes (Figure 1A). These results suggest that microfilaments are not involved in the mechanism of MBP mRNA translocation. Colchicine, on the other hand, caused a major redistribution of MBP mRNA in the cells, resulting in a localization of the MBP mRNAs within the oligodendrocyte cell bodies (Figure 1C and 1D) suggesting that microtubules are involved in the mechanism of MBP mRNA translocation.

Cytoskeletal-disrupting agents do not affect CNPase translocation in oligodendrocytes

In parallel sets of experiments we examined the effects of colchicine and cytochalasin B on the distribution of CNPase mRNA in the enriched oligodendrocytes. Unlike the MBP mRNAs, the CNPase mRNAs are localized primarily within oligodendrocyte cell bodies *in vivo* (Trapp et al., 1988; Vogel et al., 1988; Jordan et al., 1989) and in primary mixed glial cell cultures (Amur-Umarjee et al., 1990a). However, in enriched oligodendrocytes we observed CNPase mRNA to be distributed throughout the cell somas and processes in much the same fashion as MBP mRNA (see Figure 2A). In experiments identical to those described for examining MBP mRNA distribution, neither colchicine (Figure 2B) nor cytochalasin B (Figure 2C) influenced the distribution of CNPase mRNA. These data indicate that cytoskeletal elements are not involved in the transport of CNPase mRNAs, and, therefore they are translocated by a mechanism different from that of the MBP mRNAs.

Influence of astrocytes on the translocation of MBP mRNA and protein in oligodendrocytes

MBP mRNA translocation occurred in only about 25% of the oligodendrocytes in mixed culture, in marked contrast to the situation *in vivo* and in enriched oligodendrocytes. Since the major glial cell type in the mixed cultures is astrocytes, we wondered if astrocytes could be inhibiting MBP mRNA translocation in the mixed cultures. To test this directly, enriched oligodendrocytes were prepared and plated on to purified astrocytes. These oligodendrocytes grew very well and elaborated delicate networks of processes as determined by immunohistochemical studies with an anti-galactocerebroside (GC) antibody (results not shown). In these enriched oligodendrocytes cocultured with astrocytes the MBP mRNAs were sequestered almost completely within the oligodendrocyte cell bodies (Figure 3B). These results were in striking contrast to plating the same enriched oligodendrocytes on to poly-L-lysine coated plates in the absence of astrocytes (Figure 3A). In this case, MBP mRNA was distributed throughout the cell somas and within the elaborate network of processes, with the MBP mRNA in the somas exhibiting a granular appearance.

The distribution of MBP polypeptides was examined in enriched oligodendrocytes grown in the presence or absence of astrocytes. In cells plated on poly L-lysine (no astrocytes), MBP polypeptide distribution was essentially the same the MBP mRNA (Figure 3C). In the enriched oligodendrocytes plated on to astrocytes, the MBP

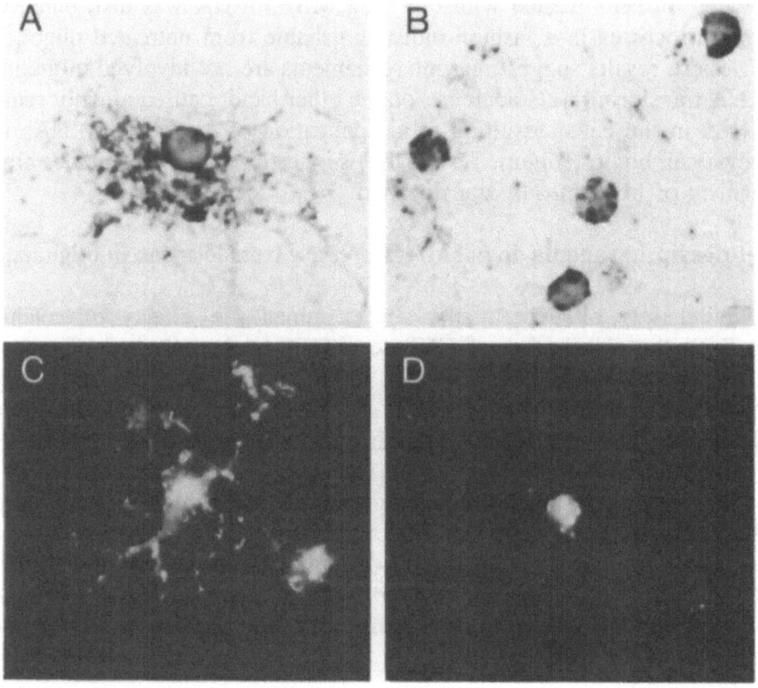

FIGURE 2. Effect of colchicine (10μg/ml) and cytochalasin B (10μg/ml) on CNPase mRNA distribution within enriched oligodendrocytes in culture detected by *in situ* hybridization. (A) distribution of CNPase mRNA in enriched oligodendrocytes grown on poly-L-lysine plates. Treatment of the cells with either cytochalasin B (B) or colchicine (C) did not affect the distribution of CNPase mRNA.

polypeptides were found to be confined to the cell somas, again consistent with the MBP mRNA localization under these conditions (Figure 3D). Thus, in enriched oligodendrocytes grown in the absence of astrocytes, MBP mRNA and polypeptides were observed in the cell bodies and processes, but when they were grown in the presence of astrocytes, MBP mRNA and polypeptides were confined to cell bodies.

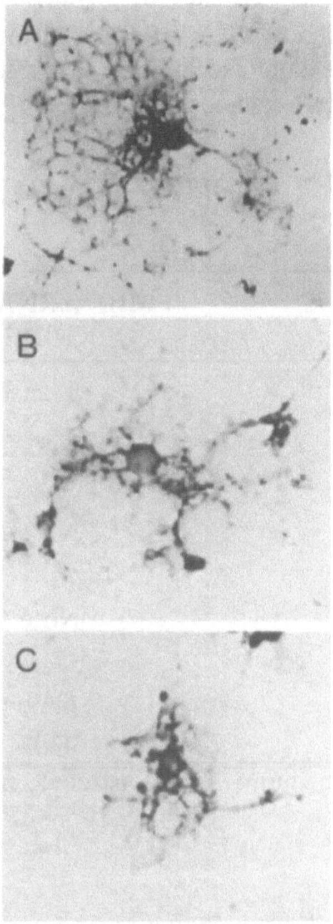

FIGURE 3. Montage illustrating the effect of astrocytes on MBP mRNA translocation within enriched oligodendrocytes. Panels A and B illustrate the distribution of MBP mRNA in enriched oligodendrocytes grown in the absence (A) or presence (B) of astrocytes. The MBP mRNA is localized primarily in the cell bodies of enriched oligodendrocytes cultured in the presence of astrocytes. Panels C and D illustrate the distribution of MBP polypeptides within enriched oligodendrocytes, detected by immunofluorescence, in the absence (C) or presence (D) of astrocytes. The MBP polypeptides, like MBP mRNA, is localized primarily in the cell bodies of enriched oligodendrocytes cultured in the presence of astrocytes.

How do the astrocytes mediate their inhibitory effect?

We investigated the possibility that the astrocytes might be secreting a soluble factor that was involved in this effect. When astrocyte conditioned medium was added to the growth medium of enriched oligodendrocytes no effect was observed on the the distribution of MBP mRNAs in the enriched oligodendrocytes (Table 1). Enriched oligodendrocytes have been reported to exhibit increased growth and survival when plated on to astrocyte matrix after being shaken off primary mixed cultures (Rome et al., 1986). We examined the effect of plating enriched oligodendrocytes onto astrocyte matrix on MBP mRNA translocation and did not observe any effect (i.e. no inhibition of MBP mRNA translocation), in either the presence or absence of astrocyte-conditioned medium.

In order to test whether live astrocytes were necessary for the effect on MBP mRNA translocation, purified oligodendrocytes were plated on to astrocytes fixed with 4% paraformaldehyde and MBP mRNA distribution examined three days later. MBP mRNA translocation within the enriched oligodendrocytes was inhibited when they were grown on fixed astrocytes. In these experiments almost all of the MBP mRNAs appeared to be confined to the cell somas. However, normal MBP mRNA translocation seemed to occur in the few oligodendrocytes that were not in physical contact with astrocytes. These findings indicate that the inhibition of MBP mRNA translocation by astrocytes is mediated through cell to cell contact. The data also indicate that a living astrocyte is not necessary to elicit the effect on MBP mRNA translocation.

TABLE 1. Influence of Growth Conditions on MBP mRNA Translocation within Enriched Oligodendrocytes

GROWTH CONDITIONS	MBP mRNA TRANSLOCATION INTO OLIGODENDROCYTE PROCESSES
Enriched oligodendrocytes	Translocation
Enriched oligodendrocytes + ACM*	No effect
Enriched oligodendrocytes + AMX*	No effect
Enriched oligodendrocytes + AMX + ACM	No effect
Enriched oligodendrocytes + "fixed" astrocytes	Inhibition of translocation

* ACM, astrocyte conditioned medium; AMX, astrocyte matrix

DISCUSSION

Cytoskeletal elements have been shown to be involved in the transport of several mRNAs such as actin, MAP2, Vg1, bicoid, etc.(see reviews by Hesketh and Pryme, 1991; Steward and Banker, 1992). Myelin basic proteins have been shown to be associated with microtubules (Dyer and Benjamins, 1989; Wilson and Brophy, 1989). These two lines of evidence suggested to us that cytoskeletal elements, possibly microtubules, might be involved in the translocation of MBP mRNA. To examine this possibility, we tested the effect of two pharmacological agents, colchicine and cytochalasin B, on the distribution of MBP mRNA in oligodendrocytes. In untreated enriched oligodendrocytes at 6 DIV after shaking off, MBP mRNA was distributed throughout the cell somas and processes, but in colchicine-treated oligodendrocytes, MBP mRNA was found to be confined almost entirely to the cell somas. The disappearance of MBP mRNA from the processes of the colchicine-treated oligodendrocytes could be due to degradation of the mRNA in the processes or to retraction of the mRNA into the cell somas. Cytochalasin B treatment did not alter the distribution of MBP mRNA in oligodendrocytes. These results suggested that microtubules, and not microfilaments, are involved in the translocation of MBP mRNA. Microtubules have been shown to be involved in the transport of vg1 mRNA (Yisraeli et al., 1990), bicoid mRNA (Pokrywka and Stephenson, 1991) and cyclin B mRNA (Raff et al., 1990). Other RNAs such as rpL32, alpha actin, gamma actin and histone H4 mRNAs have been found associated with cytoskeleton, but not with microfilaments in differentiating L6 myoblasts (Bagchi et al., 1987).

In mixed glial cultures, MBP mRNA translocation occurred in 20-25% of the oligodendrocytes (Amur-Umarjee et al., 1990a) even though the mRNA was translocated into the processes of almost all enriched oligodendrocytes (see Fig 1A). These results suggested to us that astrocytes might suppress the translocation of MBP mRNA since astrocytes are the predominant cell type in the mixed glial cultures. On adding enriched oligodendrocytes to astrocyte cultures, we observed that astrocytes did inhibit translocation of MBP mRNA in oligodendrocytes. These findings are consistent with those of Rosen et al. (1989) that astrocytes inhibit myelination by oligodendrocytes *in vitro*, since an inhibition of MBP mRNA translocation might be expected to inhibit myelination Inhibition of translocation by astrocytes was not due to factors secreted by astrocytes since astrocyte-conditioned medium did not influence the distribution of MBP mRNA. The astrocyte matrix also did not inhibit the MBP mRNA distribution, even though fixed astrocytes inhibited the movement of the mRNA. These results suggest that surface components of astrocytes are responsible for the inhibition of MBP mRNA translocation.

The essentially complete translocation of CNPase into the processes of enriched oligodendrocytes is interesting since translocation of these mRNAs occurs rarely in oligodendrocytes in mixed cultures or *in vivo*. It suggests that CNPase mRNA translocation may also be modulated by influences external to the oligodendrocyte, like the translocation of the MBP mRNAs. The effects of colchicine on the translocation of the two mRNAs, however, indicate that the mechanism by which they are translocated are quite different, as well as the external influences that modulate the translocation of these two sets of mRNAs.

ACKNOWLEDGEMENTS

This research was supported by NIH grants NS23022, NS23322 and HD25831 and National Multiple Sclerosis Society grant RG2233A1 to ATC.

REFERENCES

Amur-Umarjee, S.G., Hall, L., and Campagnoni, A.T., 1990a, Spatial distribution of mRNAs for myelin proteins in primary cultures of mouse brain. Dev. Neurosci. 12:263-272.

Amur-Umarjee, S.G., Dasu, R., and Campagnoni, A.T., 1990b, Temporal expression of myelin-specific components in neonatal mouse brain cultures: Evidence that 2,3-cyclic nucleotide 3-phosphodiesterase appears prior to galactocerebroside. Dev. Neurosci. 12:251-262.

Bagchi, T., Larson, D.E., and Sells, B.H., 1987, Cytoskeletal association of muscle-specific mRNAs in differentiating L6 rat myoblasts. Exp. Cell Res. 168:160-172.

Bernier, L., Alvarez, F., Norgard, E.M., Raible, D.W., Mentaberry, A., Schembri, I.G., Sabatini, D.D. and Colman D.R., 1987, Molecular cloning of a 2,3-cyclic nucleotide 3-phosphodiesterase: mRNAs with different 5-ends encode the same set of proteins in nervous and lymphoid tissues. J. Neurosci. 7: 2703-2710.

Dyer, C.A., and Benjamins, J.A., 1989, Organization of oligodendroglial membrane sheets. I: Association of myelin basic protein and 2,3-cyclic nucleotide 3-phosphohydrolase with cytoskeleton. J. Neurosci. Res. 24:201-211.

Hesketh, J.E., and Pryme, I.F., 1991, Interaction between mRNA, ribosomes and the cytoskeleton. Biochem. J. 277:1-10.

Jordan, C., Friedrich, V., Jr., and Dubois-Dalcq, M., 1989, *in situ* hybridization analysis of myelin gene transcripts in developing mouse spinal cord. J. Neurosci. 9:248-257.

Lawrence, J. B., and Singer, R.H., 1986, Intracellular localization of messenger RNAs for cytoskeletal proteins. Cell 45:407-415.

Lenk, R., Ransom, L., Kaufmann, Y., and Penman, S., 1977, A cytoskeletal structure with associated polyribosomes obtained from HeLa cells. Cell 10:67-78.

Miyamoto, K., Kumagai, H., and Imazawa, M., 1984, The maturation of oligodendrocytes and the accumulation of myelin basic protein in oligodendrocytes; in Tsukada Y (ed): Research report from the national center of Neurology and Psychiatry of the Ministry of Health and Welfare, Japan. 79-84.

Omlin, F.X., Webster H. de F., Palkovits, C.G., and Cohen, S.R., 1982, Immunocytochemical localization of basic protein in major dense line regions of central and peripheral myelin. J. Cell Biol. 95: 242-248.

Pokrywka, N.J., and Stephenson, E.C., 1991, Microtubules mediate the localization of bicoid mRNA during Drosophila oogenesis. Development. 113:55-66.

Raff, J.W., Whitfield, W.G.F., and Glover, D.M., 1990, Two distinct mechanisms localise cyclin B Transcripts in syncytial Drosophila embryos Development. 110:1249-1261.

Reidl, L.S., Campagnoni, C.W. and Campagnoni, A.T. (1981) Preparation and properties of an immunosorbent column specific for the myelin basic protein. J. Neurochem. 37:373-380.

Rome, L.H., Bullock, P.N., Chiappelli, F., Cardwell, M., Adinolfi, A.M., and Swanson, D., 1986, Synthesis of a myelin-like membrane by oligodendrocytes in culture. J. Neurosci. Res. 15:49-65.

Rosen, C.L., Bunge, R.P., Ard, M.D., and Wood, P.M., 1989, Type 1 astrocytes inhibit myelination by adult rat oligodendrocytes *in vitro*. J. Neurosci. 9:3371-3379.

Roth, H.J., Kronquist, K., Pretorius, P.J., Crandall, B.F., and Campagnoni, A.T., 1986, Isolation and characterization of a cDNA coding for a novel human 17.3 kDa myelin basic variant. J. Neurosci. Res. 16:227-238.

Steward, O., and Banker, G.A., 1992, Getting the message from the gene to the synapse: sorting and intracellular transport of RNA in neurons. TINS 15:180-186.

Sundell, C.L. and Singer, R.H., 1991, Requirement of microfilaments in sorting of actin messenger RNA. Science 253:1275-1277.

Suzumura, A., Bhat, S., Eccleston, P.A., Lisak, R.P., and Silberberg, D.H., 1984, The isolation and long-term culture of oligodendrocytes from newborn mouse brain. Brain Res. 324:379-383.

Trapp, B.D., Moench, T., Pulley, M. Barbosa, E., Tennekoon, G., and Griffin, J., 1987, Spatial segregation of mRNA encoding myelin-specific proteins. Proc. Natl. Acad. Sci. USA. 84:7773-7777.

Trapp, B.D., Bernier, L., Andrews, S.B., and Colman, D.R., 1988, Cellular and subcellular distribution of 2',3'-cyclic nucleotide-3'-phosphodiesterase and its mRNA in the rat central nervous system. J. Neurochem. 51:859-868.

Van Venrooij, W.J., Sillekens, P.T.G., van Eekelen, C.A.G., and Reinders, R.J., 1981, On the association of mRNA with cytoskeleton in uninfected and adenovirus-infected human cells. Exp. Cell Res. 135:79-91.

Verity, A.N., and Campagnoni, A.T., 1988, Myelination and its underlying mechanisms: Regional expression of myelin protein genes in the developing mouse brain: *in situ* hybridization studies. J.Neurosci.Res. 21:238-248.

Vogel, U.S., Reynolds, R., Thompson, R.J., and Wilkins, G.P., 1988, Expression of the 2,3-cyclic nucleotide 3-phosphohydrolase gene and immunoreactive protein in oligodendrocytes as revealed by *in situ* hybridization and immunofluorescence. Glia 1:184-190.

Wilson, R., and Brophy, P.J., 1989, Role for the oligodendrocyte cytoskeleton in myelination. J. Neurosci. Res. 22:439-448.

Yisraeli, J.K., Sokol, S., and Melton, M., 1990, A two-step model for the localization of maternal mRNA in Xenopus oocytes: Involvement of microtubules and micro-filaments in the translocation and anchoring of Vg1 mRNA. Development. 108:289-298.

THE CYTOSKELETON IN THE DIFFERENTIATION OF MYELIN-FORMING CELLS

Peter J. Brophy, C. Stewart Gillespie, Bernadette M. Kelly, and Demetrius A. Vouyiouklis

Department of Biological and Molecular Sciences
University of Stirling
Stirling FK9 4LA
Scotland, UK

INTRODUCTION

The ensheathment of nerve fibres by oligodendrocytes and Schwann cells requires a dramatic change in cell morphology. Not only must these cells extend long processes in order to insulate axons but they must also migrate considerable distances to reach their targets. Since interactions between the cytoskeleton and the plasma membrane are known to control cell shape and movement, the cellular cytoskeleton must play a fundamental role in myelination.

In most eukaryotic cells the cytoskeleton consists of microtubules (diameter 25 nm), microfilaments (diameter 5-7 nm) and intermediate filaments (IFs), (diameter 10 nm). In addition a growing number of proteins have been demonstrated to interact with and influence the structure of the cytoskeleton. For example, over 60 different actin-binding proteins alone can modulate the structure and dynamics of the microfilament system.

We are investigating how the cytoskeletons of oligodendrocytes and Schwann cells regulate the formation of myelin processes. Here we describe the developmentally regulated expression of cytoskeletal proteins in oligodendrocytes and Schwann cells, and we discuss the significance of these modifications in cytoskeletal composition to the morphological differentiation of myelin-forming cells.

THE CYTOSKELETON OF OLIGODENDROCYTES

Oligodendrocytes have a remarkable capacity for myelin membrane synthesis. A single cell can myelinate many nerve fibres simultaneously because it can assemble a complex system of processes that extends from the cell body[1]. Outgrowth of these processes coincides with the terminal differentiation of oligodendrocyte precursors to mature myelin-forming oligodendrocytes[2,3]. Unlike their progenitors, oligodendrocytes do not contain intermediate filaments[4], consequently they must rely on microtubules and microfilaments for the morphological reorganisation required for the formation and extension of myelin processes. Microtubules are abundant in the main body of these processes whereas microfilaments are concentrated in the finer processes at the periphery of the cell (Fig.1)[5]. Interestingly, this complementary distribution is reminiscent of the way the cytoskeleton is organised in growing axons.

In spite of the fact that cytoskeleton-binding proteins are known to affect the structure and dynamics of microfilaments and microtubules, little attention has been paid so far to their role in oligodendrocyte differentiation. In contrast, considerable work has been done on the role of microtubules and their associated proteins (MAPs) in the development of neuronal polarity[6]. Direct evidence for the role of MAPs in the development of neuronal morphology has been adduced from experiments in which the expression of MAP, tau and MAP2 has been attenuated by the introduction of antisense oligonucleotides and antisense RNA respectively [7,8].

MAP1B (also known as MAP5 and MAP1x) is a protein of 255kD that is closely associated with two smaller polypeptides known as light chains 1 and 3. MAP1B is found in association with microtubules at a very early stage of neurite extension, which has led to the proposal that it has a role in neurite growth[9,10]. This view is supported by the fact that MAP1B is down-regulated in all maturing neurons with the exception of those that can reinnervate in the adult nervous system, such as olfactory neurons[11]. Thus MAP1B appears to be fundamentally involved in the early stages of axonal growth. MAP1B is present in oligodendrocytes[12], therefore it was of great interest to us to determine if its expression coincided with the formation of myelin processes.

MAP1B Expression Precedes Terminal Differentiation of Progenitors into Complex Process-Bearing Oligodendrocytes

We have used a culture system in which oligodendrocyte progenitors are isolated from neonatal rat cerebral hemispheres. This population of cells is expanded with the mitogen basic fibroblast growth factor (bFGF) and the cells are then allowed to differentiate by withdrawal of bFGF[13]. Surface expression of the sphingolipid GalC marks differentiated oligodendrocytes upon their development from progenitors and these cells possess MAP1B (Fig 2). In contrast to oligodendrocytes, astrocytes, the other major CNS macroglia, have little detectable MAP1B except for perhaps in some regions of their fibrous processes. Adjacent to the astrocyte in Figure 2C is shown a rare flat fibroblast-like cell that does possess MAP1B, although at a much reduced level of expression in comparison with oligodendrocytes in the same cultures.

MAP1B is not detectable immunocytochemically in oligodendrocyte progenitors (data not shown) despite the abundance of MAP1B in oligodendrocytes. However, upon acquisition of the differentiation marker O4, which is at a stage immediately before preoligodendrocytes acquire the terminal differentiation marker GalC, MAP1B is detected

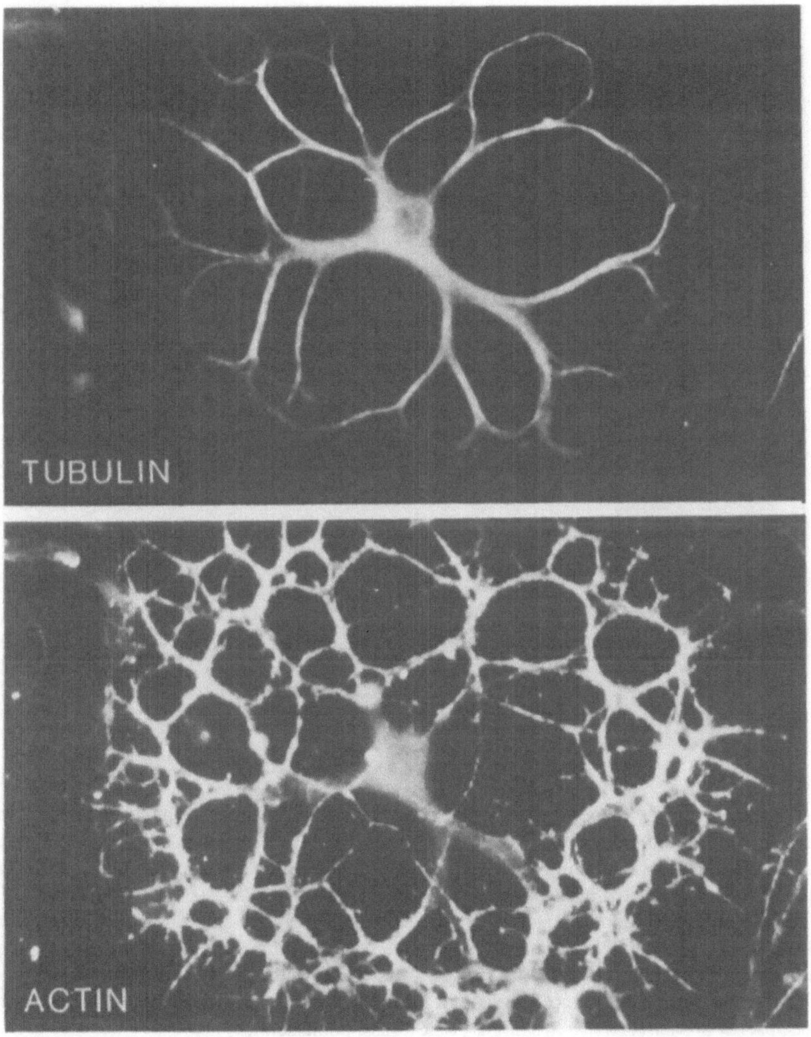

Fig 1. Actin and tubulin are localised to different regions of the myelinating oligodendrocyte. Actin is predominantly present in the peripheral processes whereas tubulin is concentrated in the major processes as they emerge from the cell body.

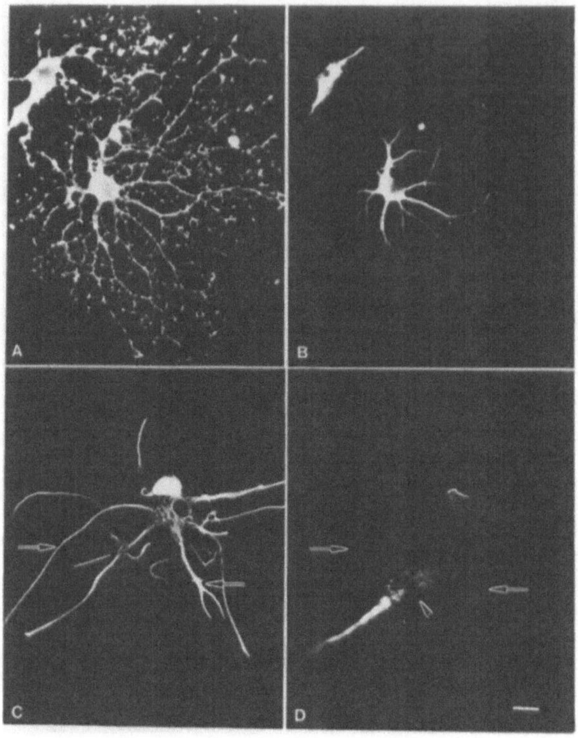

Fig. 2. MAP1B is abundantly expressed in the cell body and major processes of mature oligodendrocytes. O-2A progenitors were stimulated to divide in the presence of bFGF and then allowed to differentiate to oligodendrocytes in medium containing 1% FCS without bFGF. After six days in culture oligodendrocytes with complex, well developed processes appear, which are strongly GalC$^+$(FITC) (A); these cells express MAP1B (TRITC) in their cell bodies and major processes (B). There are small numbers of astrocytes which express GFAP (FITC) (C), however only faint MAP1B staining is detectable in some of their processes (arrows in C and D). Very rare (< 1%) flat fibroblast-like cells which are GFAP$^-$ (C) also express MAP1B (D, arrowhead). Bar, 5 μm. Reproduced by copyright permission from Vouyiouklis and Brophy, 1993, (in press) in the *Journal of Neuroscience Research,* (Wiley-Liss, publishers).

in O4$^+$ cells. From D0 to D4 of differentiation after withdrawal of bFGF there are always more MAP1B$^+$ than GalC$^+$ cells (Fig. 3), and at all stages of oligodendrocyte differentiation the GalC$^+$ cells are MAP1B$^+$. By D4 80% of the cells express MAP1B, which corresponds to the number of MBP$^+$ oligodendrocytes that ultimately develop in these cultures. As oligodendrocyte precursors mature to oligodendrocytes their major processes increase in diameter and terminate in a myriad of fine processes. This morphological transition is initiated before GalC expression and coincides with the movement of MAP1B to the major outgrowths from the cell body. Subsequently, MAP1B is consolidated in the major processes of GalC$^+$ oligodendrocytes (Fig. 4).

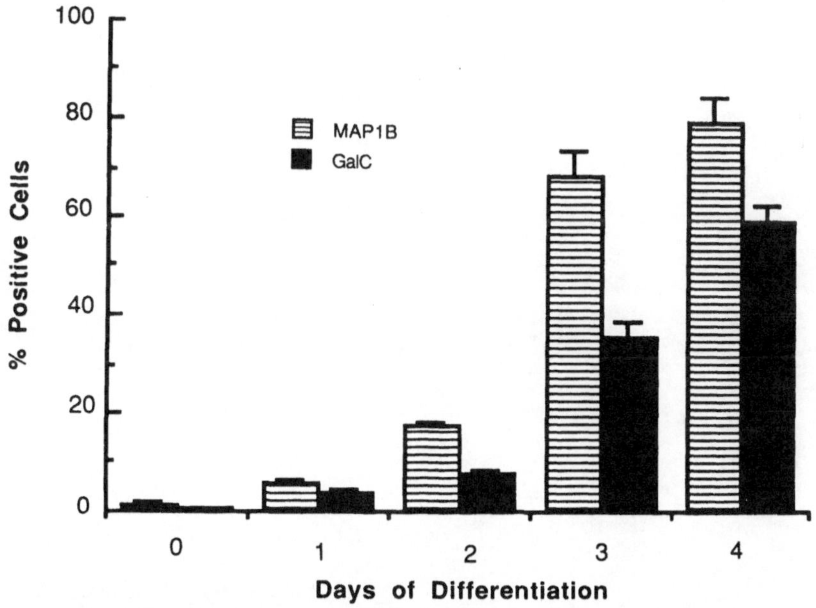

Fig. 3. Developmental expression of MAP1B in differentiating oligodendrocytes. Oligodendrocyte progenitors were allowed to differentiate and the developmental appearance of MAP1B and GalC was followed by indirect immunofluorescence using anti-IgG1-TRITC (specific for anti-MAP1B) and anti-IgM-FITC (specific for anti-GalC) secondary antibodies. Cell counts were from ten randomly selected fields of approximately 700 cells per duplicate coverslip in three different experiments, data are means ± SEM. All GalC$^+$ oligodendrocytes were MAP1B$^+$. Reproduced by copyright permission from Vouyiouklis and Brophy, 1993, (in press) in the Journal of Neuroscience Research, (Wiley-Liss, publishers).

Major reorganisation of the cytoskeletal network must play a fundamental role in preparing the myelin-forming cell for the extraordinary changes in shape necessitated by the extension of myelinating processes and the envelopment of axons. Microtubules and their MAPs are likely to have an important role in the development of this characteristic process-bearing phenotype in the vertebrate CNS. The induction of MAP1B expression coincides with the transition of progenitors to preoligodendrocytes, which is a stage in their development that immediately precedes terminal differentiation. These observations show that MAP1B expression appears to be fundamentally related to the morphological

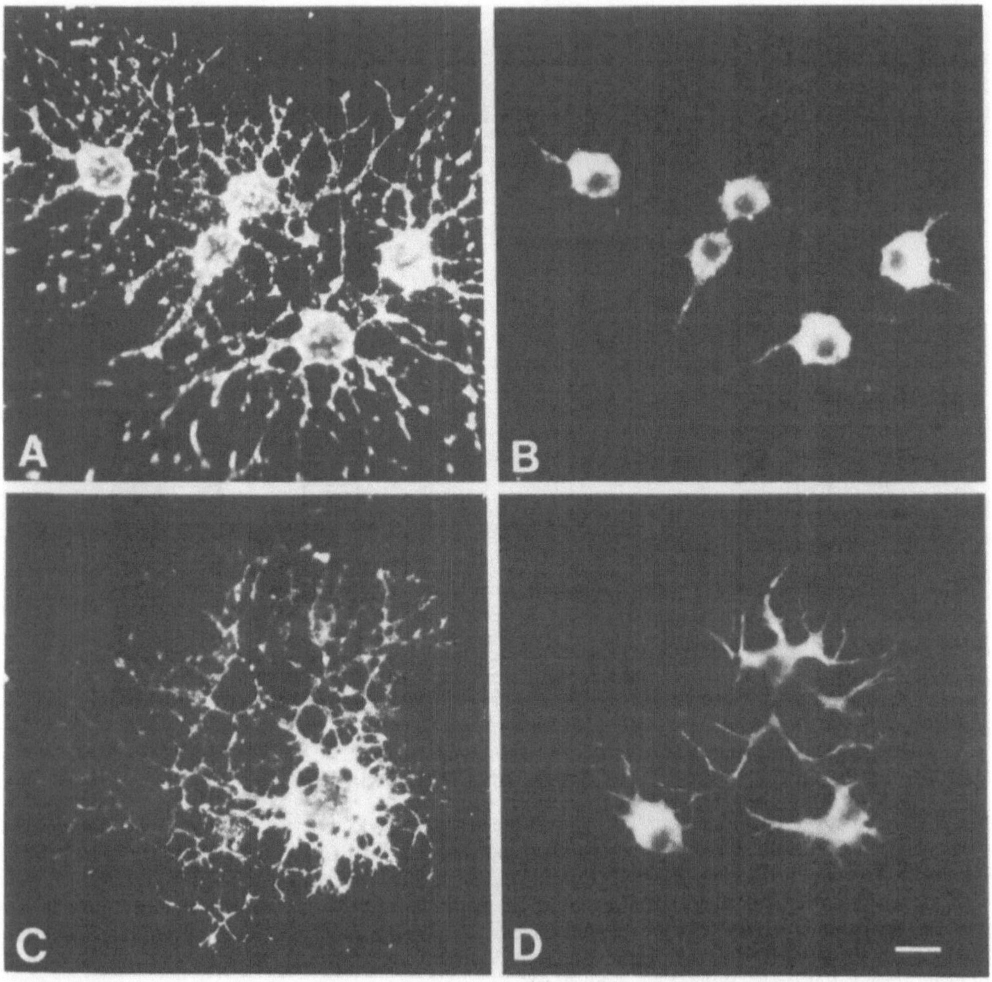

Fig. 4. O4+ preoligodendrocytes (A) express MAP1B in their cell bodies (B), whereas in more mature oligodendrocytes expressing GalC (C) MAP1B is redistributed to the major processes (D).

reorganization of the oligodendrocyte[13]. MAP1B is known to be capable of forming cross-bridges between microtubules[14] thus it could provide the structural support necessary for the stabilization of large complex outgrowths from the oligodendrocyte's cell body.

INTERMEDIATE FILAMENT PROTEIN EXPRESSION IN DIFFERENTIATING SCHWANN CELLS

IFs are morphologically similar from cell to cell but show a wide heterogeneity in their polypeptide subunits. All IF proteins share common structural features: they are rod-shaped and have a large, highly conserved central α-helical core domain of approximately 310 residues which is flanked by two nonhelical terminal domains of variable length and sequence at the amino- and carboxy- terminal regions. These proteins have been divided into six classes: types I and II (keratins in epithelial cells), type III (vimentin, desmin, glial fibrillary acidic protein (GFAP), peripherin), type IV (α-internexin and the neurofilaments NF-L, NF-M and NF-H), type V (nuclear lamins) and type VI (nestin)[15]. Since particular IF proteins are often expressed in specific cell types, they have been used as differentiation markers and as such are useful tools in the study of cell differentiation as well as in tumour identification[16].

Despite extensive knowledge of their amino acid and cDNA sequences, the physiological function of most IF proteins is unknown. Their linkage to both nucleus and plasma membrane suggests a general role in the spatial organization of the cytoplasm; alternatively, they may also be involved either in the transport of macromolecules between the nucleoplasmic and cytoplasmic compartments or in the transduction of information from the cell periphery to the nucleus[17].

IFs are commonly remodelled during cell maturation, which suggests that the IF network has a role in differentiation. Thus vimentin is exchanged for desmin in the IFs of developing muscle cells[18] and there are dramatic changes in the composition of IF networks during neuronal development. Neuroepithelial stem cells initially coexpress nestin and vimentin[19]; subsequently nestin is down-regulated and the type IV IF protein α -internexin appears [20]. α-Internexin is in turn replaced by the type IV proteins characteristic of mature neurons (NF-L, NF-M and NF-H)[15]. IFs in Schwann cells are also remodelled during development in a manner which is dependent on whether they will become myelin-forming or non-myelin-forming cells. GFAP expression is believed to be suppressed upon terminal differentiation in myelin-forming Schwann cells, whilst it is retained in non-myelin-forming Schwann cells[21]. Our interest in the IF network of peripheral nervous system glia sprang from a desire to determine if other IFs took the place of GFAP during the the development of myelinating Schwann cells.

Differentiated Schwann Cells Express NF-M

Schwann cells were extracted with a buffer containing Triton X-100[5]; the resultant cytoskeleton and soluble fractions were electrophoresed on SDS-PAGE and the separated proteins were analyzed by Western blotting. Immunoblotting with an antibody raised against a 145kD protein (p145) and rabbit anti-vimentin demonstrated that, like vimentin, all of the p145 remains associated with the cytoskeleton fraction and that the anti-p145 specifically recognizes only p145 in Schwann cell cytoskeletons; the vimentin antibody was also specific (Fig. 5). The presence of p145 in Schwann cell cytoskeletons was also demonstrated by indirect immunofluorescence (Fig. 6). The isolation and sequencing of a full-length clone from a Schwann cell cDNA library established the identity of p145 with the neurofilament protein NF-M (data not shown)[22].

When Schwann cell cytoskeletons were prepared and double-labelled with anti-NF-M and vimentin mAb followed by gold-conjugated anti-rabbit IgG and anti-mouse IgG secondary antibodies (10 nm and 5 nm respectively), the two different sized gold particles decorated the same 10 nm filaments indicating that NF-M and vimentin reside on the same IFs (Fig. 7A). In addition to containing IFs composed of both NF-M and vimentin, some Schwann cells contained IFs that were only labelled with antibodies against vimentin (Fig. 7C).

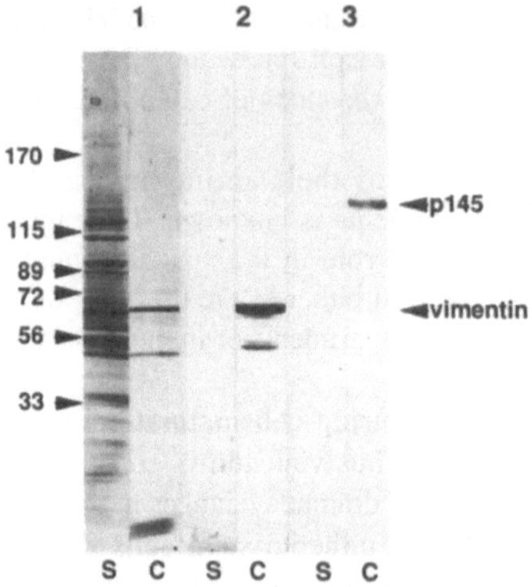

Fig. 5. Immunoblot of cytoskeleton (C) and soluble (S) fractions of Schwann cells (80 μg protein) from 5-d rat sciatic nerves as described in Materials and Methods. 1, proteins stained with Amido Black; 2, immunoblot with anti-vimentin; 3, immunoblot with anti-p145. Both vimentin and p145 were exclusively associated with the Schwann cell cytoskeleton and their respective antibodies were specific. Reproduced by copyright permission from Kelly et al., 1992, 118, pp397-410 in the *Journal of Cell Biology*, (Rockefeller University Press, publishers).

Developmental Expression of NF-M in Schwann Cells

The relationship between NF-M expression and Schwann cell development was investigated by comparing the expression of NF-M with that of well-characterized differentiation markers, O4, which appears at E16, and GalC, which is first detectable in myelin-forming Schwann cells at E18-19. After 16 h in culture, the cells were double-labelled for NF-M and GalC and for NF-M and O4. The percentage of cells in these cultures that were Schwann cells was determined by measuring the number of S100$^+$ cells in parallel cultures (87±1% at E18, 86±2% at P1 and 72±2% at P4). O4 expression increased from 69% of Schwann cells on E18 to 96% of Schwann cells on P4 (Fig. 8). The number of Schwann cells expressing NF-M also showed an increase with age from 29% on E18 to 58% on P4. The percentage of Schwann cells expressing GalC increased from 9% on E18, through 14% on P1 to 36% on P4. At P4, three GalC, NF-M phenotypes were observed: GalC$^+$,NF-M$^+$(50±3%), GalC$^+$,NF-M$^-$ (7±2%) and GalC$^-$,NF-M$^+$ (43±4%). Thus, 88% of the GalC$^+$ cells expressed NF-M, which indicates that NF-M is present in myelin-forming Schwann cells. Myelin synthesis begins in earnest on P1 and we found the expected rise in P_0 expression by Schwann cells from 31% on P1 to 56% on P4. Our values

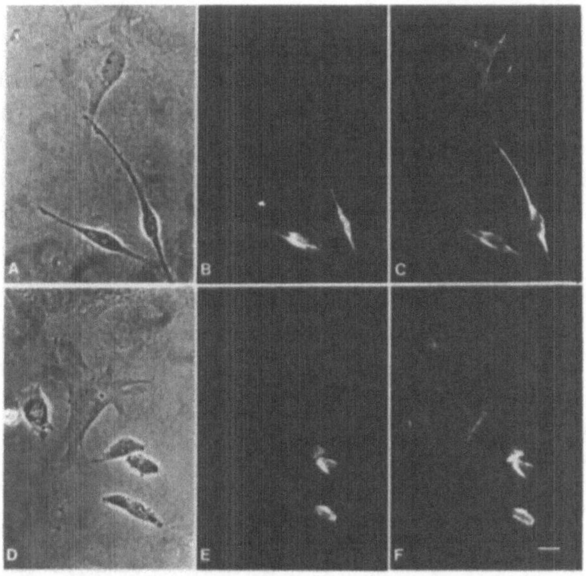

Fig. 6. Immunofluorescence localization of p145 in Schwann cell cytoskeletons. Cells were grown on poly-D-lysine coated plastic for 24 h prior to extraction. Cytoskeletons were viewed with *(A)* phase constrast optics and *(B)* rhodamine optics to visualize p145. Note the filamentous staining in the lower cell. Bar, 10 µm. Reproduced by copyright permission from Kelly et al., 1992, 118, pp397-410 in the *Journal of Cell Biology*, (Rockefeller University Press, publishers).

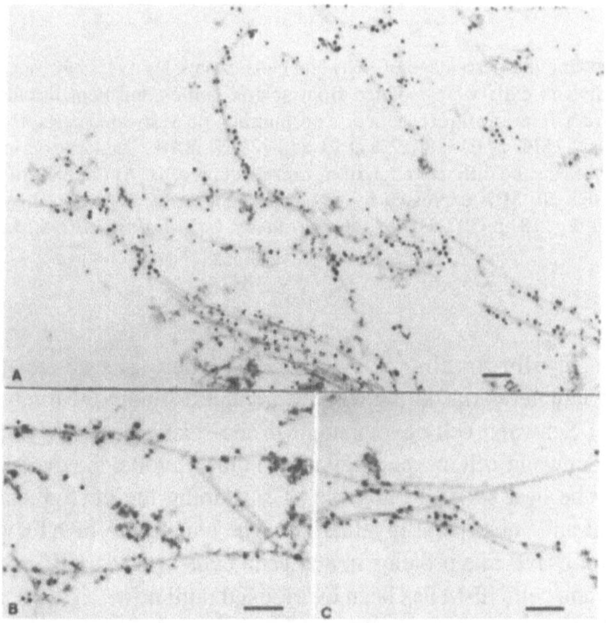

Fig. 7. Immunoelectron localization of NF-M and vimentin in Schwann cells. Cytoskeletons were immunostained with p145 antibody and vimentin mAb and then reacted with secondary antibodies that were labeled with colloidal gold particles. IFs were clearly immunostained for both NF-M (10 nm gold) and vimentin (5 nm gold) *(A* and *B)*. Schwann cell IFs composed only of vimentin which were not labelled with anti-p145 are shown in *C*. Bars 0.1µm. Reproduced by copyright permission from Kelly et al., 1992, 118, pp397-410 in the *Journal of Cell Biology*, (Rockefeller University Press, publishers).

for the observed expression of P_0 on P1 and P4 parallelled those for NF-M and support our argument that NF-M is a component of myelin-forming Schwann cells. Further work will be needed in order to delineate the expression of NF-M in non-myelin-forming cells.

The two IF proteins of Schwann cells that have been described so far are vimentin and GFAP. Vimentin is present in both differentiated and undifferentiated Schwann cells[23] and can be detected in Schwann cell cultures from E15 onwards[24]. GFAP is first detected on E18 but is believed to down-regulate as the myelinating phenotype develops[24]. It seems that NF-M

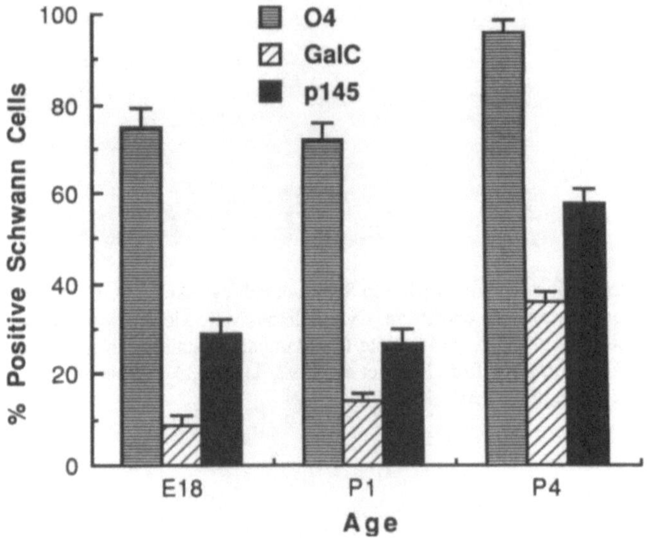

Fig. 8. Histogram showing the percentage of Schwann cells expressing O4, GalC, and p145 (NF-M) at ages E18, P1 and P4. Schwann cells were isolated from sciatic nerves and kept in culture for 16 h prior to processing for indirect immunofluorescence. The number of Schwann cells as determined by S100 immunoreactivity was: 87±1% at E18, 86±2% at P1 and 72±2% at P4. The increase in the number of $GalC^+$ cells from E18 to P4 reflects the differentiation of myelin-forming cells. At P4, 88% of $GalC^+$ cells expressed NF-M which shows that NF-M is expressed by myelin-forming cells. Reproduced by copyright permission from Kelly et al., 1992, 118, pp397-410 in the *Journal of Cell Biology*, (Rockefeller University Press, publishers).

replaces GFAP in myelin-forming Schwann cells. There are several precedents for the exchange of IF protein networks during cell differentiation and NF-M may replace GFAP at a very early stage of Schwann cell development in the sciatic nerve. Myelination necessitates major changes in Schwann cell morphology which must involve the cytoskeleton and the NF-M IF protein may be best suited to the role of sustaining the altered shape thus promoting axon ensheathment and subsequent myelination. The abundance of NFs in peripheral nerves makes the detection of NF-like proteins in Schwann cells *in vivo* a difficult task, which might explain why Schwann cell NF-M has been overlooked until now.

Although NF proteins are normally restricted to postmitotic neurons, NF-L has been detected in the the beta cells of the islets of Langerhans in the embryonic rat thus raising the possibility that these cells originate in the neural crest[25]. Furthermore, NF proteins have been incorporated into IF-like arrays in non-neuronal cells following transfection[26] and in transgenic mice[27]. Why Schwann cells should synthesize NF-M alone of the NF triplet remains to be elucidated, since these proteins are normally coordinately expressed in neurons.

ACKNOWLEDGEMENTS

This work has been supported by the Multiple Sclerosis Society of Great Britain in Scotland, the Wellcome Trust and the Science and Engineering Research Council.

REFERENCES

1. P.M. Wood and R.P.Bunge, The biology of the oligodendrocyte. In "Oligodendroglia" Norton WT ed: Plenum Press, NY pp 1-46 ,1984.
2. A.L. Gard and S.E. Pfeiffer, Oligodendrocyte progenitors isolated directly from developing telencephalon at a specific phenotypic stage: Myelinogenic potential in a defined environment. *Development* 106:119-132 1989.
3. A.L. Gard and S.E. Pfeiffer, Two proliferative stages of the oligodendrocyte lineage ($A2B5^+O4^-$ and $O4^+GalC^-$) under different mitogenic control. *Neuron* 5:615-625, 1990.
4. M.C. Raff, Glial cell diversification in the rat optic nerve. *Science (Wash)* 243:1450-1455 1989.
5. R. Wilson and P.J. Brophy, Role for the oligodendrocyte cytoskeleton in myelination. *J. Neurosci. Res.* 22: 439-448 1989.
6. A. Matus. Microtubule-associated proteins: Their potential role in determining neuronal morphology, *Annu Rev Neurosci* 11:29-44 1988.
7. A. Caceres and K.S. Kosik, Inhibition of neurite polarity by tau antisense oligonucleotides in primary cerebellar neurons. *Nature* 343:461-463 1990.
8. J.H. Dinsmore and F. Solomon, Inhibition of MAP2 expression affects both morphological and cell division phenotypes of neuronal differentiation. *Cell* 64:817-826 1991.
9. R.P. Tucker, L.I. Binder and A. Matus, Neuronal microtubule-associated proteins in the embryonic avian spinal cord. *J Comp Neurol* 271:44-55, 1988.
10. R.P. Tucker, L.I. Binder, C. Viereck, B.A. Hemmings and A.I. Matus, The sequential appearance of low- and high-molecular-weight forms of MAP2 in the developing cerebellum. *J Neurosci* 8:4503-4512, 1988.
11. C. Viereck, R.P. Tucker and A. Matus, The adult rat olfactory system expresses microtubule-associated proteins found in the developing brain. *Neuroscience* 9:3547-3557, 1989.
12. I. Fischer , Konola and E. Cochary, Microtubule associated protein (MAP1B) is present in cultured oligodendrocytes and colocalizes with tubulin. *J Neurosci Res* 27:112-124, 1989.
13. R. Sato-Yoshitake, Y. Shiomura, H. Miyasaka and N. Hirokawa, Microtubule-associated protein 1B: molecular structure, localization, and phosphorylation-dependent expression in developing neurons. *Neuron* 3:229-238, 1989.
14. D. A. Vouyiouklis and P. J. Brophy, Microtubule-Associated Protein MAP1B expression precedes the morphological differentiation of oligodendrocytes. *J. Neurosci. Res.*, (in press).
15. P.M. Steinert and R.K.H. Liem, Intermediate Filament Dynamics. *Cell.* 60: 521-523, 1990.
16. M. Osborn and K. Weber, Biology of Disease. Tumour diagnosis by intermediate filament typing: a novel tool for surgical pathology. *Lab. Invest.* 48:3372-3394, 1983.
17. B. Geiger, Intermediate filaments. Looking for a function. *Nature (Lond.).* 329: 392-393, 1987.
18. E. Lazarides, Intermediate Filaments: a chemically heterogeneous, developmentally regulated class of proteins. *Ann. Rev. Biochem.* 51: 219-250, 1982.
19. U. Lendahl, L.B. Zimmerman and R.D.G. McKay, CNS stem cells express a new class of intermediate filament protein. *Cell*, 60: 585-595, 1990.
20. K.H. Fliegner, G.Y. Ching and R.K.H. Liem, The predicted amino acid sequence of a- internexin is that of a novel neuronal intermediate filament protein. *EMBO (Eur. Mol.Biol.Organ.) J.* 9: 749-755, 1990.
21. K.R. Jessen and R. Mirsky, Schwann cell precursors and their development. *Glia* 4: 18-194, 1991.
22. B.M. Kelly, C.S. Gillespie, D.L. Sherman and P.J. Brophy, Schwann cells of the myelin-forming phenotype express neurofilament protein NF-M. *J. Cell Biol.*, 118, 397-410, 1992.
23. L.J. Autilio-Gambetti, J. Sipple, O. Sudilovsky, and P. Gambetti, Intermediate filaments of Schwann cells. *J.Neurochem.* 38: 774-780, 1982.
24. K.R. Jessen, L. Morgan, H.J.S. Stewart, and R. Mirsky,. Three markers of adult non-myelin-forming Schwann cells, 217c (Ran-1), A5E3 and GFAP: development and regulation by neuron-Schwann cell interactions. *Development.* 109: 91-103, 1990.
25. M.K. Escurat, K. Djabali, C. Huc, F. Landon, C. Bécourt, C. Boitard, F. , Gros, and M-M. Portier, Origin of the beta cells of the islets of Langerhans is further questioned by the expression of neuronal intermediate filament proteins, peripherin and NF-L, in the rat insulinoma RIN5F cell line. *Dev. Neurosci.* 13:424-432, 1991.

26. M.J. Monteiro and D.W. Cleveland, Expression of NF-L and NF-M in fibroblasts reveals coassembly of neurofilament and vimentin subunits. *J. Cell Biol.* 108: 579-593, 1989.
27. M.J. Monteiro, P.N. Hoffman, J.D. Gearhart, and D.W. Cleveland, Expression of NF-L in both neural and nonneural cells of transgenic mice: increased neurofilament density in axons without affecting caliber. *J. Cell Biol.* 111: 1543-1557, 1990.

REGULATION OF BRAIN STEAROYL CoA DESATURASE mRNA EXPRESSION DURING MYELINATION

James DeWille

The Ohio State University
Department of Veterinary Pathobiology
1925 Coffey Rd.
Columbus, OH 43210-1093

INTRODUCTION

Myelin is a unique membrane system found only in nervous tissue[1-7]. Myelin is highly enriched in lipids (75% lipids, 25% protein)[1,7]. The lipid content of myelin is composed almost entirely of structural lipids, i.e., phospholipids (45%), galactolipids (35%) and cholesterol (30%)[7]. Although myelin is composed mainly of lipid, little attention has been focused on the regulation of genes encoding lipogenic enzymes, transport proteins or receptors during the neonatal myelinating period [6]. A better understanding of the regulation of genes involved in lipid biosynthesis in myelination is important for several reasons.

Myelin is synthesized almost exclusively during early neonatal life[1-7]. The developmental signals directing the synthesis and assembly of myelin are poorly understood[2,3,6]. The coordinated expression of genes encoding myelin-specific proteins, lipid biosynthetic enzymes and transport proteins is required for the synthesis and assembly of the myelin membrane[2,3,6]. The mechanism underlying this coordination of gene expression has not been elucidated. In fact, there is little direct evidence to explain how this coordination occurs at the gene level[3,6]. Thus, fundamental questions regarding the basic mechanisms underlying myelin synthesis are unanswered.

In addition to providing answers to questions regarding developmental regulation, a better understanding of myelination is needed to design strategies to intervene in demyelinating conditions. Demyelinating conditions of genetic, viral and mechanical etiology may be more effectively treated if the basic mechanisms of myelination were better understood. Recent evidence suggests that oligodendrocytes may interfere with neuronal regeneration following spinal cord injury[8]. A better understanding of gene expression in oligodendrocytes could lead to more effective clinical management in of spinal cord injury[8].

A significant number of inherited neuropathies can be traced to disorders of lipid metabolism[9]. Phytanic acid storage diseases, lipoprotein deficiencies (Tangier and Bassen-

A Multidisciplinary Approach To Myelin Diseases II
Edited by S. Salvati, Plenum Press, New York, 1994

Korzweig diseases), Fabry's disease, the leukodystrophies (see below), Niemann-Pick disease, Gaucher's disease and the Gangliosidoses all exhibit abnormalities in lipid metabolism that result in abnormal nervous system function[9]. In many of these disorders the defective gene has been identified, but clinical management is of limited effectiveness. A better understanding of the mechanisms controlling lipid related genes in the nervous system should prove useful in defining disease mechanisms and devising treatment strategies in this unique class of disorders.

There is now evidence demonstrating that specific dietary fatty acids can slow the progression of certain demyelinating conditions[10,11]. Oils providing high levels of long chain monounsaturated fatty acids (i.e., oleic acid C18:1, erucic acid C22:1) reduce plasma C26:0 levels in individuals with X-linked adrenoleukodystrophy (ALD)[11]. This oil (Lorenzo's oil) provides about 20% of total calories from a blend that contains 18% C22:1, 76% C18:1, 3% C24:1, 2% C20:1[11].

The biochemical lesion in ALD has been traced to a defect in long chain fatty acid oxidation[10,11]. The defect has been localized to a peroxisomal long chain acyl CoA synthetase required for fatty acid activation and oxidation[10,11]. The consequence of an inability to oxidize long chain fatty acids is the accumulation of long chain saturated fatty acids (C24-30:0) in nervous tissue. Although the subsequent steps are not well understood, long chain fatty acid accumulation is followed by progressive demyelination and neurological deterioration.

These results suggest the exciting possibility that specific fatty acids may have significant therapeutic benefit in demyelinating conditions. The potential role of diet in the amelioration of demyelinating diseases, however, has not been adequately evaluated. This is due, in large part, to the lack information regarding the influence of diet on neural fatty acid metabolism in general and myelin lipid metabolism in particular.

MYELIN SYNTHESIS IS INFLUENCED BY DIETARY FAT

All aspects of development are ultimately determined by the timely expression of specific genes. The primary function of the neonatal diet is to provide the optimal ratios of essential nutrients to insure an environment for the appropriate induction of developmentally regulated genes. Lack of essential nutrients, or improper ratios of nutrients, may influence developmental gene expression with long term deleterious effects.

The neonatal diet can alter myelination[5-7,12-17]. Since myelination is restricted to a limited time in neonatal life, environmental factors inhibiting myelin formation may have long lasting effects[5].

Dietary fat is known to influence the expression of genes involved in carbohydrate and fatty acid metabolism in liver and adipose tissue[18-20]. Many of these genes are regulated, at least in part, at the level of transcription[20]. This suggests a link between dietary fat and transcriptional regulation of specific genes in the liver and adipose tissue[20]. If dietary fat exerted similar effects on genes implicated in fatty acid synthesis during neonatal myelin biosynthesis, then diet could significantly alter myelination.

Understanding the effects of dietary fat on myelin-specific and lipid-related gene expression in the developing CNS is of great practical significance. The initial neonatal source of dietary fat is breast milk. In humans, commercial infant formula often replaces breast milk and serves as a relatively sole source of nutrition until 3-6 months of life[18]. This early time in human development coincides with active myelination of the human CNS[21]. No commercial infant formula fat blend matches the human milk fatty acid profile[18,22,23]. Human milk provides long chain (>18 carbon) polyunsaturated fatty acids (PUFA) of both the n-6 and n-3 series[22,23]. Since commercial infant formulas use vegetable oils as a fat source, PUFA with >18 carbons are absent[22,23]. To compensate, the content of PUFA in formulas may be as high as 30% of calories, compared with 7-17% in human milk[22,23]. The significance of wide variations in PUFA intake on neural development and myelin synthesis at the molecular level has not been adequately investigated. However, most infants thrive on commercial infant formulas. One exception is premature infants in which elongating and desaturating enzymes may be limiting. Long chain PUFA (>18 carbons) and possibly oleic acid may be dietary essentials for these infants[23].

Although numerous reports indicate that diet can alter tissue fatty acid composition it important to note that tissues of similar fatty acid composition (i.e., brain and liver) can exhibit divergent mRNA profiles[6]. This is true even for genes that encode enzymes that carry out the same catalytic function in both tissues[6]. This indicates that tissue fatty acid profiles alone are a poor indicator of the relationship between dietary fat and tissue fatty acid metabolism and gene expression.

STEAROYL CoA DESATURASE: RATE-LIMITING STEP IN OLEIC ACID BIOSYNTHESIS

Oleic acid (C18:1, n-9) is the most prevalent fatty acid in myelin[7]. The rate limiting step in the endogenous synthesis of oleic acid from acetate is Stearoyl CoA desaturase[24]. Stearoyl CoA desaturase (SCD) catalyzes the desaturation of stearic acid (C18:0) in the 9 position to form oleic acid (C18:1, n-9)[24]. Myelin oleic acid content is about twofold brain synaptosomal oleic acid content and about 4-10 fold higher that liver phospholipid oleic acid content[25]. Oleic acid is derived from diet, or through endogenous synthesis, via the SCD reaction[6].

If SCD synthesis were induced in nervous tissue during myelination, then endogenous synthesis of oleic acid could provide a key myelin structural component. A brain biosynthetic pathway for oleic acid would necessarily be regulated in a manner that is coincident with neonatal myelination and divergent from hepatic oleic acid biosynthesis. That is, brain oleic acid biosynthesis should respond to developmental regulation, i.e. increase during myelination, whereas liver oleic acid biosynthesis would respond directly to diet.

Recent work by Lane and coworkers established the presence of two SCD structural genes in the mouse[26,27]. These two SCD genes differ markedly in promoter region sequence and tissue specific expression, but encode proteins that share >87% amino acid sequence identity. Mouse SCD1 contains a typical "TATA" box in the immediate 5' flanking region and is expressed in liver as well as other tissues. Mouse SCD2, which is expressed in brain, but not in liver, lacks a "TATA" box, but does have two "CAAT" boxes approximately 100 bp 5' to the transcription initiation site. One potential Sp1 binding site (GGGCGG) is present at position -175 relative to the transcription initiation site. The mouse SCD2 promoter contains at least one consensus CCAAT/enhancer binding protein (C/EBP) site[26,27].

Mouse brain SCD2 mRNA levels increase in a developmentally regulated fashion coincident with the neonatal myelinating period (Figure 1). It is important to recognize that the developmental increase in brain SCD2 mRNA levels from day 14 to day 21 of life occurs during a time when the nursing pup is consuming milk that is providing about 30% of fatty acids as oleic acid[22,27]. At the same time, hepatic SCD1 mRNA is undetectable (see Figure 3).

This developmentally induced increase in neonatal brain SCD2 mRNA levels, however, can be altered by neonatal fatty acid intake (Figure 2). Pups nursed by mothers fed an essential fatty acid deficiency (EFAD) diet during lactation exhibit reduced brain SCD2 mRNA levels, compared with pups nursed by mothers fed the control diet (Figure 2).

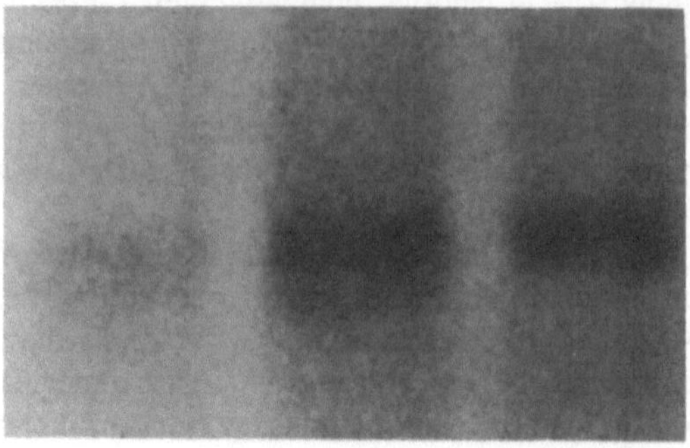

1 2 3

Figure 1. Developmental increase in mouse brain Stearoyl CoA desaturase2 (SCD2) mRNA levels. Poly A+ RNA was isolated from brains of neonatal mice (10 micrograms/lane) and probed with 32P-labelled mouse SCD2 cDNA insert. Lanes contain: (1) day 14 brain, (2) day 21 brain, (3) day 28 brain.

Brain SCD

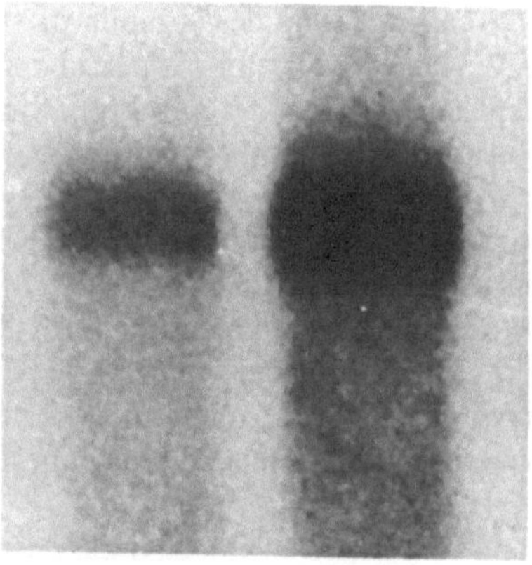

1 2

Figure 2. Limiting maternal essential fatty acid (EFA) intake reduces pup brain SCD2 mRNA levels. Poly A+ RNA was isolated from brains of 21 day old mice (10 micrograms/lane) and probed with 32P-labelled mouse SCD2 cDNA insert. Lanes contain: (1) brain, (reduced EFA intake), (2) brain (EFA adequate intake).

The hepatic response to the same diet is completely reversed. That is, the hepatic SCD1 mRNA response in the pups nursed by the EFAD mothers in markedly elevated (Figure 3). The hepatic SCD1 mRNA levels in control pups is essentially undetectable by Northern blot (Figure 3). Although dietary fat influences brain SCD2 and liver SCD1, mRNA levels, the SCD transcription initiation sites in both tissues are unaffected by diet[28].

The divergent effects of dietary fat on brain and liver SCD mRNA levels are consistent with previous reports. Reduced SCD2 mRNA levels, which suggests reduced myelination, is consistent with earlier studies demonstrating that neonatal myelination is impaired by limiting EFA intake[15,17]. Pups nursed by mothers fed EFAD diets exhibit reduced proteolipid protein (PLP) and myelin basic protein (MBP) mRNA levels and reduced brain myelin content[6,15-17]. In contrast, the marked induction of hepatic SCD1 by EFAD is also consistent with a large body of evidence demonstrating SCD1 induction and n-9 fatty acid accumulation in response to limiting n-6 fatty acid (EFA) intake[6,25-27]. These data demonstrate the unique tissue-specific regulation of the two mouse SCD structural genes by dietary fat.

The correlation between neonatal brain SCD2 mRNA expression and myelination is supported by results from quaking mice[29] (Figure 4). Reduced SCD2 mRNA levels in neonatal quaking brain parallels the reduction of proteolipid protein (PLP) and myelin basic protein (MBP) mRNA levels (Figure 4). These results support the coordinate expression of

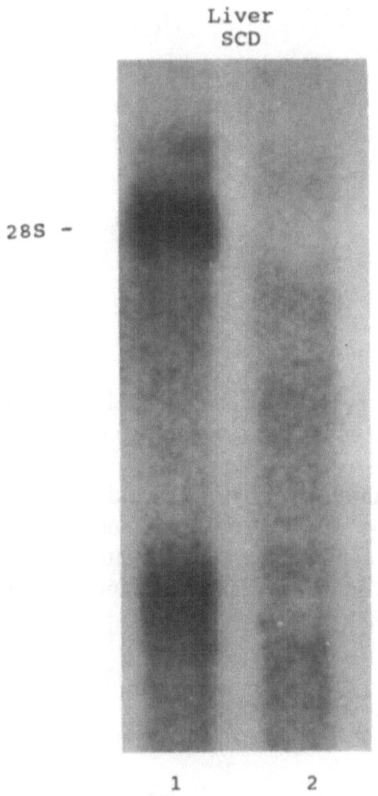

Figure 3. Limiting maternal essential fatty acid (EFA) intake induces pup hepatic SCD1 mRNA levels. Poly A+RNA was isolated from livers of 21 day old neonatal mice (10 micrograms/lane) and probed with 32P-labelled mouse SCD1 cDNA insert. Lanes contain: (1) liver, (reduced EFA intake), (2) liver, (EFA adequate intake).

myelin-specific protein encoding genes and SCD2 in neonatal brain. The mechanism by which these genes and their mRNAs are coordinately regulated during myelination, however, has not been elucidated.

One consistent feature of mylin mutants is that the absence of one structural protein (i.e. PLP or MBP) has a devastating effect on myelination. The sensitivity of neonatal myelin synthesis to EFAD is somewhat surprising as myelin is not highly enriched in EFAs[7,12-14]. It is possible that removing EFAs from the diet removes an essential myelin structural component much the same as removing an essential myelin structural protein by mutation or deletion[2,3]. Therefore, even though the EFA content of myelin may be relatively small, EFAs appear to play an essential role in the synthesis or assembly of myelin during neonatal life.

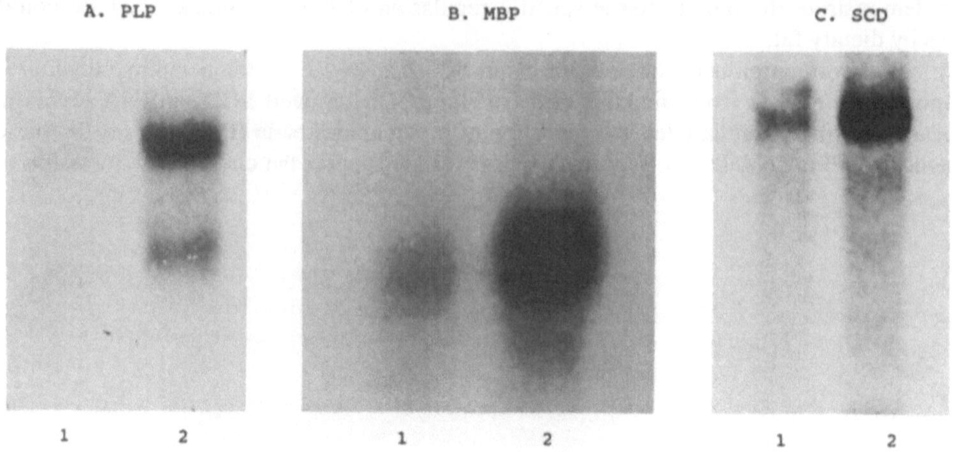

Figure 4. Influence of quaking phenotype on neonatal brain mRNAs. Poly A+ RNA (10 micrograms/lane) was isolated from brains of 23 day old quaking (lane 1) and control (lane 2) mice and probed with 32P-labelled PLP (A), MBP (B), and SCD2 (C).

Brain PLP, MBP and SCD2 mRNA levels remain relatively high in postweaning life, indicating that continuous synthesis of myelin structural components is required to maintain the myelin membrane in response to metabolic turnover .EFAD feeding reduces myelin-specific mRNAs during the neonatal myelinating period but has little impact on these mRNAs in postweaning life (Figure 5). This suggests a constitutive level of expression of genes encoding myelin structural components and biosynthetic enzymes during adult life that is relatively EFA independent. In contrast, induced expression of these genes as occurs during neonatal myelination is EFA dependent.

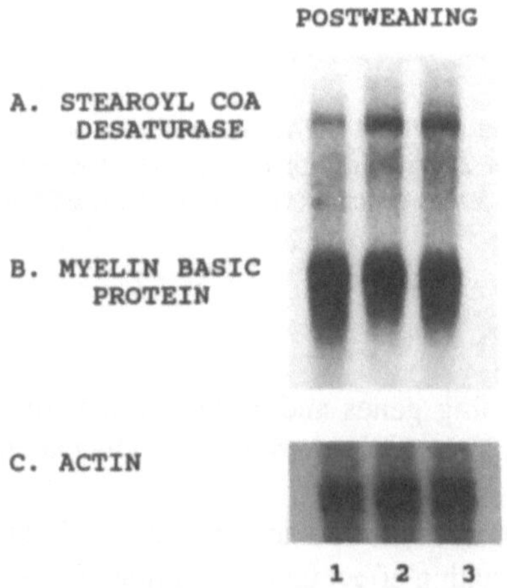

Figure 5. Postweaning essential fatty acid (EFA) intake does not influence brain SCD2 or MBP mRNA levels. Poly A+ RNA (10 micrograms/lane) was isolated from brains of 42 day old mice after 21 days of dietary treatment. The blot was probed with 32P-labelled SCD2, MBP and actin cDNA inserts. Lanes contain: (1) chow diet, (2) fat free (EFAD), (3) control (5% corn oil). Continuing the feeding an additional 35 days also had no effect on brain MBP or SCD2 mRNA levels.

MYELIN CHOLESTEROL AND NEONATAL BRAIN APOLIPOPROTEIN E (APO E) mRNA EXPRESSION DURING MYELINATION

Cholesterol comprises about 25% of myelin lipids[7]. Most of the cholesterol incorporated into brain structural lipids is synthesized within the brain, with diet playing a minor role[30]. In the systemic circulation, cholesterol is transported in lipoprotein complexes and taken up by cells via specific receptors[7]. A similar mechanism may operate in the CNS[7,31-34].

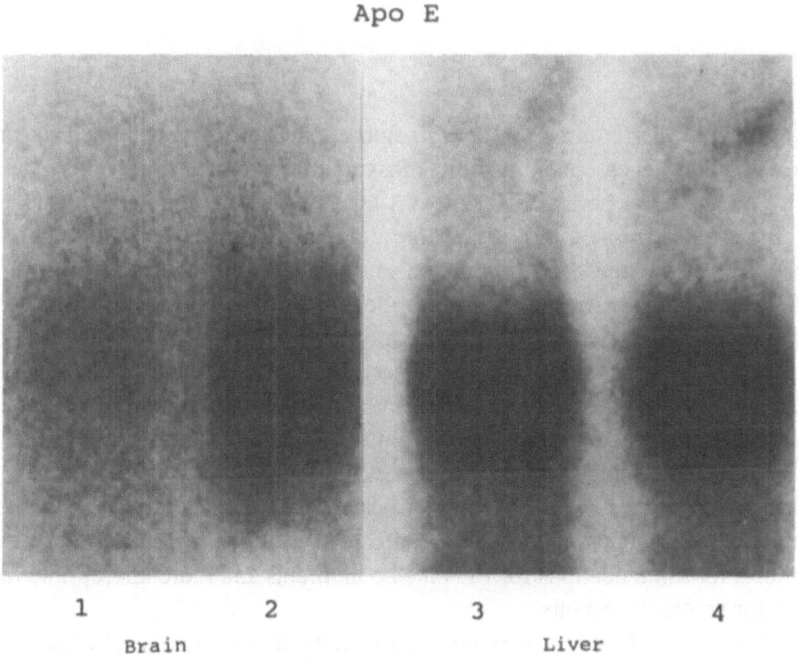

Figure 6. Influence of quaking phenotype on brain and liver Apo E mRNA levels. Poly A+ RNA (10 micrograms/lane) was isolated from brain and liver tissue of 23 day old quakers (lanes 1 & 3) and controls (lanes 2 & 4) and probed with 32P-labelled Apo E cDNA.

Apo E mRNA is expressed in neonatal brain suggesting a potential involvement in the transport of fatty acids or cholesterol during development[6,7,31-34]. The cell type synthesizing Apo E in the brain has been reported to be the astrocyte, a cell of glial origin with cholesterolgenic capacity[33,34]. Apo E synthesized by astrocytes may transport fatty acids or cholesterol via the brain interstitial fluid to lipid requiring myelinating cells expressing the LDL receptor[6,31-34]. This suggests that the CNS may utilize a lipoprotein transport system which parallels the systemic lipoprotein transport system[31-34]. Synthesis of Apo E is stimulated following mechanical, toxic, or viral injury to the central or peripheral nervous system[35]. The stimulation of neural Apo E during development and following injury suggests that Apo E may play a role in nerve or glial cell growth, repair or regeneration.

The deficit in myelination is more severe in dysmyelinating mutants than in animals consuming EFAD diets (data not shown). The extreme myelin deficiency manifested in some dysmyelinating mutants (i.e. "quaking" mice) provide a more complete perspective on the range of genes expressed during myelination. We found that dietary fat has little impact on neonatal brain Apo E mRNA levels, however, Apo E mRNA levels are significantly reduced in quaking brain[6,28] (Figure 6). It is highly significant that Apo E mRNA levels were reduced in neonatal quaking brain, but liver Apo E mRNA levels were normal (Figure 6). Since astrocytes are believed to be the source on endogenous Apo E in the brain this suggests that quaking may influence the biosynthetic capacity of astrocytes as well as oligodendrocytes[3]. The regulation of neural Apo E production is unknown, but this evidence suggests a link between Apo E production and myelination.

CONCLUSIONS

Myelin synthesis requires the coordinated expression of genes encoding myelin-specific proteins and genes encoding lipid biosynthetic enzymes, transport proteins and receptors[1-7]. The increase in brain SCD2 mRNA levels in the nursing neonate suggests that the endogenous capacity to synthesize oleic acid increases in concert with neonatal myelination[6]. Developmental SCD2 induction in neonatal brain is divergent from hepatic SCD1 induction and indicative of a unique mechanism for oleic acid biosynthesis in developing brain[6]. The developmental regulation of brain SCD2 mRNA levels is in contrast to hepatic SCD1 mRNA regulation, which is directly influenced by diet[6].

Increasing evidence indicates the developing brain exhibits the capacity to synthesize structural lipids, rather than extensively incorporating preformed dietary lipids[6,28-34]. While this insulates the brain to some degree from dietary perturbations, certain aspects of brain development, i.e. myelination, are sensitive to dietary EFA intake[6,7,12-17]. Understanding the mechanism underlying the effects of EFAs on myelin synthesis may lead to more uniform recommendations for EFA intake in infants and more appropriate fatty acid formulations for premature infants.

The developing brain possesses the capacity to synthesize cholesterol[30], the apolipoprotein to transport cholesterol (Apo E)[6,31-34], and the LDL receptor to take up the cholesterol complexed to Apo E [6,31-34]. This evidence suggests that a mechanism to transport cholesterol may be functional within the CNS and may be induced during neonatal myelination. Accumulating evidence indicates that Apo E may also play a role in lipid trafficking following nerve injury and regeneration[35]. A better understanding of the role of fatty acids and Apo E could be directly useful in the clinical management of demyelinating conditions and spinal cord injury[35].

Recent clinical studies indicate that specific fatty acids are effective in slowing the progression of ALD, a fatal demyelinating condition[10,11]. This suggests the exciting possibility that fatty acids, or synthetic analogues, may be used as dietary or pharmacological agents to slow the progression of ALD or other demyelinating conditions exhibiting abnormalities in fatty acid metabolism or accumulation.

REFERENCES

1. W.T. Norton and W. Cammer, Isolation and characterization of myelin, in: "Myelin", 2nd edition, P. Morrell, ed., Plenum Press, New York (1984).
2. G. Lemke, Unwrapping the genes of myelin, *Neuron* 1:535(1988).

3. A.T. Campagnoni and W. Macklin, Cellular and molecular aspects of myelin protein gene expression, *Molec. Neurobiol.* 2:41(1988).
4. A. Gottlieb, I. Keydar, and H.T. Epstein, Rodent brain growth stages: an analytical review, *Biol. Neonate* 32:166(1977).
5. R.C. Wiggins, Myelin development and nutritional insufficiency, *Brain Res. Rev.* 4:151(1982).
6. J.W. DeWille and S.J. Farmer, Postnatal dietary fat influences mRNAs involved in myelination, *Dev. Neurosci.* 14:61(1992).
7. J.W. DeWille and L. Horrocks, Synthesis and turnover of myelin phospholipids and cholesterol. *in*: "Myelin", 3rd edition, R.E. Martenson, ed., CRC Press, Boca Raton (1992).
8. M. Schwartz, A. Cohen, C. Stein-Izsak, and M. Belkin, Dichotomy of the glial cell response to axonal injury and regeneration, *FASEB J.* 3:2371(1989).
9. P.K. Thomas, Inherited neuropathies related to disorders of lipid metabolism, *Adv. Neurology* 48:133(1988).
10. H.W. Moser, The peroxisome: Nervous system role of a previously underrated organelle, *Neurology* 38:1617(1988).
11. W.B. Rizzo, R.T. Leshner, B.A. Odone, A.L. Dammann, B.S. Craft, M.E. Jensen, S.S. Jennings, S. Davis, R. Jaitly, and J.A. Sgro. Dietary erucic acid therapy for X-linked adrenoleukodystrophy, *Neurology* 39:1415(1989).
12. B.D. Trapp and J. Bernsohn, Essential fatty acid deficiency and CNS myelin: Biochemical and morphological observations, *J. Neurol. Sci.* 37:249(1978).
13. Y.Y. Yeh, Maternal dietary restriction causes myelin and lipid deficits in the brain of offspring, *J. Neurosci. Res.* 19:357(1988).
14. C. Galli, H.I. Trzeciak, and R. Paoletti, Effects of essential fatty acid deficiency on myelin and various subcellular structures in rat brain, *J. Neurochem.* 19:1863(1972).
15. M.C. McKenna and A.T. Campagnoni, Effects of pre- and postnatal essential fatty acid deficiency on brain development and myelination, *J. Nutr.* 109:1195(1979).
16. S.E. Berkow and A.T. Campagnoni, Essential fatty acid deficiency: Effects of cross-fostering mice at birth on brain growth and myelination, *J. Nutr.* 111:886(1981).
17. S.E. Berkow and A.T. Campagnoni, Essential fatty acid deficiency: Effects of cross-fostering mice at birth on myelin levels and composition, *J. Nutr.* 113:582(1983).
18. S.D. Clarke, M.A. Armstrong, and D.B. Jump, Dietary polyunsaturated fats uniquely suppress rat liver fatty acid synthetase and S14 mRNA content, *J. Nutr.* 120:225(1990).
19. G. Shillabeer, J. Hornford, J.M. Forden, N.C.W. Wong, and D.C.W. Lau, Hepatic and adipose tissue lipogenic enzyme mRNA levels are suppressed by high fat diets in the rat, *J. Lipid Res.* 31:623(1990).
20. W.L. Blake and S.D. Clarke, Dietary polyunsaturated fatty acids reduce hepatic fatty acid synthetase gene transcription, *FASEB J.* 5232(1990).
21. S.J. Fomon, Reflections on infant feeding in the 1970s and 1980s, Am. *J. Clin. Nutr.* 46:171(1987).
22. T. Heim, How to meet the lipid requirements of the premature infant, *Ped. Clinics N. Amer.* 32:289(1985).
23. R. Uauy and D.R. Hoffman, Essential fatty acid requirements for normal eye and brain development, *Sem. in Perinatol.* 15:449(1991).
24. R. Jeffcoat R and A.T. James, The regulation of desaturation and elongation of fatty acids in mammals, *in*: "Fatty acid metabolism and its regulation", S. Numa, ed., Elsevier Science Publishers, New York (1984).
25. C. Alling, A. Bruce, I. Karlsson, O. Sapia, and L. Svennerholm, Effect of maternal essential fatty acid supply on fatty acid composition of brain, liver, muscle and serum in 21-day-old rats, *J. Nutr.* 102:773(1972).
26. J.M. Ntambi, S.A. Buhrow, K.H. Kaestner, R.J. Christy, E. Sibley, T.J. Kelly, and M.D. Lane, Differentiation-induced gene expression in 3T3-L1 preadipocytes. Characterization of a differentially expressed gene encoding stearoyl-CoA desaturase, *J. Biol. Chem.* 263:17291(1988).
27. K.H. Kaestner, J.M. Ntambi, T.J. Kelly, M.D. Lane, Differentiation-induced gene expression in 3T3-L1 preadipocytes. A second differentially expressed gene encoding stearoyl-CoA desaturase, *J. Biol. Chem.* 264:14755(1989).
28. S. Farmer and J.W. DeWille, Dietary fat affects mRNAs in neonatal brain and liver, *FASEB J.* (in press) (1993).
29. J.W. DeWille and S.J. Farmer, Quaking phenotype influences brain lipid-related mRNAs, *Neurosci. Lett.* 141:195(1992).
30. J. Edmond, R.A. Korsak, J.W. Morrow, G. Torok-Both, and D.H. Catlin, Dietary cholesterol and the origin of cholesterol in the brain of developing rats, *J. Nutr.* 121:1323(1991).
31. S.L. Hofmann, D.W. Russell, J.L. Goldstein, M.S, Brown, mRNA for low density lipoprotein receptor in brain and spinal cord of immature and mature rabbits, *Proc. Nat. Acad.* Sci. 84:6312(1987).
32. L.W. Swanson, D.M. Simmons, S.L. Hofmann, J.L. Goldstein, and M.S. Brown, Localization of mRNA

for low density lipoprotein receptor and a cholesterol synthetic enzyme in rabbit nervous system by in situ hybridization, *Proc. Nat. Acad. Sci.* 85:9821(1988).
33. N.A. Elshourbagy, W.S. Liao, R.W. Mahley, and J.M. Taylor, Apolipoprotein E mRNA is abundant in the brain and adrenals, as well as the liver, and is present in other peripheral tissues of rats and marmosets, *Proc. Nat. Acad. Sci.* 82:203(1985).
34. R.E. Pitas, J.K. Boyles, S.H. Lee, D. Hui, and K.H. Weisgraber, Lipoproteins and their receptors in the central nervous system. Characterization of the lipoproteins in the cerebrospinal fluid and identification of apolipoprotein B,E (LDL) receptors in the brain, *J. Biol. Chem.* 262:14352(1987).
35. R.W. Mahley, Apolipoprotein E: Cholesterol transport protein with expanding role in cell biology, *Science* 240:622(1988).

DIETARY LIPIDS: EXOGENOUS CONTROL OF MYELINATION

Serafina Salvati, Lucilla Attorri, Cristina Avellino,
Antonella Di Biase, Annamaria Confaloni

Metabolism and Pathological Biochemistry Dept.
Istituto Superiore di Sanità
Rome, Italy

INTRODUCTION

Myelin formation represents a terminal phenotipic expression of oligodendrocytes and Schwann cells that must involve the expression of a myelin specific genetic programme. This process is also influenced by environmental factors and only the correct interaction of such factors makes possible the formation and the maintenance of the myelin sheath able to carry out its functional activities. Among environmental factors the important role of diet and in particular of dietary lipids is becoming more evident[1]. The importance of lipids can be understood from the fact that lipid deposition and metabolism are intimately connected with the biogenesis of myelin[2].

Myelin composition contributes to both endogenous and exogenous lipids[3]. The potential sources of lipids in the nervous system are both synthesis *in situ* and uptake from plasma as such or after modification at the level of various organs. The relative contribution of each pathway in myelin formation is not yet clear.

The importance of lipids is highlighted by the fact that if these molecules, especially in the most active phase of myelin synthesis , are not available or are metabolically blocked amyelination, dysmyelination or demyelination may occur[4-6].

DIETARY INFLUENCES

Many studies indicate that the nature of dietary lipids influence the chemical composition of myelin. The most pronounced changes are observed in the progeny born to mothers fed diets with different lipid sources during pregnancy and lactation[7,8]. Alterations in dietary lipids even in post-weaning rats can induce fatty acid changes in central (CNS) and peripheral nervous system (PNS) myelin[9,10].

Essential fatty acid (EFA) deficiency is known to have a dramatic effect on myelin development[11].

ESSENTIAL FATTY ACIDS

Two independent families of fatty acids the ω 6 and ω 3, are known to be essential for normal cell function. The ω 6 and ω 3 fatty acids accepted as essential in the human diet are linoleic (18:2 ω 6) and linolenic (18: 3 ω 3) acids, respectively.

The effect of EFA dietary deficiency on myelin formation has been extensively documented. It is well established that their deficiency causes hypomyelination associated with a delay in behavioral development[12-16]. Since the percentage of EFA in myelin is very low, the biochemical and functional alterations observed could be due not to their absence, but either to a deficiency of their metabolites (mainly C 20: 4 ω 6 and C 22: 6 ω 3) or to a rise in C20 and C22 trienoic acid. The consequent lipoacidic imbalance caused could be reflected in reduced myelin assembly. The changes caused by EFA deficiency could also be linked to the important functional role played by such acids[19,20] It was demonstrated that post natal deficiency in EFA influences the synthesis of myelin proteins. (see J. De Wille, same volume)

In 1972 Schlenck included odd-chain fatty acids of the ω 5 series with essential fatty acids. Odd-chain fatty acids occur in several natural dietary sources like fish and milk, but their concentration is very low, generally not exceeding 1-3% of the total fatty acids, with the exception of mullet in which their concentration reaches about 20%.

Odd-chain fatty acids have a double-bond structure that resembles that of the next higher homologue. In odd-chain fatty acids the double bonds are *cis* and methylene interrupted and they are, in reference to the carboxyl group, in the same position as in the next higher homologue. The physiological role of these odd-numbered fatty acids is still obscure.

ODD-CHAIN FATTY ACIDS AND BRAIN

In order to investigate the role of these unusual fatty acids on brain development, rats were fed diets containing 10% of odd-chain phospholipids obtained from yeast (*Candida lypolitica*) grown on n-alkanes, during pregnancy and lactation. The lipid fraction of the control diet constituted a mixture of margarine and phosphatidylcholine with a percentage of phospholipids equal to that of the test diet[18].

As shown in table 1 the body and the brain weight of offspring increases with age, and in both no significant differences are observed with controls. The protein content of the brain homogenate , also shown in table 1, increases with development and it is significantly higher in test animals at 7 and 14 days of age. The increase of protein content in the test group in the first weeks of life suggests an earlier ontogenesis induced by the presence of odd-chain phospholipids in the diet of the test animals.

To verify this hypothesis we focused our attention on myelin since it is one of the most sensitive membranes to variations in EFA. Firstly, we studied the 2',3' cyclic nucleotide 3'-phosphohydrolase (EC 3.1.4.37) (CNPase) activity, since its activity is closely associated to

Table 1. Body and brain weight and total protein content of control and test rats at 7, 14 and 21 days postnatal development.

Age (days)	Group	Body weight (g±SD)a	Brain weight (g±SD)a	Total protein (mg/g brain±SD)b
7	C	17.9±0.7	0.64±0.07	66.4±4.7
	T	15.4±2.3	0.64±0.10	75.0±4.3*
14	C	30.4±1.1	1.16±0.07	79.4±0.9
	T	29.2±1.5	1.12±0.16	86.0± 4.2**
21	C	49.9±1.2	1.44±0.07	94.6±5.0
	T	48.6±2.3	1.36±0.12	92.3±1.5

a; the results are the means ± SD of 40 animals; b; the results are the means ± SD of 5 pools. Each pool consisted of eight brains. Statistically different: $p<0.05$; $p<0.01$ (Student's t Test) C:control; T: test

myelin in developing rats; its increase is generally considered as an indicator of *in vivo* myelinogenesis[22,23].

The CNPase specific activity shows a marked and progressive increase with age in myelin of the rat brain in both groups (Fig. 1). At 7 and 14 days the enzyme activity in myelin of the test rats is significantly higher ($p<0.05$ and $p< 0.01$, respectively) than that of the control rats, whereas at 21 days the values are essentially the same. The earlier increase of CNPase specific activity in the test animals compared to controls indicate differences in the pattern of development of myelin biogenesis between two groups.

ODD-CHAIN FATTY ACIDS AND MYELIN

An early myelination due to dietary lipids is confirmed by immunohistochemical studies.

Myelination proceeds in a caudo-rostral direction, involving different tracts at different times. In the CNS of 7-day-old rats the highest density of myelinated fibers is present in the medulla oblongata. In the cerebellum some myelinated fibers are found in the white matter, whereas no positivity is seen in the cortex cerebelli. No myelin sheaths are detected in the forebrain regions rostral to the telencephalon[24,25]. On the contrary, in our experiments the test sections immunostained by peroxidase-antiperoxidase (PAP) method[26] show myelin sheaths throughout the brain also at this stage of development.

Positivity to myelin basic protein (MBP) and proteolipid protein (PLP) is also observed in the more rostral level such as the corpus callosum at 7 days postnatal in rat fed the test diet. In the control rats, immunopositivity in the same region is detected later in brain development (Fig.2 Ref.27).

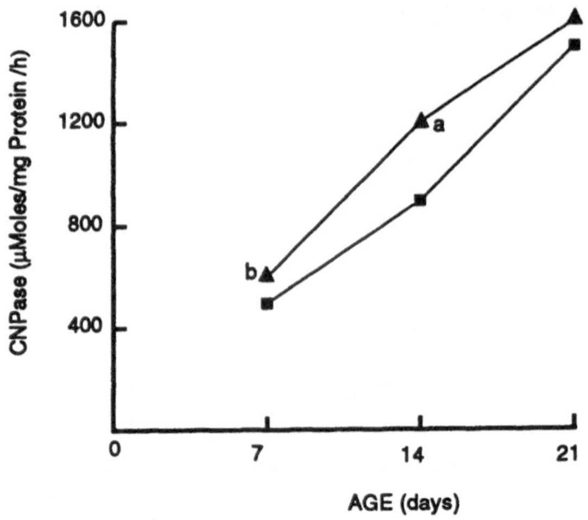

Fig. 1. CNPase specific activity in myelin of control and test animals at 7,14 and 21 days postnatal. The data represent means of at least three determinations.a: $p < 0,01$;b : $p < 0,05$ ❐ control; Δ test

The early appearance of myelin could indicate that dietary lipids interfere with brain development, most probably affecting the ontogenesis of oligodendrocytes. It can be hypothesized that lipids carry out their action influencing either oligodendrocyte progenitors and/or bipotential cells able to generate both astrocytes and oligodendrocytes. Astrocytes and oligodendrocytes could descend from a common progenitor cell that selects its lineage late in development under the influence of intrisic and extrinsic factors[27]. Consistent with this last point is the fact that in the test animals at the 7 days marked immunopositivity, not only to MBP and PLP but also glial fibrillary acidic protein (GFAP), specific marker of astrocytes, is present compared to controls.

In the cerebellum of the test rats marked positivity to GFAP is present in the white matter, while in controls no positivity is detected (Fig. 2).

In the more rostral level, such as in the telencephalon area, at this stage of development no staining to GFAP is observed in controls. In contrast, in the test animals marked immunopositivity is detected in these regions as shown in Fig. 3.

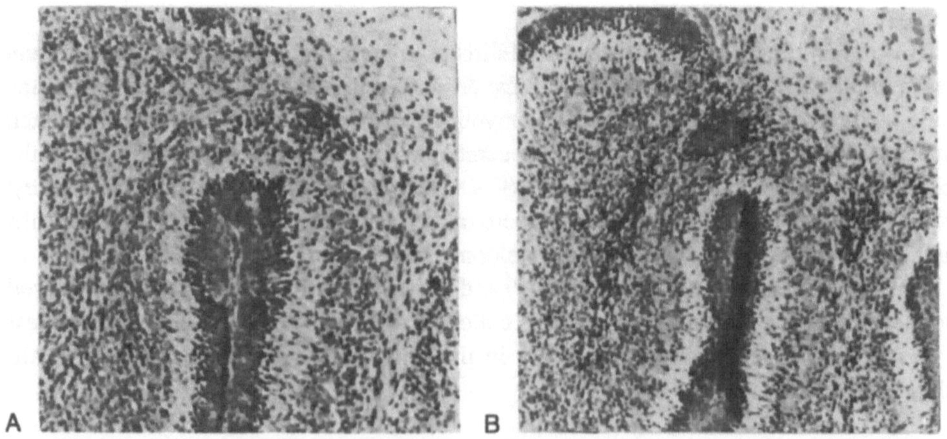

Fig.2 Sagittal sections of cerebellum immunostained with antisera to GFAP. At 7 days postnatal in the white matter of the cerebellum no positivity is present in controls(A), marked immunopositivity (asterisk) is detected in tests (B).x 140

Since the acceleration of myelinogenesis is not accompanied by any relevant changes in the body weight of the rats or of their brains, it might be ascribed to the direct influence of the composition of the lipidic fraction on brain development.

Presence of odd-chain fatty acids in brain lipids of rats is observed both at birth and at weaning (Table 2).

The highest percentage of odd-chain fatty acids in the test animals compared to controls at birth as well as their increase during development suggests that these acids cross the placenta, they are secreted in the maternal milk and cross the blood-brain- barrier.

A

B

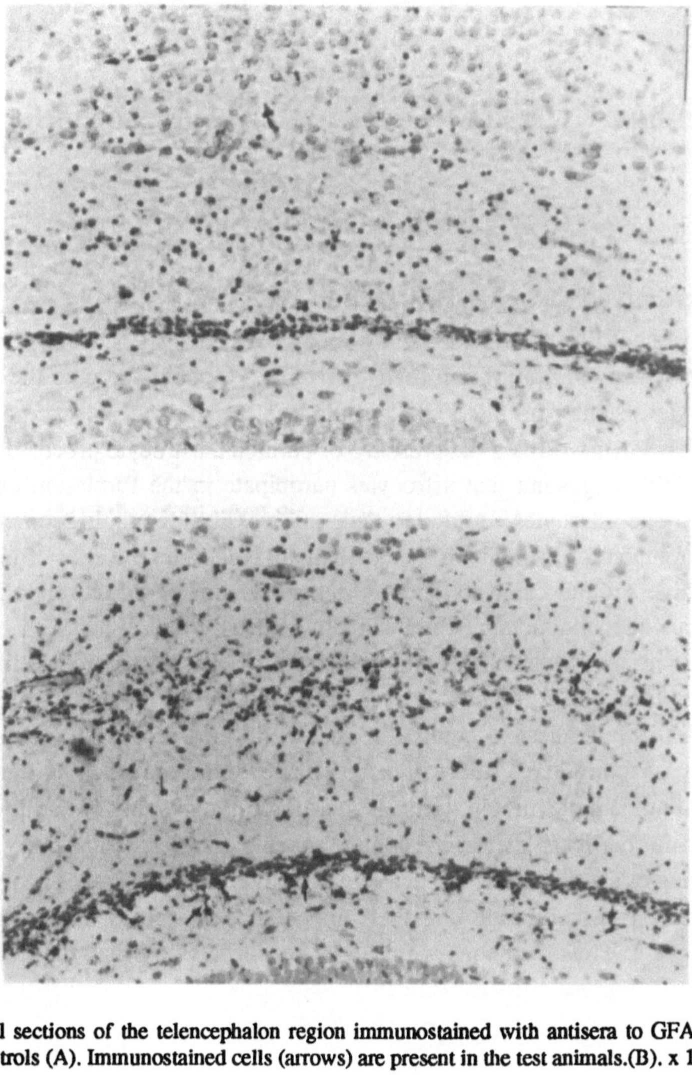

Fig.3. Sagittal sections of the telencephalon region immunostained with antisera to GFAP. No positivity is present in controls (A). Immunostained cells (arrows) are present in the test animals.(B). x 140

Table 2. Percentage of odd-chain fatty acids in brain lipids of rats at birth and at weaning.

		Fatty acid						
		C-15:0	C-17:0	C-17:1	C-17:2	C-19:1	C-19:2	Total
Birth	C	0.2±0.0	0.2±0.1	-	-	-	-	0.6
	T	0.5±0.1	0.9±0.3	1.0±0.5	-	tr	0.2±0.1	2.6
Weaning	C	0.2±0.0	0.3±0.1	tr	-	tr	-	0.5
	T	0.6±0.2	1.3±0.5	1.0±0.2	-	0.4±0.0	0.3±0.1	3.6

Myelinogenesis is a well scheduled process that in the rat occurs prevalently during the 2nd week of postnatal life. The presence of both MBP and PLP since the latter protein appears at a later stage of maturation than MBP does[28], could indicate the presence of mature myelin. Consistent with this hypothesis is also the marked presence of astrocytes. Several studies have demonstrated the presence of perinodal astrocyte processes at nodes of Ranvier in the CNS suggesting that astrocytes participate in the formation of the mature central node[29,32]. The astrocytes could play their role possibly regulating ion concentration in the microenvironment of the node or its metabolism. Nodes are the sites of high densities of voltage-sensitive sodium channels and are generally considered to support the saltory mode of action potential conduction exhibited by myelinated fibers [33].

The marked presence of astrocytes and the protein specific to mature myelin leads us to hypothesize that this myelin-membrane is already carrying out its functional role.

To establish this hypothesis a number of behavioral measures were assessed. Table 3 shows that many of the reflexes studied reached maturity more quickly in the test groups than in controls. For example, motor ability such as forelimb grasping or bar holding appear respectively six and four days later in the control animals. Similarly, sensitive reflexes such as vibrissae placing or startle appear respectively four and three days later in controls. These reflexes are related to myelination and their appearance coincides temporally with the appearance of myelin. Thus the axons wrapped by myelin already play their physiological role.

Table 3. Reflexes and behavioral measures

Behavior	Control	Test
Rooting	1.0±0.0	1.0±0.0
Cliff	3.0±0.5	3.0±0.5
Hair growth	10.5±1.5	10.5±1.5
Ears open	14.0±1.5	14.0±2.0
Righting	4.5±0.5	3.0±0.5
Forelimb placing	5.0±1.0	4.0±0.5
Forelimb grasping	12.0±1.5	6.0±0.5**
Bar holding	14.5±1.0	10.5±0.5**
Vibrissae placing	8.0±0.5	4.0±0.0**
Startle	17.0±2.5	14.5±0.5*

Mean scores on each behavioral test for each chronological age. X^2 tests were determined for each reflex in control and test rats. The p values resulting from the X^2 test for the days on which it was significant by one ($p<0.05$) or two asterisks($p<0.01$).

CONCLUSION

Exogenous lipids can positively influence myelin development. The effect of the lipid fraction is due to either odd-chain fatty acids or their metabolites. They can act like a growth-factors stimulating the synthesis and accumulation of myelin components and their assembly into myelin. *In vitro* studies have shown the mitogenic effect of phospholipids containing polyunsaturated fatty acyl groups [34].

These findings suggest at least two distinct roles for lipids in cellular regulation: a structural role as building blocks of the membrane and a functional role in transducing intra-or extracellular signals.

In vitro studies indicate that certain signal transduction pathways, in particular those involving phosphoinositide-derived second messengers, may be intimately involved in the process of oligodendrocyte differentation[35,38].

The prototype of the future adult organism is present in the fertilized ovum, but the realization of the development of the prototype is modulated by environmental factors among which nutrition is one of the most important.

This information may be relevant since it leads us towards the development of treatments which could promote remyelination

ACKNOWLEDGMENTS

Thanks are due to L. Malvezzi Campeggi and F. Pieroni for their helpful technical assistance.

REFERENCES

1. Y.Y. Yeh,Maternal dietary restriction causes myelin and lipid deficits in the brain of offspring, J.Neurosci.Res.19:357(1988)
2. P.S. Sastry,Lipids of nervous tissue: composition and metabolism, Prog.LipidRes.24:69(1985)
3. J.M. Naughton,Supply of polyenoic fatty acids to the mammalian brain,Int.J.Biochem.13:21(1981)
4. K. Toda,T. Kobayashi, I. Goto, K. Ohno, Y. Eto, K. Inui and S. Okada, Lysosulfatide (sulfogalactosylsphingosine accumulation in tissuesfrom patients with metachromatic leucodistrophy, J.Neurochem. 55:1585(1990)
5. T. Kobayashi, I. Goto, T. Yamanaka, Y. Suzuki, T. Nakano and K. Suzuki,Infantile and fetal globoid cell leucodistrophy: analysis of galactosylceramide and galactosylsphingosine,Ann.Neurol. 24:517 (1988)
6. W.T.Norton and W. Cammer,Chemical pathology of diseases involving myelin,in:"Myelin" P. Morell,ed.,Plenum Press,New York-London pp.369(1984)
7. J.M. Bourre, M. Francois, C. Weidner et al.,Brain cell and tissue recovery in rats made deficient in n-3 fatty acids by alteration of dietary fat, J.Nutr.119:15(1989)
8. S. Salvati, L. Malvezzi Campeggi, P.Corcos Benedetti, M. Di Felice, V. Gentile, M. Nardini, G. Tomassi, Effects of dietary oils on fatty acid composition and lipid peroxidation of brain membranes (myelin and synaptosomes) in rats, J.Nutr.Biochem.4:1(1993)
9. P. Divakaran, T. Pavlina, R.C. Johnson, R. Cotter, D.Madsen and R. Wiggins,Dietary supplementation of undernourished rats with soy or safflower oil:effects on myelin polyunsaturated fatty acids, Met.BrainDis.1:137(1986)
10. J.K. Yao,R.T. Holman, M.F. Lubozynski and P.J. Dyck,Changes in fatty acid composition of peripheral nerve. Myelin in essential fatty acid deficiency,Archs.Biochem.Biophys.204,175(1980)
11. R.C. Wiggins,Myelin development and nutritional insufficiency, Brain Res.Rev.,4:151(1982)
12. M.C. McKenna, A.T. Campagnoni,Effect of pre-and postnatal essential fatty acid deficiency on brain development and myelination,J.Nutr. 109:1195(1979)
13. S.E. Berkow, A.T. Campagnoni, Essential fatty acid deficiency: effect of cross-fostering mice at birth on brain growth and myelination, J.Nutr.11:886(1981)
14. S.E. Berkow, A.T. Campagnoni, Essential fatty acid deficiency: effect of cross-fostering mice at birth on myelin levels and composition, J.Nutr.113:582(1983)
15. C. Galli, P. Messeri,A. Oliverio,R. Paoletti,Deficiency of fatty acids during pregnancy and avoidance learning in the progeny, Pharmacol.Res.Commun. 7:71(1975)
16. B. D'Udine,A. Oliverio,Lipid malnutrition and early development on study of motor reflexes and electrocortical activity in the mouse,Behav. Proc.1:183(1976)
17. H. Schlenk,Odd-numbered and new essential fatty acids, Fed.Proceedings 31:1430(1972)
18 . S. Gozzo, A. Oliverio, S. Salvati, G. Serlupi Crescenzi, B. Tagliamonte and G. Tomassi, Nutritional studies on the lipid fraction of n-alkane grown yeast IV.Effect on behavioral development, Nutr.Rep.Int.17:357(1978)
19. G. Hertting and A. Seregi,Formation and function of eicosanoids in the central nervous system, Ann.N.Y.Acad.Sci.559:84(1989)
20. P. Hoffman and H.J. Meist,What about the effect of dietary lipids on endogenous prostanoid synthesis? A state-of-the art review, Biomed.Biochim.Acta 46:639(1987)
21. G.Y. Sun and R.A. MacQuarrie,Deacylation-reacylation of arachidonoyl groups in cerebral phospholipids,Ann.N.Y.Acad.Sci.559:37(1989)
22. T. Kurihara and Y. Tsukada, The regional and subcellular distribution of 2':3'-cyclic nucleotide 3'-phosphohydrolase in the central nervous system, J.Neurochem.14:1167(1967)
23. R.W. Olafson, G.I. Drummond and J.F. Lee, Studies on 2'-3'-cyclic nucleotide 3'-phosphohydrolase from brain,Can.J.Biochem. 47:961(1969)
24. S. Jacobson,Sequence of myelination in the brain of the albino rat A.Cerebral cortex,thalamus and related structures, J.Comp.Neurol. 121:5(1963)
25. B. Bjelke,A. Seiger,Morphological distribution of MBP-like immunoreactivity in the brain during development. Int.J.Devl.Neurosci.7:145(1989)
26. N.H. Sternberger,J. Itoyama, M.W. Kies, HdeF. Webster, Immunocytochemical method to identify basic protein in myelin-forming oligodendrocytes of newborn rat C.N.S.,J.Neurocytol.7:251(1978)
27. A. Confaloni, C. Avellino,L.Malvezzi Campeggi and S. Salvati, Accelerated myelinogenesis induced by dietary lipids in rats, Dev.Neurosci.(1993)in press

28. B.K. Hartman,H.C. Agrawal,D. Agrawal,S. Kalmbach, Development andmaturation of central nervous system myelin: comparison of immunohistochemical localization of proteolipid protein and basic protein in myelin and oligodendrocytes, Proc.Natl.Acad.Sci. USA 79:4217(1982)
29. C. Hildebrand,Ultrastructural and light-microscopic studies of the developing feline spinal cord white matter I. The nodes of Ranvier,ActaPhysiol.Scand.363:81(1971)
30. S.G. Waxman and J.A. Black,Freeze-fracture ultrastructure of the perinodal astrocyte and associated glial junctions,BrainRes.308:77 (1984)
31. T.J. Sims, s.g. Waxman, J.A. Black and S.A. Gilmore,Perinodal astrocitic processes at nodes of Ranvier in developing normal and glial cell deficient rat spinal cord, Brain RES.337:321(1985)
32. S.G, Waxman and J.M. Ritchie,Organization of ion channels in the myelinated nerve fiber, Science228:1502(1985)
33. J.A. Black and S.G. Waxman,The perinodal astrocyte,Glia1:169(1988)
34. W.Imagarwa ,G.K. Bandyopadhyay, D. Wallace, S. Naudi,Phospholipids containing polyunsaturated fatty acyl groups are mitogenic for normal mouse mammary epithelial cells in serum-free primary cell culture,Proc.Natl.Acad.Sci.USA 86:4122(1989)
35. M.C. Raff,R.H. Miller,M.D. Noble, A glial progenitor cell that develops in vitro into an astrocyte or an oligodendrocyte depending on the culture medium. Nature 303,390(1983)
36. M.C. Raff, L.E. Lillien, W.D. Richardson, J.F. Burne, M.D. Noble, Platelet-derived growth factor from astrocytes drives the clock that times oligodendrocyte development in culture,Nature 333:562(1988)
37. M. Noble,S.C. Bornett,O. Bogher,H. Land, G. Wolswijk, D. Wren,Control of division and differentiation in oligodendrocyte type 2-astrocyte progenitor cells, Ciba Found Symp. 150:227(1990)
38. Noble M.,J. Fok-Seang, G. Wolswijk ,D. Wren, Development and regeneration in the central nervous system, Phil.Trans.R.Soc.Lond.B.327:127(1990)

MOCH-1 CELLS: AN OLIGODENDROCYTE CELL LINE GENERATED USING A TRANSGENIC APPROACH

Brian Popko[1,2,3], Carol Hayes[1], Yan Li[1,2], Donna Kelly[1], Shigeo Murayama[4] and Kinuko Suzuki[1,4]

[1]Brain and Development Research Center, [2]Department of Biochemistry and Biophysics, [3]Program in Molecular Biology and Biotechnology, and [4]Department of Pathology, University of North Carolina at Chapel Hill, Chapel Hill, NC 27599-7250

INTRODUCTION

Myelin, the multilayered membrane sheath that surrounds axons, facilitates rapid nerve conduction velocities[1]. The multilayered structure of myelin is formed when the myelinating cell wraps several layers of its cytoplasmic membrane around the axon. Compaction occurs as the cytoplasm is extruded from the multilayered structure and the cytoplasmic faces of the myelinating cell become juxtaposed. An electron micrograph of myelin reveals two types of membrane interphases. The major dense line is formed by the apposition of the cytoplasmic sides of the cell membrane, and the intraperiod line is formed by the extracellular membrane surfaces. The myelinating cells, and the myelin sheath, are the site of insult in individuals afflicted with the debilitating disease multiple sclerosis.

Myelinating cells express a number of proteins that are unique to myelin and which are believed to participate in the formation and maintenance of the multilayered structure[2,3,4]. The most abundant proteins of central nervous system (CNS) myelin include the proteolipid proteins (PLP and DM-20) and the myelin basic protein (MBP). Although expressed by Schwann cells[5,6], PLP and DM-20 are unique to CNS myelin and are believed to participate in the formation of the intraperiod line. MBP is abundant in both CNS and PNS myelin and available evidence suggests that it plays a role in the compaction of myelin and the formation of the major dense line. The P_0 protein, which is not present in the CNS, represents about 50% of the protein content of peripheral myelin. P_0 is a transmembrane protein that is thought to play a role in the formation of the intraperiod line as well as the major dense line. The myelin associated glycoprotein (MAG) is a member of the immunoglobulin gene superfamily and is found in both PNS and CNS myelin. MAG has been localized to the periaxonal region of the myelin sheath and is believed to mediate the contact between the myelinating cell and the axon[7]. The enzyme 2',3'-cyclic nucleotide 3'-

A Multidisciplinary Approach To Myelin Diseases II
Edited by S. Salvati, Plenum Press, New York, 1994

phosphodiesterase (CNP) has been shown to be associated with PNS and CNS myelin[7]. The function of CNP in the myelination process, however, remains unclear.

CNS myelin is formed by oligodendrocytes, which appear on the day of birth in rodents [8, 9, 10]. *In vitro*, bipotential O-2A progenitor cells give rise to either oligodendrocytes or type-2 astrocytes depending on culture conditions[11]. Cultured O-2A cells differentiate constitutively into oligodendrocytes through a series of antigenically distinct precursors[12]. The *in vitro* differentiation of O-2A cells into type-2 astrocytes requires extrinsic signals that can be supplied either by high concentrations of serum or by certain populations of neuronal cells[11,13,14].

Recently, we have used a transgenic approach to selectively transform oligodendrocytes *in vivo* that express antigenic markers of oligodendrocytes (sulfatide and galactocerbroside), as well as abundant mRNA levels for MBP and PLP[15].

MATERIALS AND METHODS

The methods used to generate the data presented here are described in detail in Hayes et al.[15].

RESULTS

MBP/c-*neu* Transgenic Mice

Transgenic mice were generated, as described by Popko et al.[16], with a chimeric gene that contained the activated *neu* oncogene[17] under the transcriptional control of 1.3 kb of the upstream region of the mouse MBP gene (Figure 1). This region of the mouse MBP

Figure 1. Structure of the MBP/c-*neu* transgene. The thin line represents pUC12 vector sequences. The upstream region of the mouse MBP gene is represented by the open box. The closed box represents the c-*neu* cDNA, and the box filled with slanted lines represents the splice and polyadenylation sequences of the SV40 viral genome. The start point of transcription of the MBP gene is denoted by an arrow. Restriction enzyme sites are indicated for EcoRI (E), HindIII (H), and SalI (S). From Hayes et al.[15], used by permission from John Wiley & Sons Inc.

gene has previously been shown to be sufficient to direct high levels of brain-specific transcription in transgenic mice[18]. The splice and polyadenylation sequences of the SV40 virus were also included in the construct, downstream from the *neu* cDNA.

The MBP/c-*neu* construct shown in Figure 1 was used to generate nine transgenic lines. The steady-state levels of MBP/c-*neu* mRNA were quite low in all of the lines[15].

MBP/c-*neu* Brain Tumors

MBP/c-*neu* animals develop a low incidence of brain tumors. Most tumors have occurred at the basal aspect of the cerebrum involving the hypothalamic region and extending into the posterior fossa (Figure 2a), compressing the brain stem (Figure 2b). All tumors examined had a component of small round cells with high cellularity, large nuclear-cytoplasmic ratio and frequent mitosis, with little evidence of differentiation. In some tumors, however, astrocytic differentiation with fine cellular processes and giant cells were seen (Figure 3a). Some tumor cells in the relatively differentiated areas reacted with antibodies to MBP (Figure 3b). Most tumor cells were stained with anti-Leu 7 (data not shown). Moreover, RNA transcripts of genes encoding the myelin-specific proteins MBP and PLP are present in the tumor RNA at levels that are much higher than that of control brain RNA samples (Figure 4). These data strongly suggest that the MBP/c-*neu* brain tumors are enriched in myelinating cells.

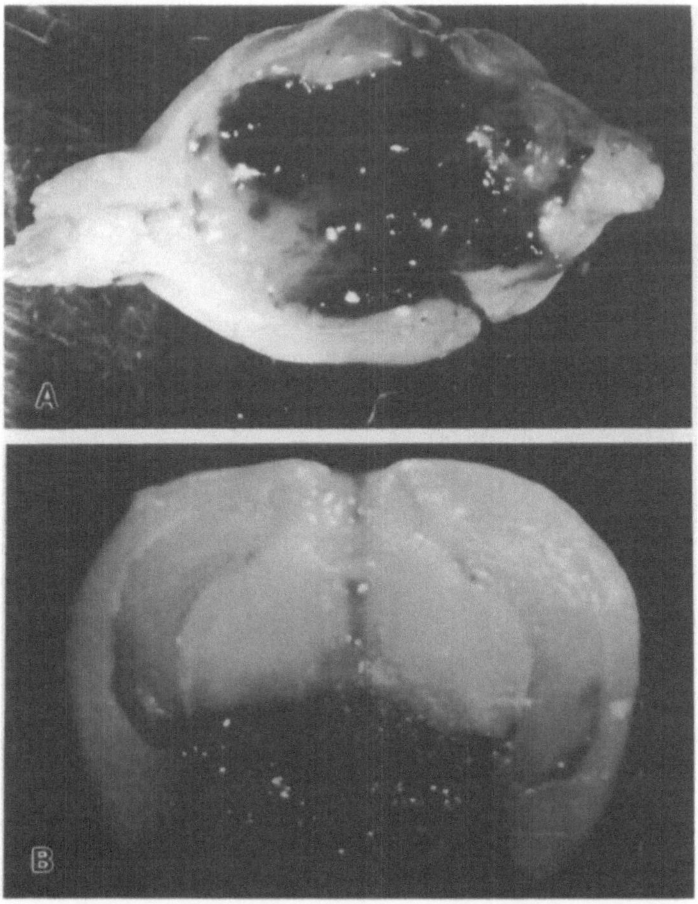

Figure 2. Gross Anatomical Side and Bottom View of MBP/c-*neu* brain tumor.

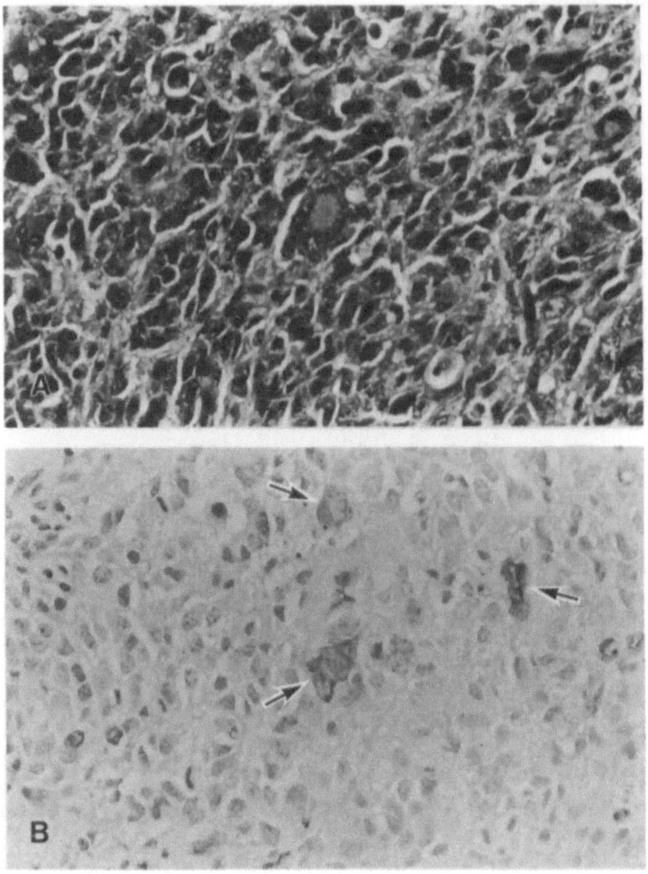

Figure 3. Histological and Immunocytochemical Properties of MBP/c-*neu* Brain Tumors. A: hematoxylin and eosin staining; astrocytic differentiation with fine cellular processes and giant cells are seen. B: avidin-biotin complex anti-MBP immunostain counterstained by methyl green; anti-MBP antibody recognizes perikaryal cytoplasm of most giant cells.

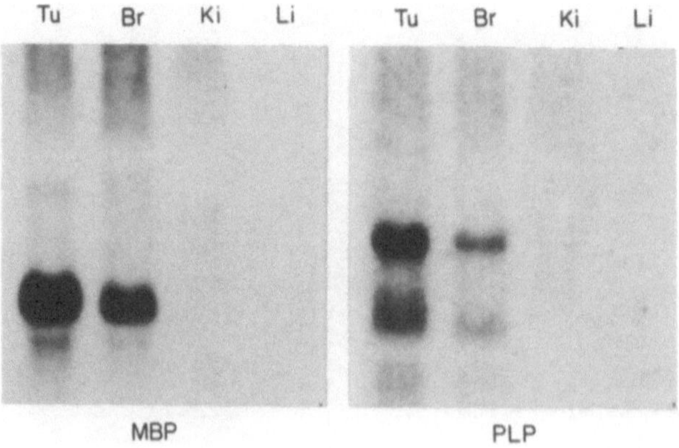

Figure 4. Northern Blot Analysis of Brain Tumor RNA. Approximately 10 ug total RNA samples from the indicated tissues were denatured, electrophoresed through agarose, transferred to nylon and hybridized with the indicated probes. Key: Tu, brain tumor #26-2; Br, normal adult mouse brain; Ki, neonatal normal rat kidney; Li, normal adult mouse liver; SN, normal adult sciatic nerve.

MBP/c-*neu* Brain Tumor Cultures

A portion of one of the MBP/c-*neu* brain tumors was plated in culture medium containing 1% fetal bovine serum (FBS). Upon initial plating many process bearing cells were present that appeared morphologically heterogeneous. From these cells a distinct population of process bearing cells arose that have been designated MOCH-1 cells. These cells have remained mitotic for over two years with a doubling time of approximately three to five days. Morphologically, MOCH-1 cells appear very similar to oligodendrocytes (Figure 5).

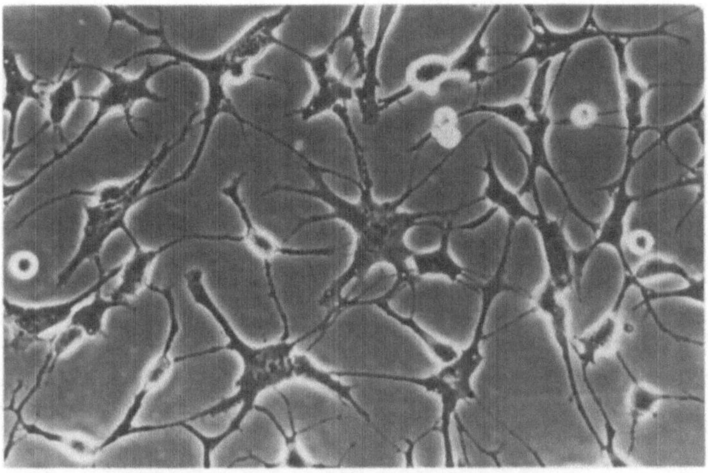

Figure 5. Phase-contrast optics view of MOCH-1 cells cultured in 1% FBS.

RNA transcripts specific for the myelin proteins MBP and PLP were detected in MOCH-1 cell RNA (Figure 6). These transcripts were abundant, but present at levels below that detected in the tumor from which the cells were derived. MOCH-1 cells also stained with antibodies to galactocerebroside, a glycolipid specific to myelinating cells[15].

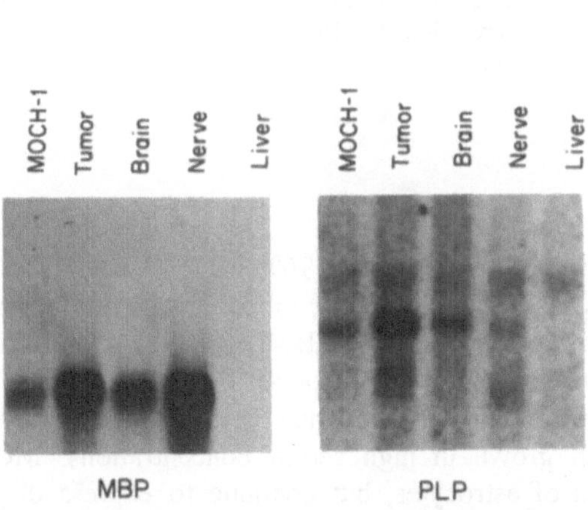

Figure 6. Northern blot analysis of MOCH-1 cell RNA. Approximately 3 ug of total RNA from MOCH-1 cells, MBP/c-*neu* brain tumor, control adult mouse brain, control adult sciatic nerve, and control adult mouse liver were electrophoresed through agarose, transferred to nylon, and hybridized with the indicated probes.

MOCH-1 cells grown in low serum conditions have phase-bright cell bodies and extend multiple thin processes (Figure 5 and Figure 7a,b). *In vitro*, O-2A progenitor cells develop into type 2 astrocytes in the presence of high concentrations of serum[11,13,14]. To determine the phenotypic effect of high serum concentrations, MOCH-1 cells were cultured in the presence of 10% FBS. Under these conditions MOCH-1 cells flatten out and lose their processes (Figure 7c,d). Almost all of these cells reacted with antibodies to GFAP

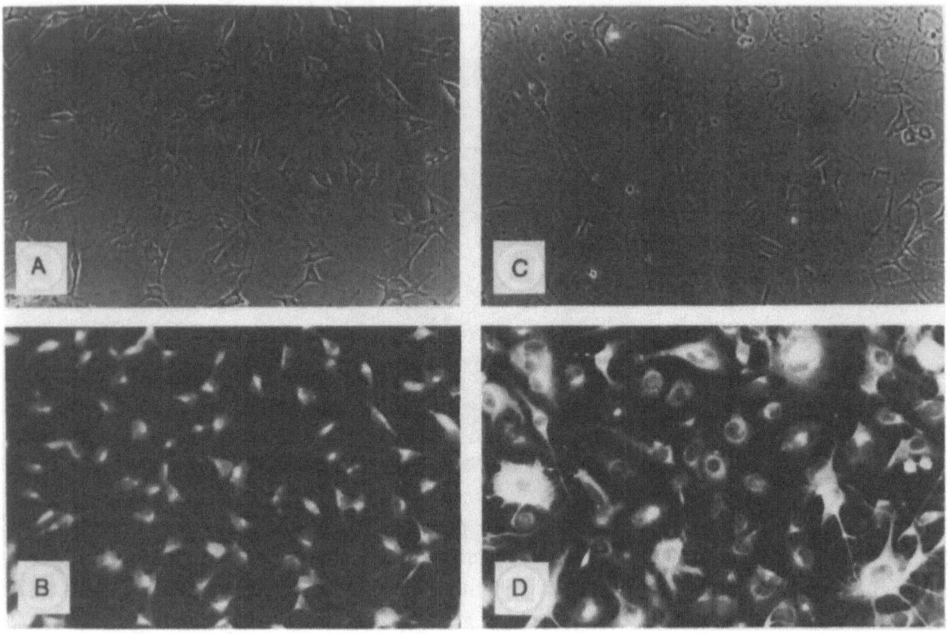

Figure 7. Immunocytochemistry of MOCH-1 Cells with GFAP antibodies. A and C, phase contrast optics; B and D, anti-GFAP antibody-fluorescein conjugate. MOCH-1 cells shown in A and B were culture in 1% FBS; and those in C and D were cultured in 10 % FBS.

(Figure 7d). Furthermore, in the presence of 10% FBS, MOCH-1 cells abundantly express GFAP mRNA, whereas, as discussed above, GFAP transcripts are undetectable in these cells when cultured in the presence of 1% FBS (Figure 8). Interestingly, MOCH-1 cells cultured in high serum concentrations express the MBP gene (Figure 8) at levels approximately two to four fold higher than when these cells are cultured in 1% FBS. These data indicate that when grown in high serum concentrations, MOCH-1 cells express molecular characteristics of astrocytes, but continue to express oligodendrocyte-specific genes.

DISCUSSION

Very little is known with regard to the molecular and biochemical events that control oligodendrocyte differentiation and function. A major obstacle to this analysis is the difficulty in obtaining large numbers of cells at a given stage of oligodendrocyte development. Protocols have been developed that allow for the isolation and culture of relatively pure populations of oligodendrocytes[19,20,21,22,23]. Nevertheless, these procedures provide limited numbers of cells that are heterogeneous with respect to their differentiated state, making a molecular or biochemical analysis of oligodendrocyte development and function difficult. For example, the procedure of Vick et al.[22], which

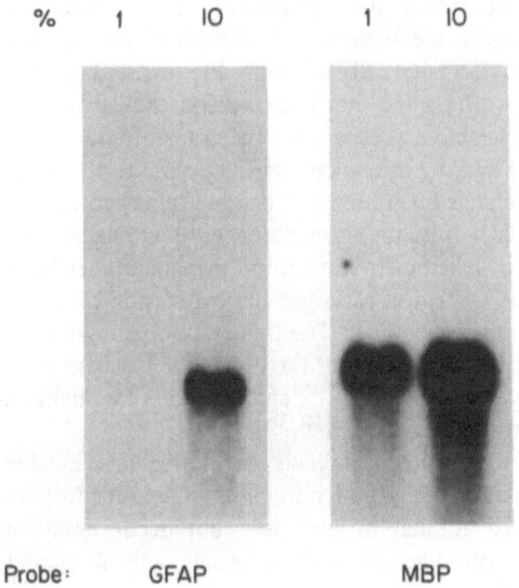

Figure 8. GFAP and MBP mRNA in MOCH-1 cells cultured in 1% and 10% FBS. Approximately 5 ug of total RNA, isolated from MOCH-1 cells cultured in either 1% FBS continuously or 10% FBS for seven days, was electrophoresed through agarose, transferred to nylon, and hybridized with either radiolabeled GFAP or MBP cDNA. From Hayes et al.[15], used by permission from John Wiley & Sons Inc.

appears to provide the largest yield of cells, produces slightly less than 10^6 oligodendrocytes per gram of starting white matter. Although this is a sufficient number of cells for many types of important experiments (e.g. patch-clamping, immunocytochemistry, etc.), the numbers are inadequate for many types of biochemical or molecular studies. For example, to generate a cDNA library, approximately 1mg of total RNA is needed for mRNA isolation and the subsequent Northern blots[24,25]. This amount of RNA can be obtained from approximately 5×10^8 cells[26], which would require at least 50 grams of starting white matter. Similarly, the isolation of sufficient numbers of oligodendrocytes for protein purification represents an arduous task, especially for proteins expressed in limited amounts (e.g. transcriptional factors).

The difficulty in obtaining large numbers of purified oligodendrocytes has motivated the search for transformed cell lines that express phenotypic properties of these specialized glial cells. Such cells would provide an unlimited source of cells with a uniform phenotype. Encouragingly, biologists interested in other systems have benefited greatly from the analysis of transformed cell lines. For example, transformed lymphocytic cell lines have been instrumental in the elucidation of molecular aspects of B-cell and T-cell development[27]. Moreover, over the last decade the neuronal PC12 cell line has played a key role in furthering our understanding of nerve cell development[28]. Unfortunately, the identification of naturally-occurring transformed glial cell lines that have more than a limited resemblance to normal oligodendrocytes has been largely unsuccessful. We have recently used a transgenic approach to target the transformation of oligodendrocytes *in vivo*[15].

By targeting the expression of oncogenes to specific cell types in transgenic animals, transformed cells of predetermined origin can be generated[29,30,31]. This approach to targeted tumorigenesis offers the opportunity to generate animal models of human neoplasias as well as a mechanism to selectively transform cells of a differentiated phenotype. By transforming cells through the expression of an oncogene under the transcriptional control of a gene that is expressed exclusively in the cells of interest, one provides an internal selective pressure on the transformed cells to maintain the ability to express the oncogene and, by inference, genes specific to the differentiated phenotype. For example, this approach has been used to generate transformed cell lines with characteristics of pancreatic beta cells[32], pituitary gonadotropes[33] and neurons[34,35].

In an effort to specifically transform myelinating cells, we have used the MBP transcriptional regulatory region to target the expression of the activated *neu* oncogene to myelinating cells in transgenic mice. The transcriptional regulatory region of the MBP gene was chosen to direct the expression of the oncogene because MBP is expressed abundantly and exclusively in myelinating cells[2,3,4], and the DNA fragment chosen has previously been successfully used in transgenic studies[18,36].

The MBP/c-*neu* transgenic animals develop a low incidence of brain tumors that express molecular markers of oligodendrocytes. Oligodendrogliomas represent about 5 to 10% of all intracranial gliomas in humans[37] but occur rarely, if at all, in mice[38]. The MBP/c-*neu* animals provide us with a small animal model with which to study oligodendrocyte neoplasia.

A brain tumor from one of the MBP/c-*neu* animals was dispersed into culture. The transformed (MOCH-1) cells that have arisen from this tumor appear very similar morphologically to oligodendrocytes. Furthermore, these cells express molecular markers of oligodendrocytes.

In vitro, O-2A progenitor cells differentiate into oligodendrocytes constitutively and into type-2 astrocytes in the presence of extrinsic signals that can be supplied by high serum concentrations[11,14]. When cultured in low serum concentrations, MOCH-1 cells abundantly express CNS myelin protein genes as well as oligodendrocyte-specific antigenic markers. In contrast, MOCH-1 cells cultured in 10% FBS abundantly express GFAP mRNA and protein. Nevertheless, these cells are not completely astrocyte-like in that they continue to express the myelin protein genes at high levels. Taken together these data suggest that the phenotype of MOCH-1 cells is not fixed, and that their plasticity can be influenced by environmental conditions. The *in-vivo* relevance of this plasticity is currently under investigation.

MOCH-1 cells provide us with an unlimited, enriched source of oligodendrocyte-like cells. These cells should facilitate the molecular analysis of the transcriptional regulatory regions of myelin protein genes, as well as provide an enriched source of the proteins that regulate these genes. These cells should also prove to be a valuable resource for the analysis of CNS glial cell lineage relationships.

Recently, others have also generated cell lines that express, or can be induced to express, properties of oligodendrocytes. Verity et al.[39] have transformed mouse oligodendrocytes *in vitro* using the temperature-sensitive T-antigen of the SV40 virus. These cells are particularly promising in that the transforming potential of this oncogene can be reduced by elevating the temperature at which the cells are grown. This ability may allow for the analysis of more differentiated properties of oligodendrocytes. Louis et al.[40] have reported the isolation of a cell line that expresses many of the properties of O-2A progenitor cells. These cells differentiate into oligodendrocytes or astrocytes depending on culture conditions, and their analysis will likely shed light on the lineage relationship of these two cell types.

The availability of oligodendrocyte cell lines should greatly further our understanding of myelinating cells. Their availability may also prove useful for the study of the response of oligodendrocytes to environmental perturbations, such as those that occur in demyelinating lesions. For example, we have recently found that interferon-gamma is capable of altering the phenotype of MOCH-1 cells in a manner that may reflect the *in-vivo* response of oligodendrocytes to immune-mediated demyelination.

ACKNOWLEDGMENTS

This work was supported in part by the National Multiple Sclerosis Society research grant RG 2089A1, the U.S Public Health Service research grants NS27336 (BP), ES 01104 (BP and KS), and NS24453 (KS). B.P. is a recipient of an NIH Research Career Development Award from NINDS and a Sloan Neuroscience Research Fellowship.

REFERENCES

1. P. Morell, R.H. Quarles, and W.T. Norton, Formation, structure, and biochemistry of myelin, *in*: "Basic Neurochemistry" 4th ed., G.J. Siegel, et al., eds., Raven Press, New York, pp. 109-136 (1989).
2. M.B. Lees and S.W. Brostoff, Proteins of myelin, *in*: "Myelin" 2nd ed., P. Morell, ed., Plenum Publishing Corp., New York/London, pp. 197-224 (1984).
3. A.T. Campagnoni and W.B. Macklin, Cellular and molecular aspects of myelin protein gene expression, Molec. Neurobiol. 2:41-89 (1988).
4. G. Lemke, Unwrapping the genes of myelin, Neuron 1:535-543 (1988).
5. C. Puckett, L. Hudson, K. Ono, V. Friedrich, J. Benecke, M. Dubois-Dalcq, and R.A. Lazzarini, Myelin-specific proteolipid protein is expressed in myelinating Schwann cells but is not incorporated into myelin sheaths, J. Neurosci. Res. 18:511-518 (1987).
6. N. Stahl, J. Harry, and B. Popko, Quantitative analysis of myelin protein gene expression during development in the rat sciatic nerve, Mol. Brain Res. 8:209-212 (1990).
7. Y. Takahashi, Gene expression in cells of the central nervous system, Prog. Neurobiol. 38:523-569, 1992.
8. E.R. Abney, P.P. Bartlett, M.C. Raff, Astrocytes, ependymal cells, and oligodendrocytes develop on schedule in dissociated cell cultures of embryonic rat brain, Dev. Biol. 83:301-310 (1981).
9. R.H. Miller, S. David, R. Patel, E.R. Abney, M.C. Raff, A quantitative immunohistochemical study of macroglial cell development in the rat optic nerve: In vivo evidence for two distinct astrocyte lineages, Dev. Biol. 111:35-41 (1985).
10. B.P. Williams, E.R. Abney, M.C. Raff, Macroglial cell development in embryonic rat brain: Studies using monoclonal antibodies, fluorescence activated cell sorting, and cell culture, Dev. Biol. 112:126-134 (1985).
11. M.C. Raff, Glial cell diversification in the rat optic nerve, Science 243:1450-1455 (1989).

12. R. Bansal, A.E. Warrington, A.L. Gard, B. Ranscht, S.E. Pfeiffer, Multiple and novel specificities of monoclonal antibodies O1, O4, and R-mAb used in the analysis of oligodendrocyte development, J. Neurosci. Res. 24:548-557 (1989).
13. M.C. Raff, R.H. Miller, and M. Noble, A glial progenitor cell that develops in vitro into an astrocyte or an oligodendrocyte depending on culture medium, Nature 303:390-396 (1983).
14. L.E. Lillien and M.C. Raff, Analysis of the cell-cell interactions that control type-2 astrocyte development in vitro, Neuron 4:525-534 (1990).
15. C. Hayes, D. Kelly, S. Murayama, A. Komiyama, K. Suzuki, and B. Popko, Expression of the *neu* oncogene under the transcriptional control of the myelin basic protein gene in transgenic mice: Generation of transformed glial cells, J. Neurosci. Res. 31:175-187 (1992).
16. B. Popko, C. Readhead, J. Dausman, and L. Hood, Transgenic mice in neurobiological research, *in*: "Neuromethods," Vol. 16, A.A. Boultan, G.B. Baker, and A. Campagnoni, eds., Humana Press, New Jersey, pp. 221-237 (1989).
17. C.I. Bargmann, M.-C. Hung, R.A. Weinberg, Multiple independent activations of the *neu* oncogene by a point mutation altering the transmembrane domain of p185, Cell 45:649-657 (1986b).
18. M. Kimura, M. Sato, A. Akatsuka, S. Nozawa-Kimura, R. Takahashi, M. Yokoyama, T. Nomura, M. Katsuki, Restoration of myelin formation by a single type of myelin basic protein in transgenic shiverer mice, Proc. Natl. Acad. Sci. USA 86:5661-5665 (1989).
19. M. Hirayama, D.H. Silberberg, R.P. Lisak, and D. Pleasure, Long-term culture of oligodendrocytes isolated from rat corpus callosum by percoll density gradient, J. Neuropathol. Exp. Neurol. 42:16-28 (1983).
20. K.D. McCarthy and J. Devellis, Preparation of separate astroglial and oligodendroglial cell cultures from rat cerebral tissue, J. Cell Biol. 85:890-902 (1980).
21. R.P. Lisak, D.E. Pleasure, D.H. Silberberg, M.C. Manning, and T. Saida, Long term culture of bovine oligodendroglia isolated with a Percoll gradient, Brain Res. 223:107-122 (1981).
22. R.S. Vick, S.-J. Chen, and G.H. DeVries, Isolation, culture, and characterization of adult rat oligodendrocytes, J. Neurosci. Res. 25:524-534 (1990).
23. P.M. Wood and R.P. Bunge, Myelination of cultured dorsal root ganglion neurons by oligodendrocytes obtained from adult rats, J. Neurol. Sci. 74:153-169 (1986).
24. T. Maniatis, E.F. Fritsch, and J. Sambrook, Molecular Cloning. A Laboratory Manual, Cold Spring Harbor, New York (1982).
25. J. Sambrook, E.F. Fritsch, and T. Maniatis, "Molecular Cloning. A Laboratory Manual," Second Edition, N. Ford, C. Nolan, and M. Ferguson, eds., Cold Spring Harbor, New York (1989).
26. R.E. Kingston, Guanidium method for total RNA preparation, *in*: "Current Protocols in Molecular Biology," Vol. 1, F.M. Ausubel, R. Brent, R.E. Kingston, D.D. Moore, J.G. Seidman, J.A. Smith, and K. Struhl, eds., Wiley Interscience, John Wiley & Sons, New York, pp. 421-425 (1990).
27. F.W. Alt, T.K. Blackwell, and G.D. Yancopoulos, Development of the primary antibody repertoire, Science 38:1079-1081 (1987).
28. U. Lendahl and R.D. McKay, The use of cell lines in neurobiology, Trends Neurosci. 13:132-137 (1990).
29. R.D. Palmiter and R.L. Brinster, Germ-line transformation of mice, Ann. Rev. Genet. 20:465-499 (1986).
30. R. Jaenisch, Transgenic animals, Science 240:1468-1474 (1988).
31. D. Hanahan, Transgenic mice as probes into complex systems, Science 246:1265-1275 (1989).
32. S. Efrats, S. Linde, H. Kofod, D. Spector, M. Delannoy, S. Grant, D. Hanahan, S. Baekkeskov, Beta-cell lines derived from transgenic mice expressing a hybrid insulin gene-oncogene, Proc. Natl. Acad. Sci. USA 85:9037-9041 (1988).
33. J.J. Windle, R.I. Weiner, and P.L. Mellon, Cell lines of the pituitary gonadotrope lineage derived by targeted oncogenesis in transgenic mice, Mol. Endocrinol. 4:597-603 (1990).
34. P.L. Mellon, J.J. Windle, P.C. Goldsmith, C.A. Padula, J.L. Roberts, R.I. Weiner, Immortalization of hypothalamic GnRH neurons by genetically targeted tumorigenesis, Neuron 5:1-10 (1990).
35. J.P. Hammang, E.E. Baetge,R.R. Behringer, R.L. Brinster, R.D. Palmiter, A. Messing, Immortalized retinal neurons derived from SV40 T-antigen-induced tumors in transgenic mice, Neuron 4:775-782 (1990).
36. M. Katsuki, M. Sato, M. Kimura, M. Yokoyama, K. Kobayashi, T. Nomura, Conversion of normal behavior to shiverer by myelin basic protein antisense cDNA in transgenic mice, Science 241:593-595 (1988).

37. D.S. Russel and L.J. Rubinstein, Tumours of Central Neuroepithelial Origin. *in*: Russell DS and Rubinstein LJ (eds.): "Pathology of Tumours of the Nervous System," Baltimore: Williams and Wilkins, pp. 83-350 (1989).
38. E.D. Murphy, Characteristic Tumors,. *in*: "Biology of the Laboratory Mouse," E.L. Green, ed., New York: Dover Publications, pp. 521-562 (1966).
39. A.N. Verity, D. Baredesen, C. Vonderscher, V.W. Handley, and A.T. Campagnoni, Expression of myelin protein genes and other myelin components in an oligodendrocytic cell line conditionally immortalized with a temperature-sensitive retrovirus, J. Neurochem. 60:577-587 (1993).
40. J.C. Louis, E. Magal, D. Muir, M. Manthorpe, and S. Varon, CG-4, a new bipotential glial cell line from rat brain, is capable of differentiating in vitro into either mature oligodendrocytes or type-2 astrocytes, J. Neurosci. Res. 31:193-204 (1992).

MOLECULAR BIOLOGY AND NEUROGENETICS OF MYELIN PROTEOLIPID PROTEIN

Klaus-Armin Nave,[1] Armin Schneider,[1]
Carol Readhead,[2] Ian Griffiths,[3]
Anja Pühlhofer,[1] Angelika Bartholomä,[1]
Susanna Graf,[1] and Beate Kiefer[1]

[1]Center for Molecular Biology (ZMBH),
University of Heidelberg, F.R.G.
[2]Cedar Sinai Medical Center, Los Angeles, U.S.A.
[3]University of Glasgow, U.K.

SUMMARY

Proteolipid protein (PLP) is a major component of myelin in the central nervous system where it constitutes approximately 50% of the total myelin protein by mass. This alone suggests an important structural function of PLP in compact myelin. More recently, the application of molecular genetic techniques by several groups has revealed an unexpected complex relationship between the PLP structure (a multitopic integral membrane protein) and its cellular function(s) which appear to exceed that of a simple molecular "strut" in the compacted myelin sheath. In particular the identification of mutations in the X chromosome-linked PLP gene of several neurological mutants (*jimpy*, *jimpy*^{msd}, and *rumpshaker* in mouse, *md* rat, *shaking* pup) as well as in human patients with *Pelizaeus-Merzbacher* disease has provided the opportunity to study an unexpected role of PLP in normal oligodendrocyte development and its involvement in genetic dysmyelinating diseases. Finally, the generation of transgenic mice has yielded a surprising demonstration that the accurate expression level of the PLP gene is a major determinant of normal myelination *in vivo*. The *Pelizaeus-Merzbacher* syndrome has also been associated with an interstitial duplication of the human X chromosome (region Xq13-23). PLP overexpressing transgenic mice prove experimentally that the dysmyelinated phenotype of this human chromosomal imbalance is most likely caused by the increased expression of a single developmentally regulated gene.

PROTEOLIPID PROTEIN IS A HIGHLY CONSERVED MULTI-TOPIC MEMBRANE PROTEIN

The PLP primary structure was determined by direct protein sequencing and cDNA analysis, predicting a 276 amino acid membrane protein (MW 30kd) which showed no homology to any previously known protein sequence[1]. The lack of an amino-terminal signal peptide cleaved off the nascent protein, suggested that PLP is translated and assembled into the membrane using internal signal and stop-transfer sequences[2]. Based largely on hydrophobicity analyses, theoretical considerations[3], and the location of acylated cystein residues[4], the PLP peptide chain is assumed to transverse the membrane bilayer four times with the N- and C-terminus oriented to the inside (Fig. 1, for a discussion of the 4-helix bundle and alternative PLP models see Popot et al.[3]) a view at variance with alternative models for PLP proposed earlier. In this respect, PLP resembles connexin, synaptophysin, and other "gap junction" forming proteins that form homooligomeric complexes in the plane of the membrane. A smaller isoform of PLP, DM-20 (MW 26kd), is generated by alternative RNA splicing and lacks 35 amino acids preceding the third transmembrane domain[5]. The functional difference between PLP and DM-20 is presently not known but using sensitive RT-PCR detection methods, DM-20 transcripts were observed earlier in brain development than full-length PLP mRNA which appears specific to myelin-forming glial cells[6]. The combined effort of several laboratories has established the primary sequence of PLP from five mammalian species (mouse, rat, dog, cow, and human). Remarkably, the protein sequences are nearly identical (100% between rodents and human, and >98% including cow and dog), a phenomenon unprecedented for any myelin protein[1]. The evolutionary pressure on the PLP primary structure is not understood but is highly suggestive for multiple critical protein-protein interactions, and supported by the identification of point mutations in the PLP gene (see below). In order to identify regions of the molecule that are tolerable to variability in the primary structure, we have determined the PLP cDNA sequence[7] of two non-mammalian vertebrates: *Gallus domesticus* and *Xenopus laevis* (in vertebrate evolution, the 30 kd PLP is first immunodetectable in myelin of tetrapodes whereas lower vertebrates contain P_0-related proteins in central myelin). In both species the amino acid encoded PLP diverges from the mammalian PLP sequence but the overall conservation is still high (chicken: 95%, frog 70%). Conservative substitutions are found throughout the molecule, but non-conservative amino acid differences in *Xenopus* map only outside the predicted transmembrane domains. Uninterrupted stretches of up to 18 residues are maintained at 100% identity between amphibia and mammals[7]. This suggests that the most important protein-protein interactions of PLP occur in the plane of the membrane, potentially in the formation of homooligomeric complexes. Other highly conserved regions include the extracellular loops of PLP whereas the positively charged cytoplasmic domain (partially deleted from DM-20) displayed most sequence divergence. Interestingly, we were unable to identify any DM-20 like isoforms in a cDNA library from adult *Xenopus* brain (in 80 clones isolated) or using a sensitive polymerase chain reaction. This suggests that the 26k isoform of PLP is not required in myelin of the adult *Xenopus* central nervous system. We note, however, that DM-20 related sequences exist in lower vertebrates such as *Torpedo californica* (D. Colman, this volume). *Xenopus* may have lost the alternative 5' splice site in exon 3 of the PLP gene.

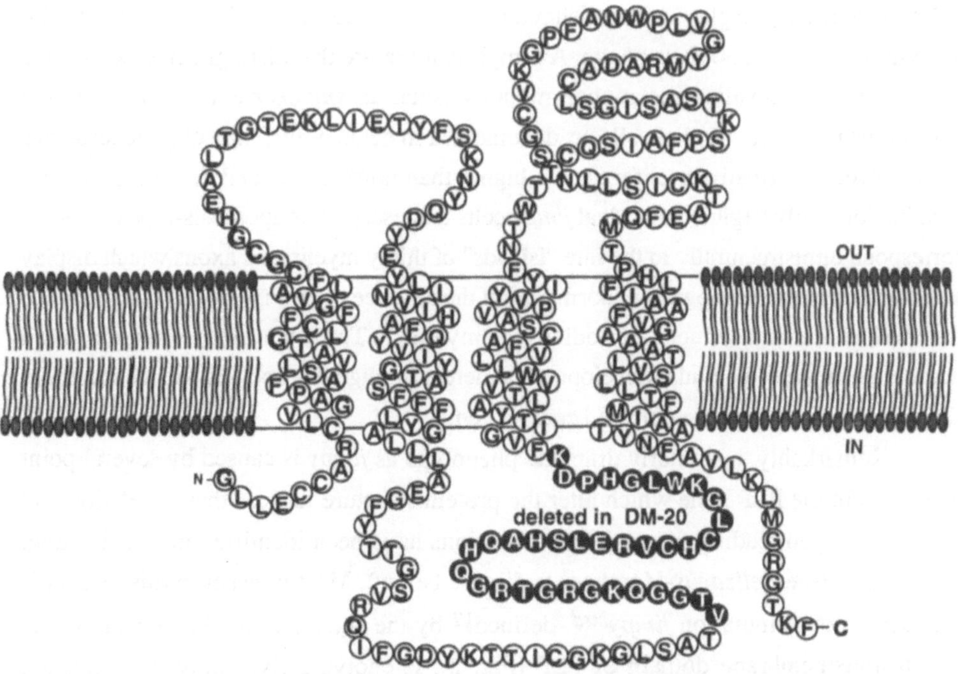

Figure 1. Hypothetical model of PLP and its alternatively spliced isoform (DM-20) depicted as a 4-helix bundel. For a discussion of the PLP topology see Popot et al.[3].

MUTATIONS OF THE PLP GENE CAUSING GLIAL CELL DEATH AND MYELIN-DEFICIENCY

The high evolutionary conservation of PLP in mammals is reflected by point mutations in the PLP gene. Some of these result in rather subtle changes of the overall protein structure whereas others cause abnormal protein folding. These mutations enable us to study the function of PLP *in vivo* and have provided an important insight into the etiology of X chromosome-linked human developmental disorders of myelin formation. More than 15 different mutations of the PLP gene have now been identified, most of them in neurological mutant mice[8-12] and in human families with *Pelizaeus-Merzbacher* disease[13-15]. The location of some amino acid substitutions and the effect of the mutations in the mouse PLP gene have been depicted in Fig. 2.

The first identified mutation of the PLP gene and the morphologically best characterized example is the *jimpy* mutation (*jp*) in the mouse. A single base change (A->G) in the splice acceptor site of intron 4 in the PLP gene (a seven exon transcription unit) results in the loss of exon 5 from the spliced *jp*-PLP mRNA, and thus in a deletion and frameshift of the normal protein coding region[8-10]. The predicted *jp*-PLP is a truncated protein (as shown in Fig. 2) in which the fourth transmembrane domain is replaced by an aberrant cystein-rich carboxyl terminus. This protein is degraded shortly after synthesis. Phenotypically, *jimpy* mice are completely myelin-deficient in the CNS (<1% of normal), display severe behavioural defects (tremors and seizures), and die prematurely at about 30 days of age. An important feature that distinguishes *jimpy* mice from other hypomyelinating mouse mutants (such as *shiverer* and *quaking*) is the developmental defect of glial cells: as documented in detail[16-18], *jimpy* oligodendrocytes fail to develop normally, proliferate at a higher than normal rate, and degenerate before myelination really begins. Individual *jimpy* cells that escape this apoptosis-type cell death correspond ultrastructurally to the rare "islands" of thinly myelinated axons which display a characteristic morphological abnormality in the absence of PLP: a "fused" single-space intraperiod line and reduced periodicity of myelin[19]. Thus, the *jimpy* mutation has a twofold phenotype, a cellular developmental defect of oligodendrocytes and a subcellular abnormality of myelin assembly and compaction.

Remarkably, a similarly dramatic phenotype as *jimpy* is caused by several point mutations in the PLP gene which alter the protein structure far less than a deletion and shift of the open reading frame. These mutations have been identified in rodents, dog, and humans (see *Pelizaeus-Merzbacher* disease below). Most unusual in this respect is the lethal mouse mutation $jimpy^{msd}$ defined[12] by the substitution Ala^{241}->Val in the fourth transmembrane domain of PLP (Fig. 2). Phenotypically, $jimpy^{msd}$ mice are nearly as severely affected as *jimpy*, with approximately twice as many oligodendrocytes escaping abnormal cell death[20].

UNCOUPLING OF PLP-DEPENDENT DYSMYELINATION FROM ABNORMAL CELL DEATH

The predicted dual function of PLP in both oligodendrocyte development and myelin assembly was also suggested by a novel neurological mutation, the *rumpshaker* mouse (*rsh*). The degree of dysmyelination in *rsh* is less severe than in other PLP mutants enabling the mice to live without seizure activity. Whereas *jimpy* mice and most rodent alleles die within 30 days after birth, *rsh* mice display a tremoring phenotype but

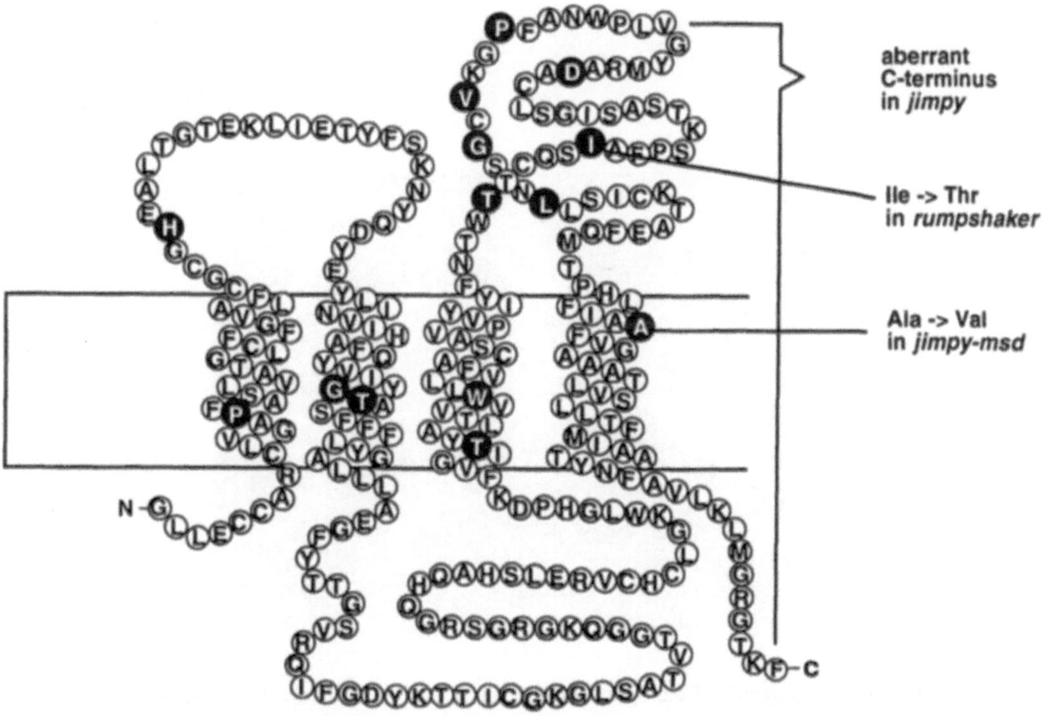

Figure 2. Summary of mutations in the mammalian PLP gene which affect the structure of the encoded protein (depicted as the 4-a helix model). Shown are the molecular defects of three allelic mouse mutants. The positions of single amino acid substitutions in other species, including human patients with *Pelizaeus-Merzbacher* disease, are indicated by black circles.

have a normal life span and breed well. The molecular defect of *rsh* was identified as a point mutation[11] in exon 4 resulting in the substitution Ile^{186}->Thr. The resulting cellular defect is less severe than in *jimpy* with most oligodendrocytes being at least ultrastructurally normal and escaping aberrant cell death. Nevertheless, a substantial fraction of the *rsh* oligodendrocytes fails to assemble normal myelin (Fig. 3) causing widespread hypomyelination of the central nervous system. The severity of dysmyelination varies significantly between brain regions with the early myelinating areas relatively better differentiated than late myelinating regions. Whereas the mutant form of PLP in *jimpy* is rapidly degraded after synthesis, the compacted myelin in *rsh* mice stains well with anti-PLP antibodies[21]. This and the presence of frequently well-compacted myelin sheaths of close to normal thickness suggests that the merely structural function of PLP is largely preserved in this mutant. The dysmyelination in *rsh* is thus a result of continued oligodendrocyte dysregulation lacking a presumably non-structural function of PLP.

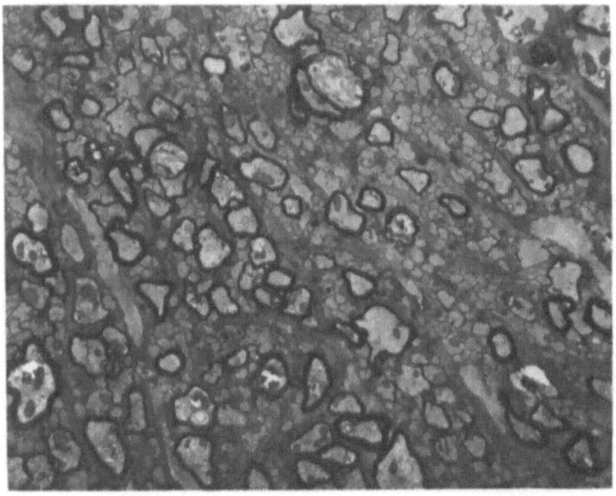

Figure 3. Electronmicrograph of the dysmyelinated spinal cord in a 90 day old *rumpshaker* mouse. Note the lack of myelin from large caliber axons next to apparently normally myelinated fibers. Compact myelin can be stained with anti-PLP antibodies and oligodendrocytes appear ultrastructurally normal (not shown).

GENERATION OF PLP-TRANSGENIC MICE: THE *JIMPY* MUTATION IS DOMINANT-NEGATIVE

By formal genetics, *jp* is a "recessive" disorder, affecting male hemizygous mice (*jp*/Y) whereas the heterozygous females (*jp*/+) are behaviourally normal. However, 50%

of oligodendrocytes in the carrier female are genotypically *jimpy* due to X-chromosome inactivation in early embryonic development. Thus, *jp*/+ mice are mosaics with respect to the PLP gene, and it is unknown whether the *jimpy* defect is truly dominant or recessive at the single cell level. To overcome X-inactivation and to generate for the first time a cellular heterozygote situation, we generated transgenic mice which carry an autosomal copy of the complete wild-type PLP gene[22]. These mice were used in a second step to transmit the PLP transgene genetically onto the *jimpy* background. The entire normal PLP gene (comprising more than 15kb of 5' flanking region, the 17kb transcription unit, and 5kb of 3' flanking sequences) was isolated from a C57Bl/6 mouse cosmid library (kindly provided by Dr. G. Evans, Salk Institute). A restriction fragment (containing a 3.8kb promotor fragment upstream of the 7 exon gene) was used for oocyte microinjections, and two transgenic lines (#66 and #72) were established (Fig. 4). In subsequent breeding experiments, mice were generated that were genotypically *jimpy* (jp/Y), male, and contained one allele of the autosomal transgene #66 or #72. These mice enabled us to quantify the transcriptional expression of the PLP transgene relative to the endogenous *jimpy* allele. By RNase protection or PCR amplification of PLP/DM-20 cDNA fragments, (utilizing the 74 nucleotide size difference of the aberrantly spliced *jimpy* transcript), we determined the expression of the wild-type PLP transgenes to be 115% and 55%, respectively.

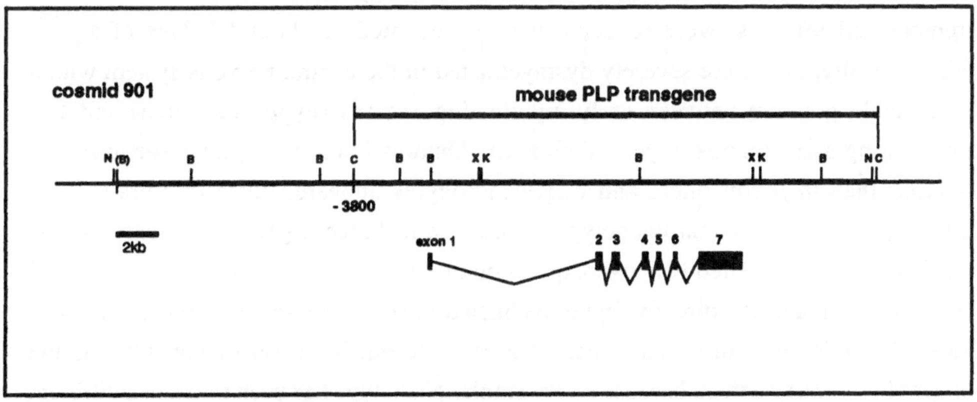

Figure 4. Structure of the mouse PLP gene (cosmid clone 901). The 7 exon transcription unit and the ClaI fragment used to generate transgenic mice are indicated.

Phenotypically, PLP-transgenic *jp*/Y mice in either line were mutants with the same clinical signs of non-transgenic *jimpy* animals, including severe tremors, seizures, and premature death before 30 days of age. Dysmyelination of the CNS was confirmed histologically and was essentially undistinguishable from the well documented myelination defect of *jimpy* mice. One difference, however, was striking: the occasional "islands" of myelin in the transgenic mutant stained well with antibodies against PLP whereas in non-transgenic *jimpy* mice, PLP is rapidly degraded in the ER and has never been detected immunologically in myelin. Thus, in the presence of a wild-type PLP transgene, a significant amount of PLP is assembled properly into the membrane and shuttled into myelin[23]. However, these amounts are insufficient to rescue the underlying *jimpy* defect. The *jimpy* mutation therefore acts as a dominant-negative allele over the wild-type PLP gene even though both alleles are about equally expressed at the RNA level. We suggest that normal PLP and the aberrantly folded mutant proteins interact with each other in the membrane of the ER, possibly in the formation of a homooligomeric PLP complex which is then subject to proteolytic degradation. It is also possible that *jp*-PLP is, in a less specific way, toxic to oligodendrocytes with its unique cystein-rich carboxy terminus[8]. These possibilities are now testable using other PLP-mutant strains of mice.

PLP OVEREXPRESSION CAUSING LETHAL DYSMYELINATION IN TRANSGENIC MICE

A very surprising observation was made when PLP-transgenic mice in line #66 were backcrossed in order to create homozygous animals. The expression of both transgene alleles (in addition to the endogenous wild-type PLP gene) caused a novel mutant phenotype in mice that strongly resembled *jimpy*. All affected mice developed tremors and seizures, were reduced in size, and died at about 60 days of age[22]. Histologically, they were severely dysmyelinated in the central nervous system with a remarkable gradient between early myelinating regions (hypomyelinated) and late myelinating areas (almost myelin-deficient). Abnormal cell death, however, was less obvious than in *jimpy* mice, and oligodendrocytes appeared ultrastructurally well differentiated. This mutant phenotype which also included a generalized astrocytosis resulted from a calculated PLP overexpression of maximal +230% at the transcriptional level. When quantified directly, hypomyelinated mice in line #66 expressed only 50% more PLP mRNA than normal mice. We can rule out insertion mutagenesis as the underlying primary defect, because a very similar phenotype has been noted in a different line of PLP transgenic mice[24].

Our observation suggests that oligodendrocytes are very sensitive to the accurate expression level of the normal PLP gene. This is surprising, since several structural genes have been overexpressed in other cell types and have never been found to affect the animal as long as transgenic overexpression was moderate (<10 fold) and non-ectopic, e.g. driven by the cognate regulatory sequences. In fact, only one (clearly regulatory) molecule, the transcription factor Hox1.4, has been shown to cause a developmental defect when overexpressed in transgenic mice[25]. The dysmyelination described here adds further support to our suggestion that PLP serves a non-structural "premyelin" function[11] in developing glial cells. It is tempting to speculate that this function is part of a signal-transduction pathway between oligodendrocytes and their neighbouring cells.

IMPLICATIONS FOR HUMAN GENETIC DISEASE

Several groups have identified mutations of the human PLP gene as the primary cause of *Pelizaeus-Merzbacher* disease (PMD), a lethal developmental disorder of myelin formation in the central nervous system. The majority of mutations differ between affected families (as expected for X-chromosomal defects) and range from conservative to non-conservative amino acid substitutions[13-15] to the loss of the entire PLP locus[26].

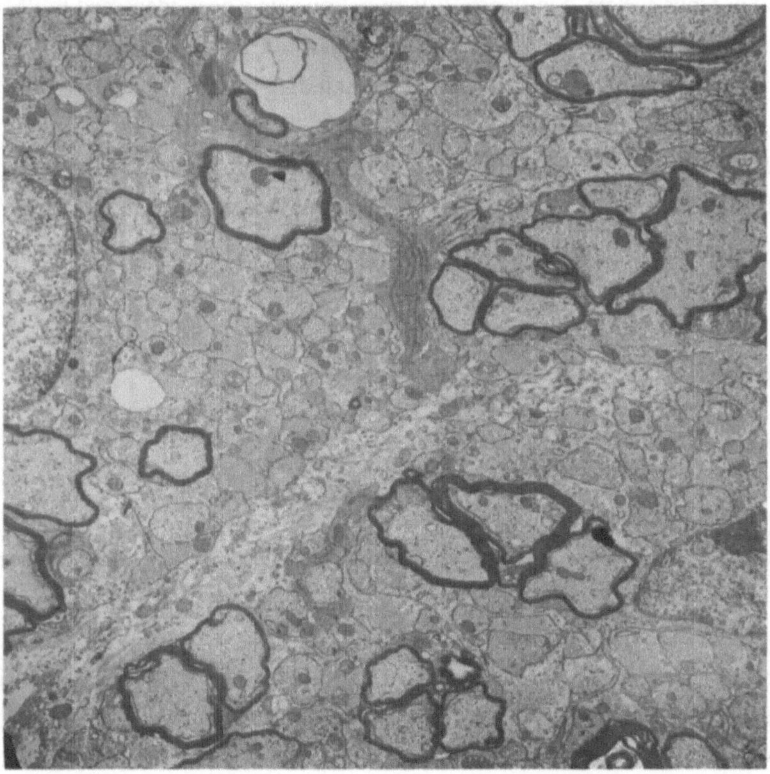

Figure 5. Dysmyelination in the optic nerve of transgenic mice (postnatal day 60) caused by moderate overexpression of the wild-type PLP gene.

No systematic correlation between severity and onset of disease and the kind of molecular defect has been identified. In general, the pathology of PMD shows more homology to *jimpy* and its lethal alleles than to the viable *rumpshaker* mouse. In several cases of typical PMD, no structural alterations of PLP have been detected by sequence analysis of the PLP coding exons in genomic DNA. Our results obtained with PLP transgenic mice suggest that mutations and chromosomal rearrangements that affect the normal transcriptional regulation of the gene may as well cause dysmyelination in humans. In fact, a report of one patient with an interstitial duplication of the X-chromosome (the q13-23 region containing the PLP gene) notes that PMD is the major clinical finding associated with this chromosomal anomalie[27]. This suggests that the twofold expression level of PLP in human is sufficient to cause dysmyelination, in good agreement with our experimental data of PLP overexpression in transgenic mice. To our knowledge, this provides the first example of a chromosomal abnormality for which the disease causing gene has been experimentally identified.

PLP shows an overall structural homology to the recently discovered myelin protein PMP-22 (see H.-W. Müller, this volume), a 4-helix bundle membrane protein of the peripheral nervous system. Similar to PLP, mutations have been identified in PMP-22 of neurological mutant mice[28] (*trembler*), and indirect evidence suggests that an interstitial duplication of human chromosome 17 (including the PMP-22 locus) underlies some forms of *Charcot-Marie-Tooth* disease type 1a[29,30]. Future studies will determine whether PLP and PMP-22 serve corresponding functions in the central and peripheral nervous system which may exceed that of strictly structural proteins of myelin assembly.

REFERENCES

1. G. Lemke, Unwrapping the genes of myelin, *Neuron* 1: 535 (1988).
2. R.J. Milner, C. Lai, K.-A. Nave, D. Lenoir, J. Ogata, and J.G. Sutcliffe, Nucleotide sequences of two mRNAs for rat brain myelin proteolipid protein, *Cell* 42:931 (1985).
3. J.-L. Popot, D. Pham-Dinh, and A. Dautigny, Major myelin proteolipid: the 4-a-helix topology, *J. Membr. Biol.* 120:233 (1991).
4. T. Weimbs and W. Stoffel, Proteolipid protein (PLP) of CNS myelin: positions of free, disulfide-bonded, and fatty acid thioester-linked cysteine residues and implications for the membrane topology of PLP, *Biochem.* 31:12289 (1992).
5. K.-A. Nave, F. E. Bloom, and R. J. Milner, A single nucleotide difference in the gene for myelin proteolipid protein defines the *jimpy* mutation in mouse. *J. Neurochem.* 49:1873 (1987).
6. K. Ikenaka, T. Kagawa, and K. Mikoshiba, Selective expression of DM-20, an alternatively spliced myelin proteolipid protein gene product, in developing nervous system and in nonglial cells, *J. Neurochem.* 58:2248 (1992).
7. B. Kiefer, A. Schneider, and K.-A. Nave, Primary structure of two proteolipid protein isoforms from *Xenopus laevis*: unusual conservation of the membrane-spanning domains, (submitted).
8. K.-A. Nave, C. Lai, F.E. Bloom, and R.J. Milner, *Jimpy* mutant mouse: a 74-base deletion in the mRNA for myelin proteolipid protein and evidence for a primary defect in RNA splicing, *Proc. Natl. Acad. Sci. USA* 83:9264 (1986).

9. L.D. Hudson, J.A. Berndt, C. Puckett, C.A. Kozak, and R.A. Lazzarini, Aberrant splicing of proteolipid protein mRNA in the dysmyelinating *jimpy* mutant mouse, *Proc. Natl. Acad. Sci. USA* 84:1454 (1987).

10. W.B. Macklin, M.V. Gardinier, K.D. King, and K. Kampf, An AG › GG transition at a splice site in the myelin proteolipid protein gene in *jimpy* mice results in the removal of an exon, *FEBS Lett.* 223:417 (1987).

11. A. Schneider, P. Montague, I. Griffiths, M. Fanarraga, P. Kennedy, P. Brophy, and K.-A. Nave, Uncoupling of hypomyelination and glial cell death by a mutation in the proteolipid protein gene, *Nature* 358:758 (1992).

12. S. Gencic and L.D. Hudson, Conservative amino acid substitution in the gene encoding myelin proteolipid protein disrupts oligodendrocyte differentiation, *J. Neurosci.* 10:117 (1990).

13. L.D. Hudson, C. Puckett, J. Berndt, J. Chan, and S. Gencic, Mutation of the proteolipid protein gene *PLP* in a human X chromosome-linked myelin disorder, *Proc. Natl. Acad. Sci. USA* 86:8128 (1989).

14. J.A. Trofatter, S.R. Dlouhy, W. DeMyer, P.M. Conneally, and M.E. Hodes, Pelizaeus-Merzbacher disease: tight linkage to proteolipid protein gene variant, *Proc. Natl. Acad. Sci. USA* 86:9427 (1989).

15. D. Pham-Dinh, J.-L. Popot, O. Boeseflug-Tanguy, P. Landrieu, J.-F. Deleuze, J. Boué, P. Jollés, and A. Dautigny, Pelizaeus-Merzbacher disease: a valine to phenylalanine point mutation in a putative extracellular loop of myelin proteolipid, *Proc. Natl. Acad. Sci. USA* 88:7562 (1991).

16. E. Farkas-Bargeton, O. Robain, and P. Mandel, Abnormal glial maturation in the white matter in *jimpy* mice, *Acta Neuropath.* 21:272 (1972).

17. R.P. Skoff, Increased proliferation of oligodendrocytes in the hypomyelinated mouse mutant *jimpy*, *Brain Res.* 248:19 (1982).

18. P.E. Knapp, R.P. Skoff, and D.W. Redstone, Oligodendroglial cell death in *jimpy* mice: an explanation for the myelin deficit, *J. Neurosci.* 6:2813 (1986).

19. I.D. Duncan, J.P. Hammang, S. Goda, and R.H. Quarles, Myelination in the *jimpy* mouse in the absence of proteolipid protein, *Glia* 2:148 (1989).

20. S. Billings-Gagliardi, L.H. Adcock, and M.K. Wolf, Hypomyelinated mutant mice: description of jp^{msd} and comparison with jp and qk on their present genetic backgrounds, *Brain Res.* 194:325 (1980).

21. M.L. Fanarraga, I.R. Griffiths, M.C. McCulloch, J.A. Barrie, P.G.E. Kennedy, and P.J. Brophy, Rumpshaker: an X-linked mutation causing hypomyelination: developmental differences in myelination and glial cells between the optic nerve and spinal cord, *Glia* 5:161 (1992).

22. C. Readhead, A. Schneider, I. Griffiths, and K.-A. Nave, Dysmyelination and astrocytosis in transgenic mice overexpressing the proteolipid protein gene, (submitted).

23. A. Schneider, I. Griffiths, C. Readhead, and K.-A. Nave, Dominant-negative action of the *jimpy* mutation in the mouse proteolipid protein gene, (submitted).

24. K. Ikenaka, (personal communication).

25. D.J. Wohlgemuth, R.R. Behringer, M.P. Mostoller, R.L. Brinster, and R.D. Palmiter, Transgenic mice overexpressing the mouse homeobox-containing gene *Hox-1.4* exhibit abnormal gut development, *Nature* 337:464 (1989).
26. W.H. Raskind, C.A. Williams, L.D. Hudson, and T.D. Bird, Complete deletion of the proteolipid protein gene (PLP) in a family with X-linked Pelizaeus-Merzbacher disease, *Am. J. Hum. Genet.* 49:1355 (1991).
27. F.P.M. Cremers, R.A. Pfeiffer, T.J.R. van de Pol, M.H. Hofker, T.A. Kruse, B. Wieringa, and H.H. Ropers, An interstitial duplication of the X chromosome in a male allows physical fine mapping of probes from the Xq13-q22 region, *Hum. Genet.* 77:23 (1987).
28. U. Suter, A.A. Welcher, T. Özcelik, G.J. Snipes, B. Kosaras, U. Francke, S. Billings-Gagliardi, R.L. Sidman, and E.M. Shooter, *Trembler* mouse carries a point mutation in a myelin gene *Nature* 356:241 (1992).
29. P.I. Patel, B.B. Roa, A.A. Welcher, R. Schoener-Scott, B.J. Trask, L. Pentao, G. Jackson Snipes, C.A. Garcia, U. Francke, E.M. Shooter, J.R. Lupski, and U. Suter, The gene for the peripheral myelin protein PMP-22 is a candidate for Charcot-Marie-Tooth disease type 1A, *Nature Genet.* 1:159 (1992).
30. V. Timmerman, E. Nelis, W. Van Hul, B.W. Nieuwenhuijsen, K.L. Chen, S. Wang, K. Ben Othman, B. Cullen, R.J. Leach, C.O. Hanemann, P. De Jonghe, P. Raeymaekers, G.-J.B. van Ommen, J.-J. Martin, H.W. Müller, J.M. Vance, K.H. Fischbeck, and C. Van Broeckhoven, Myelin gene PMP-22/gas-3 is contained within the Charcot-Marie-Tooth disease type 1a duplication, *Nature Genet.* 1:171 (1992).

MEASLES VIRUS INDUCED AUTOIMMUNE REACTIONS AGAINST BRAIN ANTIGEN

Uwe G. Liebert and Volker ter Meulen

Institut für Virologie und Immunbiologie, Universität Würzburg
Versbacher Str. 7, D-97078 Würzburg, Germany

SUMMARY

The recovery from viral infections is a result of complex interactions between specific and nonspecific host immunoreactions and the infectious agent. A variety of immune mechanisms are important factors in this event and operate together in overcoming the infectious process. Although much is known about viral defense mechanisms, it has proved remarkably difficult to assign a determinative role *in vivo* to any single immunological antiviral mechanism in recovery from a viral disease. Furthermore, the immune response to the virus itself may contribute to disease pathology. If virus induced immune responses are also directed against normal host components, autoimmune disease may develop. In this context measles virus (MV), which still causes significant morbidity and mortality despite the availablility of a live vaccine, is of interest because a well known complication of many cases of acute measles is postinfectious encephalomyelitis (acute measles encephalomyelitis). A cell-mediated immune response against myelin basic protein (MBP) has been detected in this disease and such an autoimmune response can lead to allergic encephalitis (EAE) as was shown in our animal model of MV-induced encephalitis. Further evidence for its pathogenetic role is provided by observations that (1) MV can substitute for sequences from the encephalitogenic region of MBP in the induction of EAE, and (2) some animals enter a refractory state for EAE following infection with recombinant MV nucleocapsid protein. In the following pages virological and immunological findings of MV infections in a rat model in relation to autoimmune reactions will be presented and the mechanisms by which measles virus may alter host reactivity against self-antigens discussed.

INTRODUCTION

Measles virus (MV) is a common pathogen which causes a variety of diseases. In a seronegative host, usually acute measles develops which is an extremely contageous, febrile disease characterized by rash and catarrhal inflammation of the eyes and respiratory tract. It

A Multidisciplinary Approach To Myelin Diseases II
Edited by S. Salvati, Plenum Press, New York, 1994

is principally a benign disease but depending on the socio-economic status of the patient and on the occurrence of complications during the course of measles, a high rate of mortality in particular in the developing countries is observed.[1] Characteristic features of MV are also its neurotropism and neurovirulence. Abnormal EEG pattern and elevated lymphocyte counts in the cerebro-spinal fluid (CSF) are actually present in 50 % or more of patients during the exanthematous stage of measles and 1 in 1000 patients with measles develop acute encephalitis. In addition to acute measles, MV is the etiological agent for subacute sclerosing panencephalitis (SSPE) and measles inclusion body encephalitis (MIBE). These two CNS diseases develop on the basis of a MV persistent infection of brain cells months to years after the acute MV infection.[2]

Beside the known MV diseases there are other disorders in which MV could not be unequivocally identified as the etiological agent for these disorders. In particular, in multiple sclerosis (MS) MV is still one of the leading candidate viruses since many MS patients exhibit a humoral hyperimmune response to MV in serum and cerebrospinal fluid specimens. In addition, MV RNA has been detected in some MS brains by *in situ* hybridization.[3,4,5] Similarly, in Paget's disease as well as in active otosclerosis viral RNA has been located in bone lesions by *in situ* hybridization but it is unknown whether MV plays a pathogenetic role in this disease process or represents an opportunistic pathogen in these cases.[6,7] The last group of diseases that has been associated with MV are various autoimmune syndromes, such as chronic hepatitis, glomerulonephritis, and lupus erythematosus.[8,9,10] In these illnesses either MV RNA has been detected in peripheral lymphocytes or high antibody titres have been found against MV in these patients. However, so far no additional observations exist which would indicate that this virus plays an etiological or pathogenetic role in these diseases.

MEASLES VIRUS ASSOCIATED AUTOIMMUNE ENCEPHALOMYELITIS

Of particular interest is the development of measles encephalomyelitis as a complication of acute measles. The encephalitis has an abrupt onset during its exanthematous or respiratory phase. As the most common neurological complication of measles it has an incidence of approximately 1: 1,000 cases, mostly in older children and young adults with a mortality rate of 10 - 20 %. Survivers usually reveal neurological sequelae.[11] Histologically, lymphomonocytic infiltration with perivenular demyelination occurs in brain tissue which resembles to some extent changes seen in experimental allergic encephalomyelitis (EAE). There is no evidence that MV is present in brain tissue since all attempts to isolate infectious virus or to detect viral antigen or viral RNA have failed so far.[12] These negative findings together with the neuropathological changes suggest an immunopathological basis for the development of postinfectious measles encephalitis.

This interpretation is as yet only supported by circumstantial evidence. In patients with acute measles encephalomyelitis a significant proliferative response of isolated peripheral blood lymphocytes against myelin basic protein (MBP) has been found.[11] Moreover, in cerebro-spinal fluid (CSF) specimens from such patients, MBP was detected as a result of myelin breakdown. In addition to these cell mediated immune (CMI) responses to MBP patients with measles encephalitis reveal prolonged immunological abnormalities in particular elevation of IgE as well as lower titers of soluble Il-2 receptor in comparison to patients with uncomplicated measles.[13,14]

An MBP specific lymphoproliferative response has also been seen in postinfectious encephalomyelitis following rubella, varicella, and respiratory infection as well as in patients with complications of post exposure rabies immunization.[15] The latter disorder is probably the human equivalent of EAE since the patients received rabies vaccine prepared in brain tissue. In analogy to EAE, it is conceivable, that the finding of MBP specific lymphoproliferative response in these virus infections is of pathogenetic importance.

Measles virus infection and autoimmune reactions in experimental animals

On the basis of the available evidence for a possible immunopathological process in measles encephalomyelitis animal experiments were mandatory to study in detail the mechanism by which a measles infection leads to the adverse immunological effects. We therefore developed a rat model in which a MV infection leads to CNS disease accompanied by CMI response to brain antigen. In the following sections the findings of our model will be reviewed and some of the mechanisms by which autoimmune diseases of the CNS may develop are discussed.

Measles virus infection in rats: Experimental infection of neonatal or weanling Lewis rats with the murine adapted CAM/RB strain of MV leads to an acute encephalomyelitis (AE) which is usually fatal, particularly in very young animals.[16] The lesions in these animals are inflammatory and destructive as a result of a fulminating lytic virus-cell interaction. Infectious virus can be isolated from brain tissue of these rats. The occurrence of both clinical disease and neuropathological changes is age and strain dependent. In older rats a subacute encephalitis (SAME) develops with predominantly inflammatory changes consisting of perivascular lymphomonocytic cuffing. MV persists in such animals, but infectious virus is not produced. The molecular biological definition of the virus host relationship revealed a defective MV replication cycle with close similarity to the MV infections in SSPE and MIBE.[17]
Of particular interest was the demonstration that during infection of Lewis rats with MV CMI responses against MBP were initiated. These observations provided supportive evidence for the concept that virus infection may serve as an improtant trigger for an autoimmune CNS disease.

Measles virus infection and EAE: It has been shown in animal experiments that autoimmune diseases of the CNS can be induced by injection of MBP or proteolytic apoprotein in combination with Freund complete adjuvants.[18,19] The resulting experimental allergic encephalomyelitis (EAE) is characterized by transient clinical signs including weak or paralysed hindlimbs and weight loss. Neuropathologically, inflammatory changes consisting mainly of perivascular lymphomonocytic cuffing are commonly found predominantly in the white matter of the spinal cord. In addition, in some laboratory animal species and strains spinal cord demyelination can be observed although its appearance depends on the age of animals and the inoculum used.[20,21] By intravenous injection of $CD4^+$ lymphocytes specific for MBP, EAE can be passively transferred from a diseased to a naive recipient animal. This treatment results in a CNS disorder of the host animal which is clinically and neuropathologically very similar to EAE.[18]

We observed that 4 week-old rats infected with the neurotropic MV develop a subacute measles encephalomyelitis 1 - 3 months after infection.[16] The CNS changes are very similar to those of naive rats receiving MBP specific $CD4^+$ lymphocytes. This led us to characterize the immune response to brain antigens in the infected rats. Both, during and after SAME splenic lymphocytes and superficial cervical lymph node cells were found to proliferate *in vitro* in the presence of MBP or PLP.[22] MBP-reactive class II MHC restricted T cell lines were isolated from bulk cell populations of these animals. They were shown to exhibit no cross-reactivity with dirupted measles virions or isolated MV proteins. When adoptively transferred by intravenous injection the cell lines induce EAE in naive syngeneic recipients accompanied by clinical and histopathological signs identical to T cell mediated EAE. A humoral immune response to MBP was detected only in limited numbers of rats with SAME. The observations confirmed and extended older data that showed potentiation of EAE development and severity in MV infected hamsters.[23]

Characterization of MBP-specific T cell lines from rats with SAME: As a consequence of these observations, it was of interest to analyse the fine specificity of MBP-reactive T cell lines from animals with SAME and EAE. For this purpose a panel of synthetic peptides from guinea pig MBP (Gp-MBP) representing the amino acid residues 69-84 that comprises the major encephalitogenic sequence for Lewis rats were used in proliferation assays.[24] The MBP-specific T cell lines from both, measles infected and MBP challenged rats revealed the same fine specificity and responded to *in vitro* stimulation with the encephalitogenic peptides 69-84 as well as peptide 72-84 of Gp-MBP, but not to the non-encephalitogenic peptides 75-84, 69-81.[24] This high degree of antigenic specificity is further supported by the failure of all the T cell lines to proliferate in the presence of dirupted measles virions, isolated MV proteins or other control antigens or peptide sequences. In contrast, MV-specific T cell lines only responded to MV proteins and not when MBP or its synthetic peptides were added to the cultures.

Interaction between MBP-peptides and measles virus in the induction of EAE: We also looked for possible additional epitope specificities in the polyclonal spleen cell cultures from MV-infected rats. A positive reaction was found to the non-encephalitogenic peptide 69-81 in some rats, but not to the other non-encephalitogenic peptides like 75-84. Since it has been shown that only the simultaneous immunization of rats with both of these two non-encephalitogenic peptides 69-81 and 75-84 leads to EAE-like disease and pathology, but given alone neither one is effective[25,26] it was suggestive to define the role of MV infection in inducing an autoimmune response. For these experiments, a low dose of MV infection was chosen that failed to induce a clinically recognizable subacute CNS disease in animals. Histologically, these rats did not reveal changes indicative of an active encephalitic disease process when examined 4 - 8 weeks post infection. Moreover, infectious virus could not be isolated from brain tissue of these animals and virus antigen was not demonstrable in brain cells. However, in serum as well as CSF specimens, low titers of neutralizing anti-measles antibodies were detected, documenting the interaction of the host immune system with the infecting virus. Such animals were immunized 4 - 8 weeks post infection with either of the two non-encephalitogenic peptides. While peptide 75-84 had no effect, immunization with peptide 69-81 led to clinical and histological changes in the lumbar spinal cord.[24] The lesions closely resembled EAE and the distribution of the pathological changes was different from the MV-induced CNS lesion which are mainly located in the grey and white matter of the cerebral hemispheres, midbrain and upper spinal cord. Moreover, the disease induced by peptide 69-81 is not due to activation of MV in the brain of immunized Lewis rats, because virus could not be isolated from brain material and measles antigen was not detectable. Surprisingly, the immune responses to immunization with peptide 69-81 and the infection with MV act synergistically only when at least initially MV replication in brain tissue takes place. The synergistic interaction between MBP-peptide and MV-infection was not observed when rats were infected intraperitoneally or were immunized with inactivated MV, indicating that virus replication in brain cells is probably a prerequisite for a CNS autoimmune reaction to occur in this model system.

Search for measles sequences important for cell-mediated autoimmunity: Of further interest were attempts to characterize components of MV that could be important in the substitution for sequences of the encephalitogenic region of MBP in the induction of EAE. The treatment of rats with different combinations of antibodies and T cells reactive with either MV or its structural components or MBP-peptides showed that the transfer peptide 69-81 specific T cells into MV-infected animals resulted in the development of EAE but no other treatment combination.[27] Particularly the of simultaneous immunisation with peptide 69-81 and inactivated MV or isolated MV structural proteins had no effect on the development of EAE supporting our hypothesis that only after an active viral infection the vulnerability of the brain to autoimmune aggression is enhanced in our model. In contrast to these experiments we succeeded to prevent the development of EAE following immunization with MBP in approximately one third of rats which were infected with low doses of recombinant vaccinia virus (VVR) expressing the nucleocapsid protein.[28] All attempts with VVR expressing any other MV structural protein had no effect on EAE induction with either MBP or 69-81 peptide (unpublished observation). The results provide further evidence for our hypothesis and futher wake expectations that it might be possible to define a region or an epitope on the MV nucleocapsid protein important for virus induced autoimmunity.

Rat strain differences in susceptibility to measles encephalitis and autoimmunity: Successful induction of EAE in rats or mice is dependent on the genetic background of animals used.[29] Lewis ($RT1^l$) are highly susceptible while Brown Norway (BN) rats ($RT1^n$) are usually resistent.[30] Similarly, if autoimmune mechanisms participate in the pathogenesis of virally induced encephalomyelitis, BN rats should not develop SAME. Indeed, although susceptible to MV replication in the CNS, BN rats were resistent to measles encephalitis and did not develop a subacute clinical disease.[16] Neuropathologically the lesions were less prominent than in Lewis rats and consisted mainly of astroglial and microglial proliferation. Although the proliferative response to MBP of lymphocytes taken from Lewis rats with measles induced SAME appears to be variable it nevertheless represents a significant and reproducible phenomenon and is quite distinct from that observed with lymphocytes from BN rats that were unable to proliferate in the presence of MBP.[24] Obviously, there may be many factors involved in the susceptibility of these two rat strains to the development of MV induced CNS changes and disease. Nonetheless, the development of a MBP-specific CMI response cannot be ignored as a major factor .

POSSIBLE MECHANISMS OF VIRUS-INDUCED AUTOIMMUNE REACTION

Studies in EAE have revealed that immunization of susceptible animals with MBP only leads to an autoimmune disease process of the CNS in conjunction with complete Freund's adjuvants. MBP immunization alone is ineffective which raises the question as to how MV may force the host to mount a strong CMI response to brain antigen. Although in our rat model we could show that immunization against MV and a non-encephalitogenic MBP peptide leads to EAE there are certainly other ways by which to brake the immune tolerance against self antigens.

1. Target cell changes: The fact that viruses only multiply in living cells has major consequences for the cell as well as for the host. It has been proposed that during replication, the virus may incorporate host antigens into its envelope, insert, modify or expose cellular antigen on the cell surface. Such newly exposed antigens could be recognized as foreign by the host and could elicit a reaction in the same manner as any other previously unencountered protein.[31] So far, definite proof of this hypothesis has not been presented although in the context of certain virus infections this mechanism may induce autoimmune reactions.[32,33]

2. Interaction with lymphoid tissue: Measles virus has a strong tropism towards lymphocytes and can infect *in vitro* T and B lymphocytes as well as macrophages. Such an interaction with the immune regulatory system *in vivo* could lead to destruction of lymphocyte subpopulations or could stimulate generation and/or expression of autoreactive lymphocyte clones. One of the prime examples of lymphotropic viruses is Epstein-Barr virus which infects and transforms human B lymphocytes. Such immortalized cells may secrete under certain conditions autoantibodies that can react to cellular constituants.[34,35] In our rat model, the neurotropic MV is apparently not lymphotropic since neither B or T lymphocytes of rats can be infected. Therefore infection of lymphocytes does not form the basis of the cell-mediated autoimmune response in the model. In man it is, however, long documented and well known that measles infection alters host immune functions including T cell responses of the delayed type hypersensitivity (DTH) and antibody production. Furthermore, measles is sometimes followed by serious complications such as exacerbations of tuberculosis.[35]

3. Molecular mimicry: Molecular mimicry could be another mechanism by which an immune response is raised against certain viral antigens which may cross-react with normal host cell antigens.[36] Computer analysis of a variety of viral sequences revealed that several viruses contain part of the human MBP sequence in their genome.[37,38] Moreover, the immunization of rabbits with a synthetic peptide from such a sequence from hepatitis B virus polymerase led to the induction of EAE lesions in these animals.[39] So far, encephalitogenic rat MBP sequences have not been found in the viral genome of MV and measles-specific T cell lines isolated from infected rats do not proliferate in the presence of MBP.[22] However, one cannot exclude that this mechanism could play an important role in the development of autoimmune reactions if similar structures of an infectious agent and a host cell protein share antigenic sites or interact with identical T cell receptors.

4. Does virus-mediated MHC class II induction trigger autoimmunity ?

Until recently the CNS has been considered immunologically a privileged site because it lacks a lymphatic drainage, and most lymphoid cells are excluded from free entrance due to the presence of a blood brain barrier. Additionally, major histocompatibility (MHC) antigens are not expressed on most CNS parenchymal cells.[40,41] However, recent experiments suggest that the CNS is probably effectively controlled by the immune system. Although the precise conditions under which the immune system carries out the function is still unknown, the observation of MHC class II induction on glia cells by T lymphocyte factors, the presence of perivascular microglia cells with the *in vivo* potential to present antigens, and the observation that activated T cells may non-specifically cross the blood-brain barrier shed some light on the immunological mechanisms involved in CNS immune surveillance.[40,42,43,44]

Of particular interest for the development of an immunopathological reaction in the course of a CNS viral infection are the events of MHC class II induction on brain cells and a subsequent pathological response such as a DTH reaction in genetically susceptible hosts. It has been established that MHC antigens are expressed only at low levels or not at all on the bulk of cells in the CNS.[45] There is evidence, however, that a number of glial cells may be induced to express MHC antigens by treatment with agents such as interferon-γ *in vitro*[42] or as a result of an inflammatory reaction in vivo, presumably due to the release of interferon-g by infiltrating T cells.[46] The problem, however, remains as to which factor(s) initiate(s) such a reaction in the CNS, since without the presence of MHC antigens, it is undoubtedly very difficult for T cells to recognize antigen, to become activated and to release lymphokines. Therefore, in a viral infection of the CNS, additional mechanisms probably operate to induce MHC antigen expression which would allow the T lymphocyte to find its target cell.

With respect to MV, it has been shown that exposure of astrocytes in culture may lead to the expression of MHC class II antigen that can be further enhanced by the addition of tumor necrosis factor.[47,48] Moreover, it has been shown that astrocytes expressing class II antigen can present MBP to CD4+ T lymphocytes.[49] It is tempting to extrapolate these data obtained in *in vitro* studies to the *in vivo* situation, but this may conceivably be premature. It is still unknown which role astrocytes and microglia may play in the development of immune responses. Astrocytes are effective antigen presenting cells (APC) *in vitro*, at least for secondary CD4+ T cell responses [50,51] while in the case of inflammatory changes in human or animal brain tissue microglial cells are the most abundant MHC class II expressing population.[51,52] Although MHC expression on glia cells in brain tissue is always taken as an indicator of lymphocyte/glial cell interaction leading to further T cell activation, there is no unequivocal evidence to suggest that this actually occurs *in vivo* despite the overwhelming *in vitro* data supporting the role of glia cells as effective APC for proliferative T cell response.[50] Furthermore, it was recently observed that in BN rats a high proportion of microglia are constitutively MHC class II positive in comparison to Lewis rats suggesting that the mere expression of MHC class II on microglia is not indicative of an increased susceptibility to inflammatory T cell responses in the CNS as documented by the resistance to EAE and MV associated encephalomyelitis.[53] The constitutive expression of MHC class II on microglia of BN rats and perhaps more important the low levels of expression in Lewis rats may suggest that microglial MHC expression imparts on the cells the ability to interact with and may down-regulate T cell responses rather than amplify them. This could be one of a determinating factor for the protection or the development of an autoimmune disease process.

In conclusion, the occurrence of cell-mediated autoimmune reactions to brain antigen as a result of MV infection is probably of pathogenetic importance for the development of acute measles encephalomyelitis. At present, one cannot point to any one factor which leads to this adverse immune reaction. Indeed, it is more conceivable that this disease results from the sequelae of a number of different virus-induced changes, each one a relatively common event.

ACKNOWLEDGMENT

The experiments cited in this publication have been supported by the Deutsche Forschungsgemeinschaft, Bundesministerium für Forschung und Technologie and Hertie-Stiftung.

REFERENCES

1. Mitchell CD, Balfour HH: Measles control: so near and yet so far. Progr Med Virol 1985: 31: 1-42.
2. Schneider-Schaulies S, ter Meulen V: Molecular aspects of measles virus-induced central nervous system diseases. in: Roos RP (ed): Molecular Neurovirology, Humana Press, Totowa, New Jersey, USA, 1992, 419-448.
3. ter Meulen V, Stephenson JR: The possible role of viral infections in MS and other related demyelinating diseases. in Hallpike JF, Adams CWM, Tourtellotte WW (eds): Multiple Sclerosis, Chapman and Hall, London, 1983, pp. 241-274.
4. Cosby SL, McQuaid S, Taylor MJ, Bailey M, Rima BK, Martin SJ, and Allen IV: Examination of eight cases of multiple sclerosis and 56 neurological and non-neurological controls for genomic sequences of measles virus. J Gen Virol 1989: 70: 2027-2036.
5. Haase AT, Ventura P, Gibbs CJ Jr, Tourtellotte WW: Measles virus nucleotide sequences: detection by hybridization in situ. Science 1981: 212: 672-675.
6. Basle MF, Fournier JG, Rozenblatt S, Rebel A, Bouteille M: Measles virus RNA detected in Paget's disease bone tissue by in situ hybridization. J Gen Virol 1986: 67: 907-913.
7. McKenna MJ, Mills BG: Immunohistochemical evidence of measles virus antigens in active otosclerosis. Otolaryngol. Head Neck Surg 1989: 101: 415-421.
8. Andjaparidze OG, Chaplygina NM, Bogomolova NN, Koptyaeva IB, Nevryaeva EG, Filimova RG, Tareeva IE: Detection of measles virus genome in blood leucocytes of patients with certain autoimmune diseases. Arch Virol 1989: 105: 287-291.
9. Robertson DAF, Guy EC, Zhang SL, Wright R: Persistent measles virus genome in autoimmune chronic active hepatitis. Lancet 1987: ii: 9-11.
10. Black FL, Persistent measles virus genome in autoimmune chronic active hepatitis: cause or coincidence? Hepatology 1988: 8: 186-187.
11. Johnson RT, Griffin DE, Hirsch RL, Wolinsky JS, Roedenbeck S, de Soriano IL, Vaisberg A: Measles Encephalomyelitis - clinical and immunologic studies. N Engl J Med 1984: 310: 137-141.
12. Gendelman H, Wolinsky JS, Johnson RT, Pressman NJ, Pezeshkpour GH, Boisset GF: Measles encephalitis: Lack of evidence of viral invasion of the central nervous system and quantitative study of the nature of demyelination. Ann Neurol 1984: 15: 353-360.
13. Griffin DE, Ward BJ, Jauregui E, Johnson RT, Vaisberg A: Immune activation during measles. N Engl J Med 1989: 320: 1667-1672.
14. Griffin DE, Cooper SJ, Hirsch RL, Johnson RT, de Soriano IL, Roedenbeck S, Vaisberg A: Changes in plasma IgE levels during complicated and uncomplicated measles virus infections. J Allergy Clin Immunol 1985: 76: 206-213.
15. Johnson RT, Griffin D: Virus-induced autoimmune demyelinating disease of the CNS. in Notkins AL, Oldstone MBA (eds): Concepts in Viral Pathogenesis II, Springer Verlag, 1986, New York, pp. 203-209.
16. Liebert UG, Meulen ter V: Virological aspects of measles virus induced encephalomyelitis in Lewis and BN rats. J gen Virol 1987: 68: 1715-1722.
17. Schneider-Schaulies S, Liebert UG, Baczko K, Cattaneo R, Billeter M, Meulen ter V: Restriction of measles virus gene expression in acute and subacute encephalitis of Lewis rats. Virol 1989: 171: 525-534.
18. Raine CS: Biology of disease. Analysis of autoimmune demyelination: its impact upon Multiple Sclerosis. Lab Invest 1984: 50: 608-635.
19. Yamamura T, Namikawa T, Endoh M, Kunishita T, Tabira T: Experimental allergic encephalomyelitis induced by proteolipid apoprotein in Lewis rats. J Neuroimmunol 1986: 12: 143-153.
20. Itoyama Y, Webster H de F.: Immunocytochemical study of myelin-associated glycoprotein (MAG) and basic protein (BP) in acute experimental allergic encephalomyelitis (EAE). J Neuroimmunol 1982: 3: 351-364.
21. Vandenbark AA, Gill T, Offner H: A myelin basic protein-specific T cell line that mediates experimental allergic autoimmune encephalomyelitis. J Immunol 1985: 135: 223-228.
22. Liebert UG, Linington C, ter Meulen V: Induction of autoimmune reactions to myelin basic protein in measles virus encephalitis in Lewis rats. J Neuroimmunol 1988: 17: 103-118.
23. Massanari RM, Paterson PY, Lipton HL: Potentiation of experimental allergic encephalomyelitis in hamsters with persistent encephalitis due to measles virus. J Infect Dis 1978: 139: 297-303.
24. Liebert UG, Hashim GA, Meulen ter V: Characterization of measles virus-induced cellular autoimmune reactions against myelin basic protein in Lewis rats. J Neuroimmunol 1990: 29: 139-147.
25. Offner H, Hashim G, Vandenbark AA: Response of rat encephalitogenic T lymphocyte lines to synthetic peptides of myelin basic protein. J Neurosci Res 1987: 17: 344-348.
26. Hashim GA, Day ED: Role of antibodies in T cell-mediated experimental allergic encephalomyelitis. J Neurosci Res 1988: 21: 1-5.

27. Liebert UG, ter Meulen V: Synergistic interaction between measles virus infection and MBP peptide-specific T cells in the induction of EAE in Lewis rats. J. Neuroimmunol. 1993: 46,217-224.
28. Brinckmann UG, Bankamp B, Reich A, ter Meulen V, Liebert UG: Efficacy of single measles virus structural proteins in the protection of rats from measles encephalitis. J. gen. Virol. 1991: 72, 2491-2500.
29. Levine S, Sowinski R: Experimental allergic encephalomyelitis in inbred and outbred mice. J Immunol 1973: 110: 139-143.
30. Levine S, Sowinski R: Allergic encephalomyelitis in the reputedly resistant Brown Norway strain of rats. J Immunol 1975: 114: 597-601.
31. Hirsch MS, Proffitt MR: Autoimmunity in viral infection; in Notkins AL (ed): Viral Immunology and Immunopathology, Academic Press, New York, 1975, pp.419-434.
32. Notkins AL, Onodera T, Prabhaker BS: Virus-induced autoimmunity; in Notkins AL, Oldstone MBA (eds): Concepts in Viral Pathogenesis, Springer Verlag, New York, 1984, pp. 210-215.
33. Schattner A, Rager-Zisman B: Virus-induced autoimmunity. Rev Infect Dis 1990: 12: No. 2, 204-222.
34. Rosen A, Gergely P, Jondal M, Klein G, Britton S: Polyclonal Ig production after Epstein-Barr virus infection of human lymphocytes in vitro. Nature 1977: 267: 52-54.
35. Cherry JD: Viral infections: measles; in Feigin RD, Cherry JD (eds): Textbook of Pediatric Infectious Diseases, Vol. 2, Saunders, Philadelphia, London, Toronto, Mexico City, Rio de Janeiro, Sydney, Tokyo, 1987, pp. 1607-1635.
36. Oldstone MBA, Notkins AL: Molecular mimicry; in Notkins AL, Oldstone MBA (eds): Concepts in Viral Pathogenesis, Springer Verlag, New York, 1986, pp. 195-202.
37. Jahnke U, Fischer EH, Alvord EC: Sequence homology between certain viral proteins and proteins related to encephalomyelitis and neuritis. Science 1985: 229: 282-284.
38. Weise MJ, Carnegie PR: An approach to searching protein sequences for superfamiliy relationships or chance similarities relevant to the molecular mimicry hypothesis: application to the basis proteins of myelin. J Neurochem 1988: 51: 1267-1273.
39. Fujinami RS, Oldstone MBA: Amino acid homology and immune responses between the encephalitogenic site of myelin basic protein and virus: A mechanism for autoimmunity. Science 1985: 230: 1043-1045.
40. Wekerle H, Linington C, Lassmann H, Meyermann R: Cellular immune reactivity within the CNS. Trends Neurosci 1986: 9: 271-275.
41. Sedgwick JD, Dörries R: The immune system response to viral infection of the CNS. Sem Neurosci 1991: 3: 93-100.
42. Wong GHW, Bartlett PF, Clark-Lewis I, McKimm-Breschkin JL, Schrader JW: Interferon-g induces the expression of H-2 and Ia antigens on brain cells. J Neuroimmunol 1985: 7: 255-278.
43. Hickey WF, Kimura H: Perivascular microglial cells of the CNS are bone marrow-derived and present antigen in vivo. Science 1988: 239: 290-292.
44. Hickey WF, Hsu BL, Kimura H: T-lymphocyte entry into the central nervous system. J Neurosci Res 1991: 28: 254-260.
45. Hart DNJ, Fabre JW: Demonstration and characterization of Ia-positive dendritic cells in the institial tissues of rat heart and other tissues, but not brain. J Exp Med 1981: 154: 347-361.
46. Traugott U, Scheinberg LC, Raine CS: On the presence of Ia-positive endothelial cells and astrocytes in multiple sclerosis lesions and its relevance to antigen presentation. J Neuroimmunol 1985: 8: 1-14.
47. Massa PT, Dörries R, ter Meulen V: Viral particles induce Ia antigen expression on astrocytes. Nature 1986: 320: 543-546.
48. Massa PT, Schimpl A, Wecker E, ter Meulen V: Tumor necrosis factor amplifies measles virus-mediated Ia induction on astrocytes. Proc Natl Acad Sci USA, 1987: 84: 7242-7245.
49. Fontana A, Fierz W, Wekerle H: Astrocytes present myelin basic protein to encephalitogenic T cell lines. Nature 1984: 307: 273-276.
50. Sedgwick JD, Mößner R, Schwender S, ter Meulen V: MHC-expressing non-hematopoietic astroglial cells prime only $CD8^+$ T lymphocytes: Astroglial cells as perpetuators but not initiators of $CD4^+$ T cell responses in the central nervous system. J Exp Med 1991: 173: 1235-1246.
51. Matsumoto Y, Hara N, Tanaka R, Fujiwara M: Immunohistochemical analysis of the rat central nervous system during experimental allergic encephalomyelitis, with special reference to Ia-positive cells with dendritic morphology. J Immunol 1986: 136: 3668-3676
52. Hayes GM, Woodroofe MN, Cuzner ML, Microglia are the major cell type expressing MHC class II in human white matter. J Neurol Sci 1987: 80: 25-37.
53. Sedgwick JD, Schwender S, Gregersen R, ter Meulen V: Constitutive MHC class II expression on ramified microglia from BN strain rats resistant to experimental autoimmune encephalomyelitis. J Exp Med 1993: in press.

MECHANISMS OF DE - AND REMYELINATION IN AUTOIMMUNE ENCEPHALOMYELITIS AND MULTIPLE SCLEROSIS

Hans Lassmann,[1,2] Gerda Suchanek,[1] and Mascha Schmied[1]

[1]Research Unit for Experimental Neuropathology, Austrian Academy of Sciences
[2]Neurological Institute, University Vienna, Austria

INTRODUCTION

Several different immunopathological mechanisms have been suggested to be involved in the destruction of myelin in inflammatory demyelinating diseases. They include destruction of oligodendrocytes by virus infection[1] as well as immune mediated damage by macrophage products[2], specific antibodies[3] or T-lymphocytes[4]. To evaluate the validity of these concepts it is of critical importance to determine the fate of oligodendrocytes within the lesions[5,6,7,8]. This, however, was difficult and inconclusive because of the lack of specific and sensitive markers that allow the identification of oligodendrocytes in demyelinated plaques. Recently, new techniques became available to address this question, which have been applied in the present study.

MATERIALS AND METHODS

Our study was performed on three different inflammatory demyelinating diseases. Subacute demyelinating encephalomyelitis was induced in rats by intracerebral infection with corona virus[9]. The model of chronic relapsing autoimmune encephalomyelitis was established by repeated passive cotransfer of encephalitogenic T-lymphocytes directed against myelin basic protein with demyelinating monoclonal antibodies against myelin oligodendroglia glycoprotein[10]. In addition inflammatory demyelinating plaques were studied in biopsy and autopsy material from 25 cases of multiple sclerosis with a disease duration ranging from few weeks to several years.

Appropriate lesions were selected from a large number of sections and screened by routine neuropathology. Immunocytochemistry was performed with a biotin avidin technique[11] with monoclonal and polyclonal antibodies against the following myelin antigens: MOG (8-18C5, Y1 and Y12; a gift from Dr. S. Piddlesden, University Cardiff, UK), CNPase (Affinity

A Multidisciplinary Approach To Myelin Diseases II
Edited by S. Salvati, Plenum Press, New York, 1994

Research Products, Ilkeston, UK), MBP (a gift from Dr. J.M. Matthieu, University Lausanne, Switzerland), PLP (a gift from Dr. S. Piddlesden, University Cardiff, UK) and HNK1 (Leu7, Becton and Dickinson, Vienna, Austria). The protocol for *in situ* hybridization with specific probes for MBP, PLP, CNPase, MAG and MOG has been described in detail before[12]. A method for *in situ* nick translation was developed for the detection of DNA strand breaks in tissue sections, which allows the identification of degenerating cells within the lesions[13].

RESULTS AND DISCUSSION

Detection of oligodendrocytes in normal adult and developing rat central nervous system tissue

As described in detail earlier oligodendrocytes can be detected in adult and developing brain tissue by immunocytochemistry for MOG and CNPase. MOG is expressed on the surface of oligodendrocytes and myelin sheaths, whereas CNPase is located in the cytoplasm of oligodendrocytes and their processes[14]. The mRNAs for MBP, PLP, MAG and CNPase were found in the cytoplasm of oligodendrocytes during active myelination (Figure 1a,b) and also in adult animals[12]. MOG mRNA appeared later during the process of myelination, but was prominently detectable in a subpopulation of oligodendrocytes in adult rats. Furthermore, MOG is expressed on the surface of oligodendrocytes, that survive in demyelinated lesions, such as those found during Wallerian degeneration[15].

Oligodendrocytes in corona virus induced subacute demyelinating encephalomyelitis

In this model during the acute phase of demyelination numerous oligodendrocytes could be identified, which were double labelled by *in situ* nick translation and immunocytochemistry for MOG and CNPase. Only a small fraction of these degenerating oligodendrocytes also contained detectable virus antigen by immunocytochemistry. This suggests that also uninfected oligodendrocytes can be destroyed within the lesions in the course of the immune reaction. In more chronic plaques, no oligodendrocytes were detected by either immuno-cytochemistry for MOG or CNPase or by *in situ* hybridization for PLP mRNA in the center of the lesions (Figure 1c,d). Increased mRNA for myelin proteins was detected at the edge of the lesions, indicating central remyelination from the lesional borders.

Oligodendrocytes in chronic relapsing autoimmune encephalomyelitis

As described before, cotransfer of MBP reactive T-lymphocytes with demyelinating antibodies leads to the formation of confluent demyelinated lesions in the central nervous system[3].

Furthermore, repeated cotransfers resulted in chronic relapsing clinical disease and persistently demyelinated lesions with little remyelination[10]. In the early stages of such chronic lesions, some cells, which showed DNA fragmentation by *in situ* nick translation, were labelled by anti-MOG and anti-CNPase antibodies. Furthermore, even 2 days after the last transfer, only very few cells, expressing oligodendroglia markers by immunocytochemistry or *in situ* hybridization were found in the areas of demyelination. In

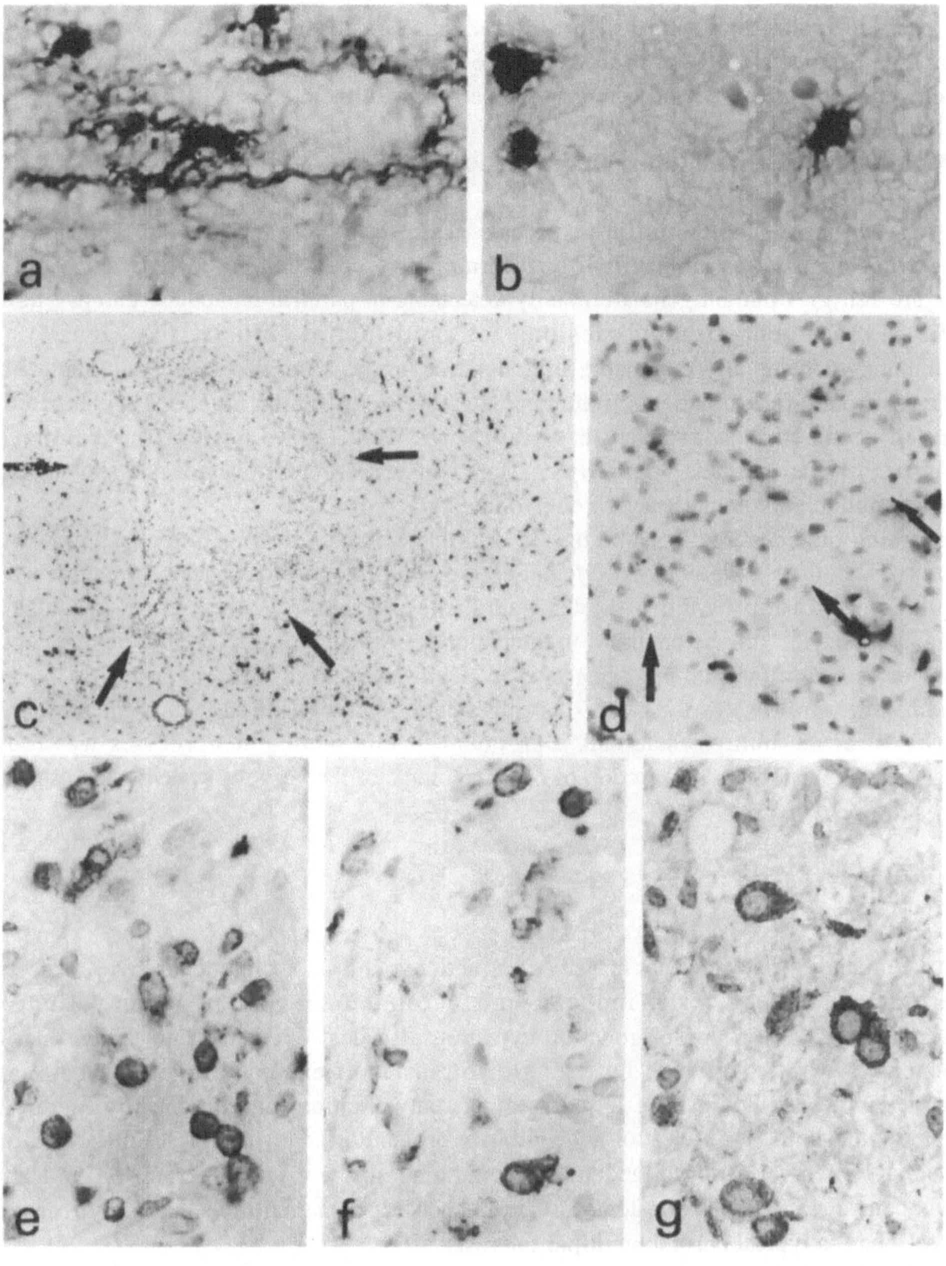

Figure 1. Oligodendrocytes in normal and diseased central nervous system tissue. a.) 7 day old rat, normal brain; *in situ* hybridization for MBPmRNA labels oligodendrocyte cell bodies and processes. x750 b.) 7 day old rat, normal brain; *in situ* hybrydization for PLPmRNA; only the perinuclear cytoplasm is stained. x750 c.) Adult rat with corona virus induced subacute demyelinating encephalomyelitis. A demyelinated plaque (arrows) contains no oligodendrocytes. Numerous oligodendrocytes are labelled in the periplaque white matter; *in situ* hybridization for PLPmRNA; x75 d.) Higher magnifications of figure 1c, showing the plaque border (arrows); *in situ* hybridization for PLPmRNA; x300e.) Inactive multiple sclerosis plaque; immuno-cytochemistry with anti-MOG antibodies shows numerous small (inactive) oligodendrocytes. x750 f.) Inactive multiple sclerosis plaque; 2 large, activated oligodendrocytes were labelled by immunocytochemistry for MOG. g.) Inactive multiple sclerosis plaque; immunocytochemistry with anti-CNPase antibodies mainly labels large activated oligodendrocytes. x750

the periphery of the lesions, however, an increased number of cells were identified that expressed mRNAs for myelin proteins. Thus, in chronic cotransfer autoimmune encephalomyelitis oligodendrocytes are destroyed during active demyelination. Remyelination is limited and occurs at the periphery of the plaques.

Oligodendrocytes in multiple sclerosis lesions

Since most of the multiple sclerosis brain tissue available came from autopsy material, our studies mainly involved immunocytochemistry for MOG and CNPase (figure 1e,f,g). Both antigens are readily detectable in paraffin embedded autopsy tissue. Within the demyelinated lesions of multiple sclerosis the density of immunocytochemically detectable oligodendrocytes was very variable, ranging within plaques from different patients from complete absence to similar numbers as compared to the adjacent white matter. Interestingly, within a single patient the number of oligodendrocytes per area was fairly constant between different plaques and showed little variation in relation to the stage of demyelinating activity.
Regarding oligodendroglia pathology in multiple sclerosis three different patterns were observed. In the vast majority of typical chronic multiple sclerosis cases, in patients with a disease duration of more than 3 years, the plaques revealed extensive loss of oligodendrocytes. On the contrary, in the majority of plaques, derived from patients at the first or second attack of the disease, oligodendrocytes were present in nearly normal numbers, regardless of the stage of demyelinating activity. In patients with Marburg's type[16] of acute multiple sclerosis a pronounced loss of oligodendrocytes was found together with extensive destruction of other elements of the nervous system such as axons or astrocytes.

Conclusions

Our studies revealed that massive destruction and loss of oligodendrocytes occurs in chronic models of virus induced and autoimmune mediated demyelinating encephalomyelitis. Since oligodendrocytes are also destroyed in acute demyelinating encephalomyelitis, induced by a single cotransfer of encephalitogenic T-cells and demyelinating antibodies, the impairment of remyelination may be due to progressive depletion of the pool of adult oligodendrocyte progenitor cells, capable of differentiating into mature remyelinating oligodendrocytes in a similar way as suggested for models of toxic demyelination[17]. The lesions formed in such experimental models closely resemble those found in typical chronic multiple sclerosis.

In addition, however, our studies indicate that in multiple sclerosis the mechanisms of demyelination may be different in the lesions formed during early exacerbations of the disease. In these plaques we found a pronounced preservation of oligodendrocytes throug-hout all stages of demyelinating activity. This suggests that during the first or second relapse, at least in a subgroup of multiple sclerosis patients, not the oligodendrocytes but the myelin itself is the primary target of the immune attack. Therefore it has to be considered that the immunopathogenesis of the lesions may differ in individual patients between early and late exacerbations in the course of their disease.

REFERENCES

1. R. Johnson, Viral aspects of multiple sclerosis, in: "Handbook of Clinical Neurology. Demyelinating Diseases," J. C. Koetsier, ed., Elsevier Science Publishers, Amsterdam (1985).
2. C. Griot, T. Bürge, M. Vanvelde, E. Peterhans, Antibody induced generation of reactive oxygen radicals by brain macrophages in canine distemper encephalitis: a mechanism for bystander demyelination, Acta Neuropathol. 78:396 (1989).
3. C. Linington, M. Bradl, H. Lassmann, C. Brunner, K. Vass, Augmentation of demyelination in rat acute allergic encephalomyelitis by circulating mouse monoclonal antibodies directed against a myelin/oligodendrocyte glycoprotein. Amer. J. Pathol. 130:443 (1988).
4. K. Selmaj, C.F. Brosnan, C.S. Raine, Expression of heat shock protein-65 by oligodendrocytes in vivo and in vitro: implications for multiple sclerosis, Neurology 42:795 (1992) .
5. C.S. Raine, Multiple sclerosis. Oligodendrocyte survival and proliferation in an active lesion, Lab. Invest. 45:534 (1981).
6. H. Lassmann. "Comparative Neuropathology of Chronic Experimental Allergic Encephalomyelitis and Multiple Sclerosis," Springer-Verlag, Berlin Heidelberg New York Tokyo (1983).
7. J.W. Prineas, The neuropathology of multiple sclerosis, in: "Handbook of Clinical Neurology. Demyelinating Diseases," J.C. Koetsier, ed., Elsevier Science Publishers, Amsterdam (1985).
8. J.W. Prineas, E.E. Kwon, P.Z. Goldenberg, A.A. Ilyas, R.H. Quarles, J.A. Benjamins, T.J. Sprinkle, Multiple sclerosis: oligodendrocyte proliferation and fresh lesions. Lab. Invest. 61:489 (1989).
9. F. Zimprich, J. Winter, H. Wege, H. Lassmann, Corona virus induced primary demyelination: indications for the involvement of humoral immune response, Neuropath. Appl. Neurobiol. 17:469 (1991).
10. C. Linington, B. Engelhardt, G. Kapocs, H. Lassmann, Induction of persistently demyelinated lesions in the rat following the repeated adoptive transfer of encephalitogenic T cells and demyelinating antibody, J. Neuroimmunol. 40:219 (1992).
11. K. Vass, H. Lassmann, H. Wekerle, H.M. Wisniewski, The distribution of Ia-antigen in the lesions of rat acute experimental allergic encephalomyelitis, Acta Neuropathol. 70:149 (1986).
12. H. Breitschopf, G. Suchanek, R.M. Gould, D.R. Colman, H. Lassmann, In situ hybridization with digoxigenin-labeled probes: sensitive and reliable detection method applied to myelinating rat brain. Acta Neuropathol. 84:581 (1992).
13. R. Gold, M. Schmied, G. Rothe, H. Zischler, H. Breitschopf, H. Wekerle, H. Lassmann, Detection of DNA fragmentation at the cellular level by in situ nick translation: application to in vitro systems and tissue sections. J. Histochem. Cytochem. 41:1023 (1993).
14. C. Brunner, H. Lassmann, Th. V. Waehnelt, J.M. Matthieu, C. Linington, Differential ultrastructural localization of myelin basic protein, myelin/oligodendroglia glycoprotein and 2'3'-cyclic nucleotide 3'-phosphodiesterase in the CNS of adult rats. J. Neurochem. 52:296 (1989).
15. S.K. Ludwin, Oligodendrocyte survival in Wallerian degeneration, Acta Neuropath. 80:184 (1990).
16. O. Marburg, Die sogenannte "akute Multiple Sklerose", Jahrb. Psychiatrie 27:211 (1906).
17. S.K. Ludwin, Chronic demyelination inhibits remyelination in the central nervous system. An analysis of contributing factors, Lab. Invest. 43:382 (1980).

THE IMMUNOLOGIC RESPONSE OF THE OLIGODENDROCYTE IN THE ACTIVE MULTIPLE SCLEROSIS LESION

Cedric S. Raine, Elizabeth Wu and Celia F. Brosnan

Department of Pathology (Neuropathology)
Albert Einstein College of Medicine
Bronx, N.Y. 10461, U.S.A.

INTRODUCTION

There is a strong belief in the field of multiple sclerosis (MS) that central nervous system (CNS) demyelination is integrally related to the early selective depletion of the CNS myelinating cell, the oligodendrocyte. Subscribers to this theory are numerous and while still a hypothetical scenario, "oligodendrogliolysis" was first proposed by Lumsden[1] in the 1950s. Thus far, however, investigators have been unable to provide evidence *in situ* for widespread lysis of oligodendrocytes in MS[2], although from studies *in vitro,* it has been known for years that the oligodendrocyte is highly susceptible to damage by a number of soluble immune mediators. Included among the latter are anti-myelin immunoglobulins[3], cytokines[4] and complement[5]. The lack of compelling evidence for selective oligodendrocyte destruction notwithstanding, it is clear from the literature on the subject that their depletion in MS must occur sometime between the initial acute inflammatory, actively demyelinating phase when oligodendrocytes are abundant, and the chronic, silent fibrous astrogliotic MS plaque where few, if any, oligodendrocytes can be located[2]. Complicating the analysis of factors underlying the demise of the oligodendrocyte in MS, are reports documenting that during periods of disease activity, oligodendrocyte proliferation occurs at the margins of established lesions. These observations, together with subsequent confirmatory reports, have led to considerations of the possibility that immune-mediated destruction of myelin may actually exert a stimulatory effect upon the oligodendrocyte. This possibility is underscored by the presence of sometimes substantial amounts of CNS remyelination within and around acute and actively expanding MS lesions[6,8,9].

The central theme of the following paragraphs will be that immune system-based mechanisms underlie the unusual proclivity of the oligodendrocyte to survive, proliferate and even synthesize new myelin in the face of ongoing, immune-mediated demyelination in MS. In view of the potential therapeutic significance of a fuller understanding of regenerative phenomena in MS, particularly from the standpoint of the oligodendrocyte, emphasis will be placed upon the immunopathology of the active lesion and upon comparisons with other CNS degenerative conditions in an attempt to identify MS-specific phenomena.

BACKGROUND

Historically, the concept of oligodendrocyte survival and proliferation (the converse of the scenario promulgated by Lumsden[1]), was probably first suggested by the histochemical studies of Ibrahim and Adams[10,11]. There was little or no follow-up to this line of thinking until 1981 when a detailed study of the edge of an actively demyelinating, chronic established lesion revealed the presence of surviving, proliferating oligodendrocytes and remyelination[6]. Remyelination had previously been documented ultrastructurally in MS[12-14] but had never been interpreted against a backcloth of oligodendrocyte proliferation. Suggestions have since been made that the proliferative activity of the oligodendrocyte in MS may be cytokine-related in that cells bearing IL-2 receptors can be localized at the margin of the MS lesion[15] and that IL-2 might induce oligodendrocyte stimulation *in vitro*[16], a view contested by others[17] who reported an opposite (inhibitory) effect of IL-2 upon oligodendrocytes in culture. On the other hand, cytokines like TNFα and TNFß have been shown to be unequivocally cytotoxic for oligodendrocytes *in vitro*[4,18,19]. This might suggest that there exists during the formative stages of the MS plaque a delicate balance between some soluble mediators with potentially beneficial and others with toxic effects upon the oligodendrocyte. Such factors have been shown to be present locally in both infiltrating and resident cell populations[20]. While suspected from works showing up- and down-regulation of oligodendrocyte behavior by a number of cytokines (IL-2, TNFα, TNFß, IFNγ), receptors for cytokines have yet to be demonstrated on these cells.

IMMUNOPATHOLOGY

Recent neuropathologic studies of active MS lesions have revealed interesting interactions between oligodendrocytes and astrocytes in MS lesions of all ages, relationships which may also have as their basis, a triggering event by locally secreted cytokines. In this regard, it was demonstrated that hypertrophic astrocytes in acute and chronic active MS lesions are capable of associating with and internalizing proliferated oligodendrocytes in a manner akin to what immunologists refer to as "emperipolesis".[21-24] No function has yet been attributed to these unusual cellular interactions. However, a phagocytic role on the part of the astrocyte has been suggested[22] while other investigators have claimed that this internalization of oligodendrocytes by hypertrophic astrocytes (Figure 1) might represent a protective phenomenon[23,24]. Subsequent studies[20], with an antibody to clusterin (SP-40,40; SGP-2), a complement inhibitory molecule implicated in protection against complement-mediated destruction (also a molecule believed to play a role in the heat shock response, and in cell adhesion and recognition), failed to support a role for such a protective mechanism against complement-mediated lysis. It remains to be shown, therefore, whether the survival and proliferation of these cells in the acute MS lesion (Figure 2), and their association with astrocytes, holds any functional significance for MS. The same phenomenon is now known to occur in a variety of different

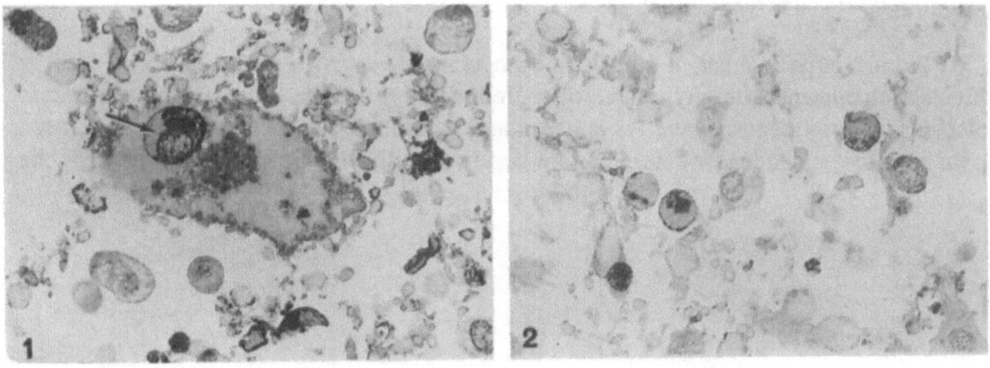

Figure 1. A hypertrophic astrocyte in the center of an acute MS lesion is stained positively for glial fibrillary acidic protein (GFAP) and contains an internalized HNK-1$^+$ oligodendrocyte (arrow). One micron epoxy section double stained for HNK-1 and GFAP x 650.

Figure 2. The edematous center of a demyelinated acute MS lesion contains several surviving HNK-1$^+$ oligodendrocytes. Elsewhere, unstained foamy macrophages and a few GFAP$^+$ astrocyte processes are seen. Similar preparation to Figure 1. x 650.

degenerative conditions of the CNS, many of which had an inflammatory component to the lesion,[20,24] leaving open a role for locally produced soluble immune mediators in the phenomenon.

The lack of evidence for the glial associations representing a protective phenomenon notwithstanding[20], it is now clear from a number of studies that the oligodendrocyte in the acute and the chronic active MS lesion is not a primary target and that by all accounts, it displays regenerative, as opposed to degenerative, activity in the form of an increase in cell number at the lesion edge, the presence of markers associated with immature or recently-derived oligodendrocytes (Figures 3 and 4), and an ability to remyelinate fibers in the face of ongoing demyelination[2,6,9] (Figures 5 and 6).

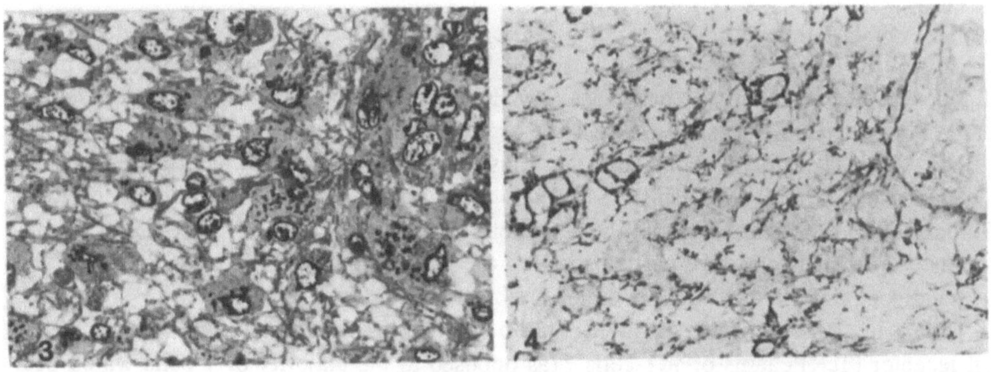

Figure 3. The edge of an acutely demyelinated MS lesion displays beginning fibrous astrogliosis and a perivascular cuff (upper right) containing large mononuclear cells. Numerous foamy macrophages containing osmiophilic myelin debris and several smaller rounded cells are also seen. Toluidine blue-stained, one micron epoxy section. x 650.

Figure 4. Adjacent epoxy section to Figure 3, immunostained for HNK-1 (oligodendrocytes). Note the same unstained perivascular cuff (upper right). Numerous HNK-1$^+$ oligodendrocytes, some arranged in rows, can be discerned. x 650.

IMMUNE RESPONSE OF OLIGODENDROCYTES IN MS

From a large number of studies both *in vitro* and *in vivo*, it has been exceedingly difficult to document with any degree of reproducibility, the presence of immune system molecules on oligodendrocytes. As stated above, although several cytokines have been shown to affect oligodendrocyte behavior *in vitro*, there has been no evidence for the

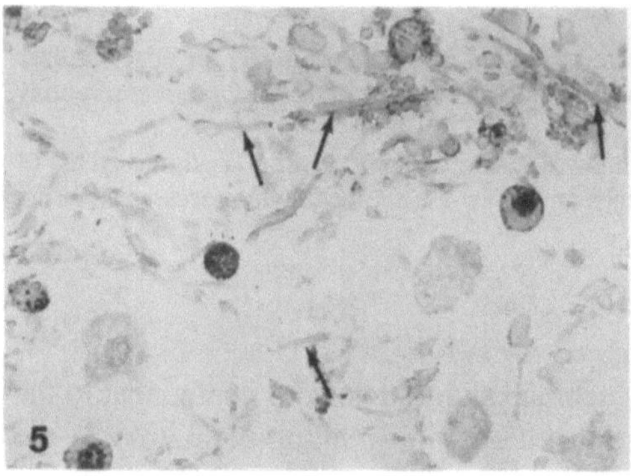

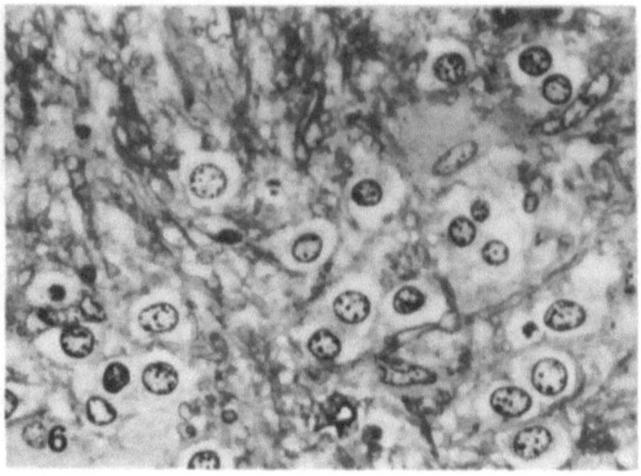

Figure 5. The center of an acute MS lesion displays several HNK-1^+ oligodendrocytes and many wisps of thinly remyelinated axons (arrows). Foamy macrophages are present but are unstained. Some pale GFAP$^+$ large astrocyte processes are present (top center). Double stained, one micron epoxy section. x 650.
Figure 6. The edge of an acute MS lesion displays widespread, thinly remyelinated fibers and numerous proliferated oligodendrocytes. A hypertrophic astrocyte is seen to the upper right. Toluidine blue-stained, one micron epoxy section. x 650.

presence of the appropriate receptors on the oligodendrocyte or for the production of cytokines by these cells *in situ*. Class I MHC has been shown to occur on the oligodendrocyte in culture, as has Class II MHC (albeit rarely), but in general, it is agreed that within the CNS, no such expression occurs. Indeed, in one detailed study of MS lesions[25], it was definitively concluded that Class II MHC did not occur on oligodendrocytes and that many of the cells displayed markers more typical of immature oligodendrocytes (Figures 7 and 8). Therefore, from the standpoint of classical CD4/CD8 T cell interactions, evidence is lacking for active participation by the oligodendrocyte in the MS lesion, either as an antigen presenting cell or as a target.

During a series of recent studies on the possible involvement of T cells bearing $\gamma\delta$T cell receptor (TcR) chains or $\gamma\delta$T cells[26] (as opposed to the majority of T cells which bear $\alpha\beta$ TcR chains and which are CD4$^+$ or CD8$^+$, we encountered an unexpected association which might be of significance to the pathogenesis of the MS lesion vis à vis the immunologic involvement of the oligodendrocyte. These findings, reported in detail elsewhere,[27,28] first concerned our ability to localize $\gamma\delta$T cells in MS lesions, particularly chronic active lesions. In our search for a functional role for these T cells (which normally constitute 5-10% of the circulating pool and which are usually CD4$^-$/CD8$^-$), we decided to examine the material for stress or heat shock proteins (HSP) and were rewarded with the finding of abundant evidence of HSP expression in MS tissue.

The heat shock response has been defined as a general homeostatic mechanism that protects cells from the deleterious effects of environmental stress[29]. HSP are among the most highly conserved and abundant proteins in nature and are among the most dominant antigens recognized in immune responses to a wide variety of pathogens[30]. They are

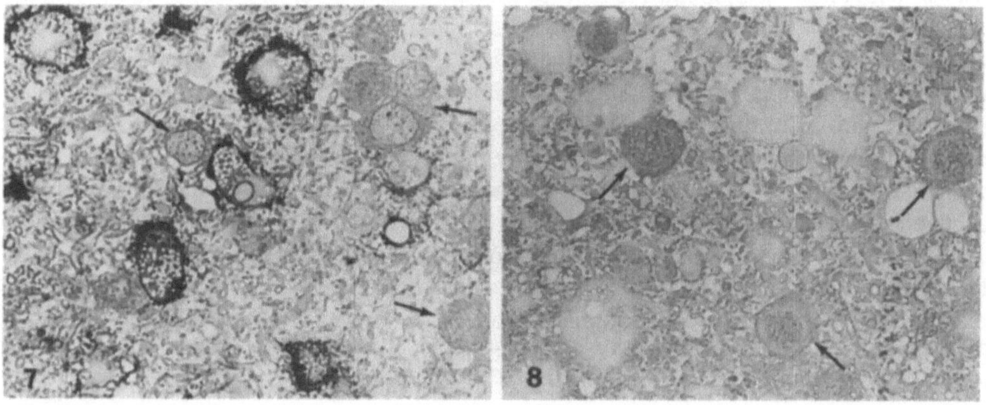

Figure 7. Foamy macrophages (ameboid microglia) stain positively for HLA.DR (Class II MHC) while oligodendrocytes (arrows) display no immunoreactivity in this one micron epoxy section. Acute MS lesion. x 900.

Figure 8. In an adjacent section from the same biopsy sample, reacted with an antibody to myelin associated glycoprotein (MAG), note how the oligodendrocytes (arrows) display positive staining while the foamy macrophages (pale ghosts) are unreactive. x 900.

classified according to approximate molecular mass in kilodaltons into three main families - HSP 60, HSP 70 and HSP 90 (although smaller forms are also recognized), and are commonly expressed in cells undergoing stress caused by, for example, heat-shock, infection, trauma, nutrient deprivation and metabolic dysfunction. Moreover, of relevance to our own studies,[27,28] is the implication that HSP play a role in target recognition by a distinct group of MHC-unrestricted T lymphocytes -γδ T cells[31]. HSP have been shown to stimulate γδ T cells to produce IL-2 and to serve as self antigens.[30] They are frequently associated with viral, bacterial and autoimmune inflammatory responses but prior to our studies[27,28] had not been examined in MS. Among the many functions of HSP are those associated with the assembly and disassembly of protein complexes, the translocation of certain proteins across cell membranes and the ability to bind to folded proteins, acting as catalysts. In their latter capacity, stress proteins are sometimes known as "chaperonins".[30,32]

For the purposes of our own investigations, we probed by immunocytochemistry for the presence (among others) of a 65kD stress protein (belonging to the HSP 60 family) with an antibody raised against the 65kD antigen of *M leprae* and revealed reactivity to be present on large cells at the margins of lesions, in particular chronic active lesions. Double-staining showed the HSP 65^+ cells to stain positively for myelin basic protein (MBP), a marker for oligodendrocytes.[27,28] We also showed that oligodendrocytes (but not astrocytes) constitutively expressed HSP 65 *in vitro*[28]. Astrocytes have been shown in the CNS of experimental animals to display induced expression of HSP 70.[33] Subsequent to our work on MS, Freedman et al[34] have reported the ability of human γδ T cell clones to lyse selectively oligodendrocytes *in vitro*. In addition, the same workers have documented differential expression after heat shock of constitutively expressed HSP 70 on oligodendrocytes, but the same cells showed the same level of HSP 60 expressed under stressed and non-stressed conditions[35]. Similar observations with HSP 65 were made on oligodendrocytes *in vitro* in work by Satoh et al.[36]

Our recent efforts to explain the possible relevance of HSP 65 in MS have included the examination of a large spectrum of CNS lesions of different ages from MS and non-MS lesions. With different antibodies to HSP 65 (in particular ML 30, an antibody to a M. leprae 65kD antigen; the kind gift of Dr. J. Ivanyi, London, U.K.) and with appropriate second labelling for glial cell phenotype, we found HSP 65 reactivity to be more widespread in acute MS lesions than previously reported[27], occurring on endothelial cells, hypertrophic astrocytes and infiltrating cells, as well as on oligodendrocytes.[37] The localization was cytoplasmic and punctate, the latter probably due to the reactivity of mitochondria.[38] As lesion activity waned, HSP 65 reactivity became more restricted to oligodendrocytes and reactive astrocytes. Reactive oligodendrocytes at the margins of chronic active MS lesions, as seen by immunocytochemistry on one micron epoxy sections, showed intense staining of the cytoplasm and cell processes (Figure 9), while macrophages and most other elements (except for the occasional hypertrophic astrocyte) displayed minimal immunoreactivity. These reactive oligodendrocytes were essentially identical to and occurred in the same location as those depicted in our original paper on this subject.[27] Interestingly, in apparently normal white matter remote from lesion areas, rows of small interfascicular oligodendrocytes also stained positively (Figure 10), as did some astrocytes. Thus, glial reactivity to HSP 65 appeared to be widespread throughout the white matter in MS. Similar analysis of MS tissue for HSP 70 reactivity, on the other hand, shows that this antigen displays considerable affinity for reactive astrocytes (Aquino et al, in preparation), observations in keeping with previous work on the CNS of rats with experimental allergic encephalomyelitis[33].

Detailed analysis of CNS tissue of a number of non-MS conditions (cerebrovascular disorders and AIDS) has also revealed HSP 65 reactivity, albeit to a lower degree, on

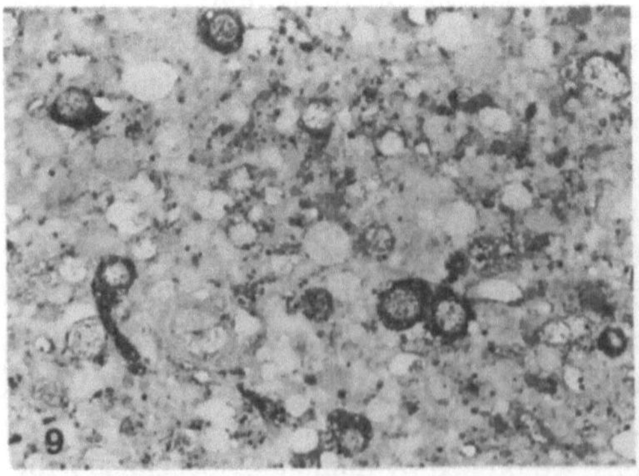

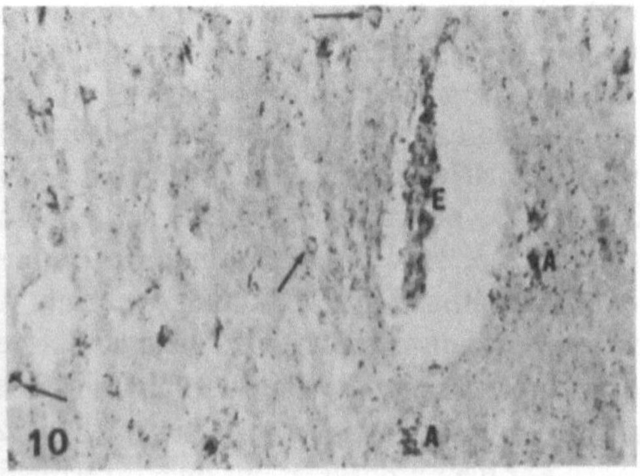

Figure 9. A one micron epoxy section from the perimeter of an acute MS lesion displays numerous HSP 65$^+$ oligodendrocytes, some with extensive cytoplasmic processes. The punctate staining seen elsewhere corresponds to HSP 65 reactivity on mitochondria in most cell types. One micron epoxy section reacted with ML 30.[38] x 650.

Figure 10. Paraffin section from an uninvolved area of white matter from a chronic silent MS lesion from another MS patient, reacted with ML 30 for HSP 65 reactivity. Note the granular staining on oligodendrocytes (arrows) and on astroglial (A) and endothelial (E) cells. x 260.

oligodendrocytes and reactive astrocytes. On the other hand, normal CNS tissue has been shown to possess little immunoreactivity for HSP 65. Our findings thus far have demonstrated in MS and a number of unrelated CNS degenerative conditions, that marked induction of HSP 65 is detectable, in particular upon oligodendrocytes and astrocytes in white matter. Whether this expression serves as a protective mechanism from the cytotoxic effects of monocytes or other infiltrating cells, as suggested in other non-CNS systems[39], as a self-antigen in the ultimate demise of the involved cells (in particular, the oligodendrocyte in MS), or plays an immunoregulatory role important in the perpetuation of an immunologic response in the CNS, perhaps via $\gamma\delta$ T cell interactions, remain to be clarified but afford interesting avenues for future work in the area.

CONCLUSIONS

Although centrally implicated in the pathogenesis of the early MS lesion, there is still no compelling evidence that the oligodendrocyte is selectively destroyed and indeed, most studies *in vivo* point to the inflammatory phase of lesion formation exerting a proliferative effect upon the cell, apparently driving it towards remyelination. Interestingly, results from recent work *in vitro* also support a high regenerative and remyelinative potential for this cell type.[40] However, the permanence of this remyelination has been questioned.[7,9] The lack of expression of "conventional" immune system molecules by the oligodendrocyte in MS, like MHC, Fc, cytokine and complement molecules reveal the cell to be relatively inert immunologically although recent findings of widespread HSP expression by these cells in white matter *in vivo* raise a number of interesting possibilities regarding their fate in MS.

ACKNOWLEDGEMENTS

We thank Drs. B. Cannella, Y.-L. Gao, D. Aquino, M.B. Bornstein and K. Selmaj for helpful discussion; Drs. V.J. Mehra, W. Welch and J. Ivanyi for providing valuable antibodies; Earl Swanson, Miriam Pakingan and Howard Finch for skilled technical assistance; and Adele Baserap for careful preparation of the manuscript.

Supported in part by USPHS grants NS 08952, NS 11920 and NS 07098; and by National MS Society Grants RG 1001-G-7 and RG 1089-F-9.

REFERENCES

1. C.E. Lumsden. The neuropathology of multiple sclerosis, in: "Multiple Sclerosis", D. McAlpine, N.D. Compston and C.E. Lumsden, ed., Livingstone, Edinburgh (1955a).
2. C.S. Raine. Demyelinating diseases, in: "Textbook of Neuropathology", R.L. Davis and D.M. Robertson, ed., Williams and Wilkins, Baltimore (1990).
3. C.S. Raine and M.B. Bornstein. Experimental allergic encephalomyelitis: An ultrastructural study of experimental demyelination *in vitro*, J. Neuropath. Exp. Neurol 29:177 (1970).
4. K. Selmaj and C.S. Raine. Tumor necrosis factor mediates myelin and oligodendrocyte damage *in vitro*, Ann. Neurol. 23:339 (1988).
5. A. Compston, N. Scolding, D. Wren and M. Noble. The pathogenesis of demyelinating disease: Insights from cell biology, Trends Neurosci. 14:175 (1991).
6. C.S. Raine, L.C. Scheinberg and J.M. Waltz. Multiple sclerosis: Oligodendrocyte survival and proliferation in an active, established lesion, Lab. Invest. 45:534 (1981).
7 J.W. Prineas, E.E. Kwon, P.Z. Goldenberg, A.A. Ilyas, R.H. Quarles, J.A. Benjamins and T.J. Sprinkle. Multiple sclerosis: Oligodendrocyte proliferation and differentiation in fresh lesions, Lab. Invest. 61:489 (1989).

8. J.W. Prineas, E.E. Kwon, E.S. Cho and L.R. Sharer. Continual breakdown and regeneration of myelin in progressive multiple sclerosis plaques, in: "Multiple Sclerosis: Experimental and Clinical Aspects", L.C. Scheinberg and C.S. Raine, ed. Ann. N.Y. Acad. Sci. 436:11 (1984).
9. C.S. Raine and E. Wu. Multiple sclerosis: Remyelination in acute lesions, J. Neuropath. Exp. Neurol. 52: 199 (1993).
10. M.Z.M. Ibrahim and C.W.M. Adams. The relationship between enzyme activity and neuroglia in plaques of multiple sclerosis, J. Neurol. Neurosurg. Psychiat. 26:101 (1963).
11. M.Z.M. Ibrahim and C.W.M. Adams. The relationship between enzyme activity and neuroglia in early plaques of multiple sclerosis, J. Path. Bact. 90:239 (1965).
12. O. Périer and A. Grégoire. Electron microscopic features of multiple sclerosis lesions. Brain 88:937 (1965).
13. K. Suzuki, J.M. Andrews, J.M. Waltz and R.D. Terry. Ultrastructural studies of multiple sclerosis, Lab. Invest. 20:444 (1969)
14. J.W. Prineas and F. Connell. Remyelination in multiple sclerosis, Ann. Neurol. 5:22 (1979).
15. F.M. Hofman, R.I. von Hanwehr, C.A. Dinarello, S.B. Mizel, D. Hinton and J.E. Merrill. Immunoregulatory molecules and IL-2 receptors identified in multiple sclerosis brain, J. Immunol. 136:3239 (1986).
16. E.N. Benveniste and J.E. Merrill. Stimulation of oligodendroglial proliferation and maturation by interleukin-2, Nature (London) 321:610 (1986).
17. R.P. Saneto, A. Altman, R.L. Knobler, H.M. Johnson and J. DeVellis. Interleukin-2 mediates the inhibition of oligodendrocyte progenitor cell proliferation *in vitro*, Proc. Nat. Acad. Sci. 83:9221 (1986).
18. D.S. Robbins, Y. Shirazi, B.E. Drysdale, A. Lieberman, H.S. Shin and M.L. Shin. Production of cytotoxic factor for oligodendrocytes by stimulated astrocytes, J. Immunol. 139:2593 (1987).
19. K. Selmaj, C.S. Raine, M. Farooq, W.T. Norton and C.F. Brosnan. Cytokine cytotoxicity against oligodendrocytes: Apoptosis induced by lymphotoxin, J. Immunol. 147:1522 (1991).
20. E. Wu, C.F. Brosnan and C.S. Raine. SP-40,40 immunoreactivity in inflammatory CNS lesions displaying astrocyte/oligodendrocyte interactions, J. Neuropath. Exp. Neurol. 52: 129 (1993).
21. N.R. Ghatak, R.T. Leshner, A.C. Price and W.L. Felten. Remyelination in the human central nervous system, J. Neuropath. Exp. Neurol. 48:507 (1989).
22. J.W. Prineas, E.E. Kwon, P.Z. Goldenberg, E.S. Cho and L.R. Sharer Interaction of astrocytes and newly formed oligodendrocytes in resolving multiple sclerosis lesions, Lab. Invest. 63:624 (1990).
23. N.K. Ghatak. Occurrence of oligodendrocytes within astrocytes in demyelinating lesions, J. Neuropath. Exp. Neurol. 51:40 (1992).
24. E. Wu and C.S. Raine. Multiple sclerosis: Interactions between oligodendrocytes and hypertrophic astrocytes and their occurrence in other, non-demyelinating conditions, Lab. Invest. 67:88 (1992).
25. S.C. Lee and C.S. Raine. Multiple sclerosis: Oligodendrocytes do not express class II major histocompatibility molecules in active lesions, J. Neuroimmunol. 25:261 (1989).
26. J. Holoshitz. Potential role of $\gamma\delta$ T cells in autoimmune diseases, Res. Immunol. 141:651 (1990).
27. K. Selmaj, C.F. Brosnan and C.S. Raine. Colocalizàtion of TCR gamma-delta lymphocytes and hsp-65+ oligodendrocytes in multiple sclerosis, Proc. Natl. Acad. Sci. 88:6452 (1991).
28. K. Selmaj, C.F. Brosnan and C.S. Raine. Expression of heat shock protein-65 by oligodendrocytes *in vivo* and *in vitro*: Implications for multiple sclerosis, Neurology 42:795 (1992).
29. B. Maresca and L. Carratu. The biology of the heat shock response in parasites, Parasitol. Today 8:260 (1992).
30. R.A. Young. Stress proteins and immunology, Annu. Rev. Immunol. 8:401 (1990).
31. R.L. O'Brien, M.P. Happ, A. Dallas, e. Palmer, R. Kubo and W. Born. Stimulation of a major subset of lymphocytes expressing T cell receptor $\gamma\delta$ by an antigen derived from mycobacterium tuberculosis, Cell 57:667 (1989).
32. M.J. Shlesinger. Heat shock proteins, J. Biol. Chem. 265:12111 (1990).
33. D. Aquino and C.F. Brosnan. Heat-shock proteins and immunopathology, Chem.Immunol. 53:1 (1992).
34. M.S. Freedman, T.C.G. Ruijs, S.K. Selin and J.P. Antel. Peripheral blood $\gamma\delta$T cells lyse fresh human brain-derived oligodendrocytes, Ann. Neurol. 30:794 (1991).
35. M.S. Freedman, N.N. Buu, T.C.G. Ruijs, K. Williams and J.P. Antel. Differential expression of heat shock proteins by human glial cells, J. Neuroimmunol. 41:231 (1992).
36. J. Satoh, H. Nomaguchi and T. Tabira. Constitutive expression of 65KDa heat shock protein (HSP 65)-like immunoreactivity in cultured mouse oligodendrocytes, Brain Res. 595:281 (1992).
37. C.S. Raine, E. Wu and C.F. Brosnan. Heat shock protein 65 (HSP 65) expression in multiple sclerosis (MS) lesions, Abstract, J. Neuropath. Exp. Neurol. 52: 311 (1993).

38. D.J. Evans, P. Norton and J. Ivanyi. Distribution in tissue sections of the human groEL stress-protein homologue, APMIS 98:437 (1990).
39. M. Jaatela and W. Wissing. Heat shock proteins protect cells from monocyte cytotoxicity: Possible mechanism of self-protection, J. Exp. Med. 177:231 (1993).
40. P.E. Knapp, R.P. Skoff and C.S. Booth. Oligodendrocytes possess essential prerequisites for remyelination, Adv. Neurol. 59:105 (1993).

AUTOIMMUNE RESPONSES TO THE MYELIN OLIGODENDROCYTE GLYCOPROTEIN (MOG) IN THE PATHOGENESIS OF INFLAMMATORY DEMYELINATING DISEASES OF THE CENTRAL NERVOUS SYSTEM

Christopher Linington, Martin Adelmann,
and Roxana Popovici

Max-Planck-Institut fr Psychiatrie
Am Klopferspitz 18A
8033 Planegg-Martinsried
Germany

INTRODUCTION

A T cell mediated autoimmune response is believed to trigger the characteristic inflammatory demyelinating pathology of multiple sclerosis (MS)[1]. However the immune effector mechanisms responsible for the selective loss of myelin in MS have still to be defined. Several authors have suggested that demyelination in MS is simply a consequence of the local inflammatory response in the CNS, "bystander demyelination". *In vitro* studies demonstrating that oligodendrocytes and myelin are highly susceptible to damage by a wide variety agents released by monocytes and T cells during an inflammatory response [2,3]. However, ultrastructural and immunocytochemical studies indicate that primary demyelination in MS may be mediated by a specific humoral response directed against the myelin membrane. Electron microscopy reveals that macrophages attack, phagocytose and degrade apparently normal myelin in MS [4], introducing processes between myelin lamellae and actively stripping sheets of membrane from the myelin sheath, which is then phagocytosed by a process resembling receptor mediated endocytosis. Interestingly the phagocytosis of myelin is closely associated with the capping of IgG on the macrophage surface suggesting that a specific receptor/ligand interaction is involved in this process[5]. The ligands involved in this interaction are unknown, but possible candidates are immunoglobulin or complement activation products (C3bi) deposited on the myelin surface. In MS there is certainly ample indirect evidence for the activation of complement on myelin surface, in particular, the intrathecal consumption of complement [6] and the presence of myelin membrane fragments coated with terminal complement complexes the cerebrospinal fluid [7]. The deposition of terminal complement components and immunoglobulins can also be demonstrated in MS lesions[8,9]. However, as yet no myelin-specific autoantibody response has been identified that can account for the observed intrathecal activation of complement in MS.

The identification of an antigen-specific autoimmune response associated with MS may well provide the key to explain the etiology and pathogenesis of this disease. Unfortunately, detection of primary events in MS is complicated by the very nature of the disease. The development of a clinically significant neurological deficit in MS is normally an insidious process. The lesions are often clinically silent so that substantial tissue damage will have occurred by the time that a definite diagnosis of MS can be established. By the time the patient is available for study, any primary pathogenic autoimmune response that is responsible for the induction of the disease process must then be distinguished from a background of secondary autoimmune responses induced following tissue damage. Clearly, some of these secondary autoimmune responses are also likely to be pathogenic and then contribute to the pathogenesis of the disease, however many such as anti-tubulin or anti-GFAP antibody responses are unlikely to play an active role in the subsequent course of the disease[10]. One approach circumventing this problem is to first identify immunodominant myelin autoantigens in a range of animal species. The pathogenic consequences of the autoimmune response to these antigens can then be correlated with the pathological findings in MS, and clinical studies performed to determine whether a similar pathogenic, autoimmune response is associated with the human disease.

Pathogenic autoantibody responses to myelin antigens

Although the identification of demyelinating autoantibody responses in MS has proved difficult, the involvement of anti-myelin antibodies in the immunopathogenesis of demyelination is well documented in an animal model, experimental allergic encephalomyelitis (EAE). As initially described, EAE induced by immunisation with whole CNS tissue in adjuvant is an inflammatory demyelinating disease of the CNS, which has many clinical and histopathological similarities with MS [11,12,13]. The key pathogenic event responsible for the induction of EAE is an autoimmune T cell response directed against protein components of the myelin sheath, in particular the myelin basic protein (MBP), but also the proteolipid protein (PLP)[14]. The adoptive transfer of syngeneic, MBP-specific T cells can therefore induce EAE in naive recipients[15]. A direct consequence of this observation has been an enormous research effort directed towards determining the role of autoimmune MBP-specific T cells in MS. However in the Lewis rat, EAE induced by the adoptive transfer of MBP-specific T cells (tEAE) differs considerably from MS, in terms of both its pathology and clinical course. MS is an inflammatory, *demyelinating* disease, in which the lesions are often clinicaly silent. In contrast, tEAE is characterised by an acute, disseminated inflammatory response in the CNS, in which demyelination is minimal and clinical disease severe[15]. Other immune effector mechanisms and/or autoantigens must therefore be responsible for the demyelination seen in models of EAE induced by active immunisation with total spinal cord in adjuvant.

It was already established some twenty five years ago that sera from animals with EAE induced by immunisation with spinal cord tissue in adjuvant contain antibodies which can demyelinate both organotypic cultures of CNS tissue *in vitro*[16] and also, providing the blood brain barrier is circumvented, the CNS *in vivo*. This was demonstrated experimentally by intra-occular injection in the rabbit and injection into the lumbral-sacral space in the rat[17,18].

The two immunodominant autoantigens responsible for this demyelinating autoantibody response in EAE are galactosyl ceramide (GC)19 and M2/myelin oligodendrocyte glycoprotein (MOG)[20,21,22]. GC is a major myelin glycoplipid and acts as a hapten requiring either a carrier protein or insertion into a membrane to induce a high titer antibody response. However, whereas rabbits are high responders to GC, extensive studies have failed to detect significant anti-GC antibody responses in association with any human inflammatory demyelinating disease. In contrast, MOG is a quantitatively minor myelin component, but one which is highly immunogenic in a variaty of species including mice, guinea pigs and man[22,23,24,25]. The identification of MOG as the first protein antigen that can initiate autoantibody mediated demyelination in the CNS has provided a series of paradigms in which the roles played by antibody, complement and T cells in the pathogenesis of autoimmune mediated demyelination have been clearly defined.

The myelin oligodendrocyte glycoprotein (MOG)

MOG was first characterised by a mouse monoclonal antibody (mAb), 8-18C5, which was raised against a preparation of rat cerebeller glycoproteins[23]. This mAb recognised a novel glycoprotein that was specific for CNS myelin. Immunohistochemical studies subsequently demonstrated that the epitope recognised by the mAb was expressed on the outer surface of the oligodendrocyte plasmamembrane and was enriched in the outermost lamellae of the compact myelin sheath[26,27]. MOG cannot be detected in the peripheral nervous system (PNS).

Immunoprecipitation of radiolabelled CNS glycoproteins or Western blotting of myelin proteins after SDS-PAGE identifies three components that bind mAb 8-18C5. The major component is a doublet with an apparent molecular weight of approximately 25-27 kD, whilst the minor component migrates with an apparent molecular weight of 55 kD [23,28]. This latter band is now believed to represent a MOG dimer. The lower molecular weight doublet is a consequence of differential glycosylation of the MOG monomer. Deglycosylation experiments clearly show that the mature form of MOG is N-glycosylated and the carbohydrate component has a molecular weight of approximately 1.5kD[28].

The deduced amino acid sequence for rat MOG is given in Figure 1. The predicted molecular weight of MOG (24,962 Daltons) is very close to that obtained by SDS-PAGE of the deglycosylated protein, and as the protein contains only one canonical N-glycosylation site at Asn-31, this is assumed to be glycosylated *in vivo* [29]. The function of MOG is obscure but its localisation at the outermost surface of the myelin sheath suggests that it may be involved in cell - cell or cell - extracellular matrix interactions, or alternatively function as a cell surface receptor. The identification of MOG as a member of the IgG gene superfamily supports these suggestions. However, unlike the other members of the IgG family which have either a single transmembrane domain, or are attached to the membrane surface by a glycolipid anchor, MOG has two hydrophobic domains (Figure 1), both of which are large enough to cross the membrane[29]. As discussed later the N-terminal IgG V-like domain of MOG is clearly located at the extracellular face of the myelin membrane, but the orientation of the carboxyl terminal tail of MOG remains to be established. Three possible orientations are illustrated in Figure 2. With respect to antibody mediated demyelination *in vivo* it is critical to determine whether or not the carboxyl terminal sequence of the protein is exposed at the external face of the membrane.

Pathogenicity of the autoimmune response to MOG

Antibody responses to MOG in EAE: The ability of MOG to act as a target for antibody mediated demyelination was first demonstrated *in vivo* following injection of the mAb 8-18C5 into the cerebrospinal fluid of rats[30,31]. Subsequently, the pathogenic effects of the mAb were also demonstrated *in vitro* using cultured rat oligodendrocytes[32] and myelinated aggregating CNS tissue cultures[33]. The pathological significance of these observations became apparent when MOG was shown to be identical to the M2 autoantigen[34]. The existence of M2, an immunodominant myelin protein autoantigen responsible for the complement dependent demyelinating activity of guinea pig EAE sera was deduced almost twenty years ago[31]. This autoantigen was shown to be distinct from MBP, PLP and galactosyl ceramide, but its identity and biochemical characteristics remained unresolved for over a decade. The role of M2/MOG in autoimmune demyelination in EAE in the guinea pig was further underlined by the demonstration that the titer of anti-M2/MOG antibodies in the sera of guinea pigs with chronic relapsing EAE is directly proportional their demyelinating activity *in vivo*[22].

The availability of the anti-MOG mAb 8-18C5 allowed, for the first time, the pathological consequences of antibody mediated demyelination to be studied *in vivo* in the absence of a concommitant T cell response. As stated previously, in the Lewis rat, MBP-specific T cell mediated EAE (tEAE) is an acute inflammatory disease without extensive primary demyelination. Intravenous injection of the antibody into rats with tEAE results in a rapid worsening of their clinical status associated with an enhanced inflammatory response in the CNS and widespread demyelination[15,26,35]. These effects are not seen following i.v. injection of irrevelant mouse IgG antibodies into rats with tEAE, or in naive controls injected with mAb 8-18C5. In the latter case the blood brain barrier effectively excludes the mAb from the CNS. However, in animals with tEAE the T cell mediated inflammatory response induces blood brain barrier dysfunction, the mAb can gains access to the CNS and triggers ADCC mediated demyelination. Activation of complement also occurs as the mAb binds to the myelin surface, but although local complement deposition is a specific feature of antibody mediated demyelination[36], demyelination in this model is itself complement independent[37].

This apparent dichotomy was analysed in rats rendered complement deficient by treatment with cobra venom factor. In the absence of complement the mAb is unable to lysis oligodendrocytes *in vitro*, and *in vivo* fails to trigger the deposition of complement component C9 within the CNS[32,37]. Yet *in vivo* the absence of complement has virtually no influence on the final extent of demyelination[37]. The mAb in fact mediates demyelination by targeting an ADCC response against the sheath, rather than by direct complement mediated lysis. Complement plays a more important role in EAE induced by the adoptive transfer of small numbers of encephalitogenic T cells[32]. In this case, antibody-independent complement activation within the CNS releases complement derived pro-inflammatory factors that stimulate migration of inflammatory cells into the CNS and thereby enhance the local inflammatory response.

The epitope recognised by the 8-18C5 mAb has now been localised to the IgG V-like domain of MOG. The mAb binds strongly to a renatured, soluble fusion protein rM-MOG which consists of the extracellular domain of rat MOG amino acid residues 1 to 125, together with the additional N-terminal sequence Met-Arg-Arg-Ser- and at the C-terminus the additional sequence -Arg-Ser-His6. Epitope mapping using a panel of overlapping peptides failed to identify any specific antibody binding site suggesting that the epitope may be conformationally dependent (manuscript in preparation).

T cell responses to MOG in EAE: The observation that MOG induces a demyelinating antibody response in the mouse and guinea pig, immediately raised the question whether or not the autoimmune T cell response to MOG is also pathogenic. The very low concentration of MOG in CNS myelin effectively precluded the direct testing of the encephalitogenicity of the native protein. However this problem was overcome by synthesising a panel of rat MOG peptides based on the deduced amino acid sequence of the protein. Using this panel of

```
                 1             *                *                *                *
Rat MOG          G-Q-F-R-V-I-G-P-G-H-P-I-R-A-L-V-G-D-E-A-E-L-P-C-R-I-S-P-G-K-N-A-T-G-M-E-V-G-W-Y
Butyrophilin     A-P-F-D-V-I-G-P-Q-E-P-I-L-A-V-V-G-E-D-A-E-L-P-C-R-L-S-P-N-V-S-A-K-G-M-E-L-R-W-F

                               50               60               70               80
                 R-S-P-F-S-R-V-V-H-L-Y-R-N-G-K-D-Q-D-A-E-Q-A-P-E-Y-R-G-R-T-E-L-L-K-E-S-I-G-E-G-K
                 R-E-K-V-S-P-A-V-F-V-S-R-E-G-Q-E-Q-E-G-E-E-M-A-E-Y-R-G-R-V-S-L-V-E-D-H-I-A-E-G-S

                               90              100              110              120
                 V-A-L-R-I-Q-N-V-R-F-S-D-E-G-G-Y-T-C-F-F-R-D-H-S-Y-Q-E-E-A-A-V-E-L-K-V-E-D-P-F-Y
                 V-A-V-R-I-Q-E-V-K-A-S-D-D-G-E-Y-R-C-F-F-R-Q-D-E-N-Y-E-E-A-I-V-H-L-K-V

                              130              140              150              160
                 W-I-N-P-G-V-L-A-L-I-A-L-V-P-M-L-L-L-Q-V-S-V-G-L-V-F-L-F-L-Q-H-R-L-R-G-K-L-R-A-E

                              170              180              190              200
                 V-E-N-L-H-R-T-F-D-P-H-F-L-R-V-P-C-W-K-I-T-L-F-V-I-V-P-V-L-G-P-L-V-A-L-I-I-C-Y-N

                              210
                 W-L-H-R-R-L-A-G-Q-F-L-E-E-L-R-N-P-F-cooh
```

The primary amino acid sequences of rat MOG[29] and the N-terminal domain of bovine butyrophilin[43] are depicted. Identical residues are underlined and the hydrophobic domains of MOG which are predicted to interact with the membrane bilayer are in bold type face.

Figure 1. The primary amino acid sequence of rat myelin oligodendrocyte gylcoprotein and its sequence homology with the N-terminal sequence of bovine butyrophilin.

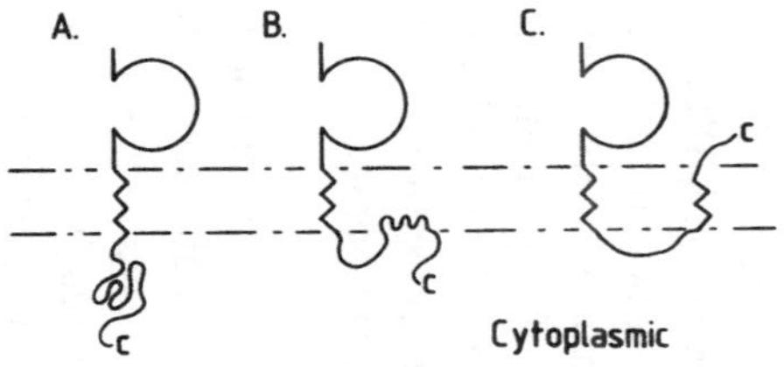

Figure 2.

synthetic peptides it was possible to identify an encephalitogenic T cell epitope of MOG for the Lewis rat[38]. Interestingly, like the demyelinating B cell epitope, this encephalitogenic T cell epitope is also located within the extracellular IgG V-like domain of the MOG.

The adoptive transfer of CD4+ T cells specific for the MOG amino acid sequence 35 to 55 (MEVGWYRSPFSRVVHLYRNGK) induces an intense, dose dependent inflammatory response in the CNS of naive Lewis rats. However, this model of autoimmune encephalomyelitis differs dramatically from the conventional model of MBP-specific T cell mediated EAE. Unlike the inflammatory response induced by MBP-specific T cell lines, MOG-peptide specific T cells does not induce a severe neurological deficit. This is not due to a reduction in the overall inflammatory response in the CNS, but is specifically associated with a decrease in the extent of parenchymal, as opposed to perivascular inflammation and a selective decrease in the number of ED1+ macrophages infiltrating the CNS. In MBP-induced tEAE the ratio of ED1+ to T cells is in the range of approximately 7 to 1. In the lesions induced by the MOG peptide-specific T cell line this is reduced to approximately 1 to 1. The decreased recruitment of macrophages into the CNS can not be ascribed to deficiency in the ability of the T cells to synthesise gamma-interferon, TNF-alpha, IL6 or IL-2. Moreover, the sub-clinical inflammatory response induced by the MOG-peptide specific T cell lines is associated with severe blood-brain barrier dysfunction, as demonstrated by the induction of severe demyelinating disease following intravenous injection of the 8-18C5 MOG-specific monoclonal antibody.

This new model T cell mediated EAE is perhaps somewhat closer to the MS, than the traditional models of MBP-induced disease in that the lesions are clinically silent and consist of large number of cells tightly packed around the perivascular space. Moreover, the induction of a neurological deficit in MOG-peptide induced tEAE is completely dependent on synergy between the encephalitogenic T cell response and the pathogenic demyelinating antibody response.

MOG specific autoimmune responses in man: At present very little is known about the presence or absence of MOG-specific autoimmune responses in MS patients and appropriate controls. Two papers document that the frequency of MOG-specific autoimmune responses is increased in some patients with MS, and that this response segregates into the CNS compartment [24,25]. However, such studies are critically dependent on the purity of the antigen preparation. Hopefully, molecular biological techniques will soon provide large quantities of standardised recombinant MOG fusion proteins thereby circumventing the problems of contamination associated with purifying MOG from human post-mortum tissue. Such reagents will enable multicenter studies to be carried out to determine the role of MOG-specific autoreactivity in MS. However, even if MOG specific autoimmune responses are found in patients with MS the pathogenicity of the response has still to be demonstrated. Our experience with the mAb 8-18C5 demonstrates that at least one pathogenic MOG B cell epitope is conformation dependent. Analysis of the human response should therefore use renatured recombinant MOG fusion proteins or native MOG expressed in mammalian cell lines, rather than rely on short linear peptides.

Myelin oligodendrocyte protein and molecular mimicry: A potential trigger for autoimmune demyelination?

The etiology of MS is even less well understood than the immune effector mechanisms involved in lesion development. The clinical and histopathological similarities that MS shares with EAE suggest that the human disease is autoimmune mediated, whilst disease susceptiblity is determined by the interplay of multiple gene products and the environment[39]. Infectious agents may provide a mechanism that links genetic susceptibility and the generation of a pathogenic autoimmune response. Viral infections involving the CNS may trigger an autoimmune response in several ways: (1) polyclonal activation of the immune system; (2) the induction of neoantigens within cells, which would then be recognised as foriegn; (3) viral lysis of cells may release cellular antigens to which no tolerance has been developed; (4) viral infection may upregulate MHC antigens within the CNS; (5) the immune response to the virus may cross-react with a structurally similar epitope within the CNS[40].

The last mechanism, molecular mimicry, requires that structure of the pathogen derived epitope is sufficiently similar to a host determinant that immune response directed against the pathogen not only cross-reacts with "self", but is also significantly different to break self-tolerance. This concept was first explored in EAE in 1985[41,42]. It was established that an amino acid sequence homology exists between the encephalitogenic site of MBP in the rabbit and the hepatitis B virus polymerase. Sensitisation of rabbits with the homologous viral peptide induced both a T cell and B cell response to MBP and, in addition, inflammatory lesions in the CNS similar to those seen in EAE[42].

In view of these experiments that established the principal that a viral epitope can indeed induce a cross-reactive and pathogenic autoimmune response to CNS antigens we screened the available protein sequence databases to identify amino acid sequence homologies between MOG and other proteins. This computer search identified three amino acid sequence homologies involving MOG that could be relevant to the immunopathogenesis of MS.

The first two amino acid sequence homologies involve the N-terminal IgG-like domain of MOG which has already been shown to be a target for both T and B cell mediated autoimmune responses in EAE (Figure 1). The first 115 amino acids of MOG share an extensive (52% identity) sequence homology[29] with the N-terminal sequence of bovine butyrophilin, the major protein of the milk fat globule membrane[43]. The significance of this observation lies in the recent report that an antibody response to another dietary antigen, bovine serum albumin, cross-reacts with a pancreatic a-cell surface protein, p69. This cross-reactive antibody response is believed to mediate a-cell distruction in autoimmune (type I) diabete mellitus[44]. In addition to the sequence homology with butyrophilin the same domain of MOG has a 39% sequence identity with the monomorphic N-terminal domain of the chicken B-G antigens[29]. Intriguingly, although the mammalian homologues of the B-G antigens have still to be identified, "natural antibodies" recognising chicken B-G antigens are present in sera of many species, including man[45]. The question whether these homologies between the extracellular domain of MOG, butyrophilin and the B-G antigen of chicken are sufficient for the latter two proteins to initiate a MOG-specific demyelinating antibody response is completely open. However, in the Lewis rat we have recently found that immunisation with a peptide corresponding to the N-terminal twenty amino acids of MOG induces an antibody response that cross-reacts with both MOG and bovine butyrophilin as determined by both ELISA and Western blotting (manuscript in preparation). Although the pathogenicity of this cross-reactive antibody response is still to be determined.

The second homology of interest involves an eleven amino acid homology with the L proteins of the Paramyxoviridae[46] and the MOG sequence 198-208 (Table 1). This group of RNA viruses includes measles virus which for many years has been implicated in the etiology of MS, although as yet no immunological cross-reactivity has been demonstrated between measles antigens and CNS determinants in man.

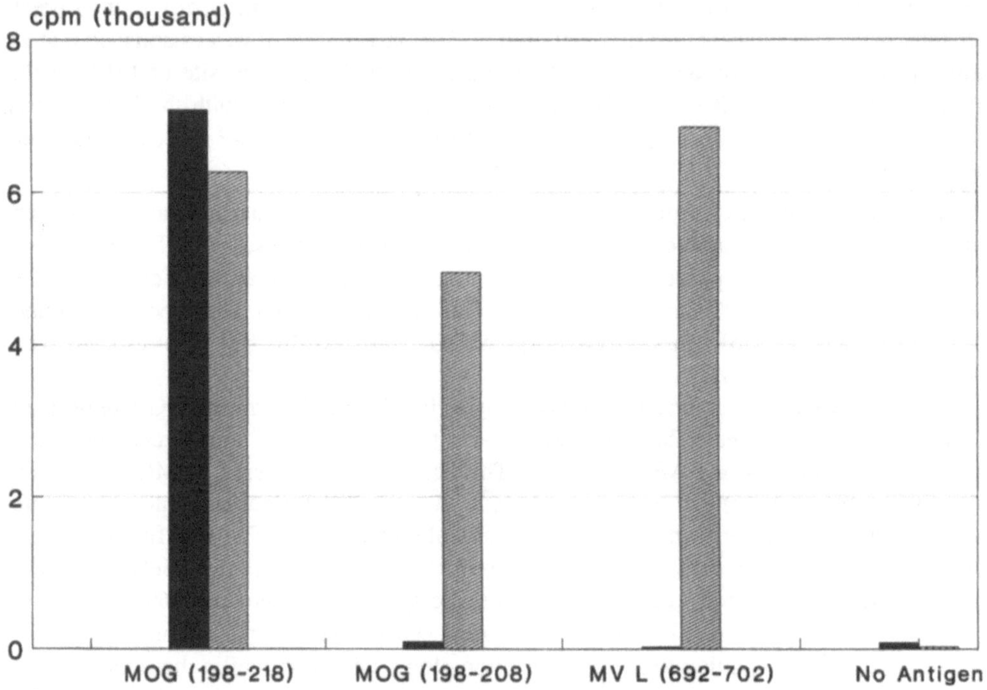

Figure 3. Antigen specificity of T cell lines CL/F5 and CL/D4.

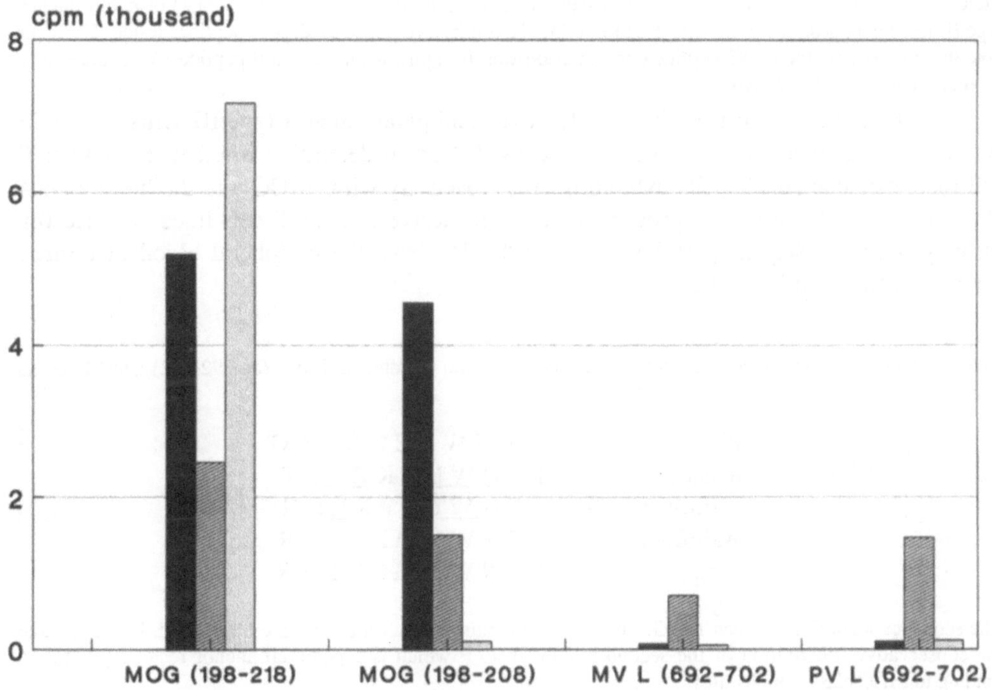

Figure 4. Antigen specificity of T cell lines MA4, MA8 and MA16.

The T cell lines CL/F5 (filled bars) and CL/D4 (hatched bars) were restimulated with irradiated autologous peripheral blood lyphocytes in the presence or absence of antigen (10 ug/ml) for 72 hours. Cells were labelled with ^{3}H-thymidine for the last 16 hours of culture, harvested and 3H-thymidine incorporation determined. The T cell line CLfF5 is completely specific for the MOG sequence 198-218 and does not respond to eithei MOG198-208, or the homologous measles L protein peptide (MV L (692-702)). In contrast the T cell line CL/D4 which was also selected using the MOG198-218 sequence recognises an epitope within the MOG198-208-sequence and responds equally well with the measles peptide. This line did not however respond to the corresponding parainfluenza type III peptide.

The T cell lines MA4 (filled bars), MA8 (hatched bars) and MA16 (stippled bars) were restimulated with irradiated autologous peripheral blood lyphocytes in the presence or absence of antigen for 72 hours. T cell proliferation was determined as described in Figure 3. The three T cell lines exhibit very different patterns of reactivity with the antigens tested. The T cell line MA16 is completely specific for the MOG sequence 198-218 and does not respond to any other antigen, whereas the cell line MA4 recognises an epitope within the MOG sequence a.a. 198-208, but does not respond to either the homologous measles (MV) or parainfluenza type III (PV) peptides. In contrast the T cell line MA8 exhibits a response to all four peptides tested. However note that in this case the MOG peptide concentration was 10 ug/ml whilst the viral peptides were added to a final concentration of 100 ug/ml.

We therefore synthesised the measles virus and parainfluenza type III virus L protein amino acid sequences listed in Table 1 and used them to determine whether the human T cell repertoire contains T cells exhibiting cross-reactivity with MOG and the homologous viral peptides. Our initial approach has been to derive human T cell lines specific for carboxyl terminal sequence of MOG (a.a. 198-218) from the peripheral blood of normal healthy controls and MS patients.

Table 1: Sequence homologies between the L proteins of the Paramxyoviridae (aa 692-702) and MOG (aa 198-208).

MOG	C Y N W L H R R L A G
Measles	F F Q W L H K R L E T
Parainfluenza III	L F N W L H P R L E G
Parainfluenza	F F N W M H P V L E R
Mumps	L F N W M H P V L E R

The sequence identities between MOG, measles and parainfluenza sequences are underlined. However note the conservative substitution of the second residue Y to F which is a potential anchor residue promoting binding to MHC class II molecules

The epitope specificity of these lines was then analysed using smaller overlapping MOG peptides. At least three epitopes are recognised within this twenty-one amino acid long domain of MOG. Approximately one third to one half of the T cell lines responding to an epitope within MOG sequence aa 198-208. The cross-reactivity of these T cell lines was then determined in a proliferation assay using the measles and parainfluenza type III L protein peptides.

Much to our suprise approximately 10% of the MOG$_{198\text{-}218}$ specific T cell lines we have so far isolated from three controls and two MS patients exhibit cross-reactivity with the measles derived peptide. Moreover the characteristics of this cross-reactive response differs dramatically between the different T cell lines examined (Figures 3 and 4). One T cell line, CLD4, proliferates equally as well to the measles derived peptide as it does to the MOG$_{198\text{-}218}$ sequence. Incontrast, a second T cell line (Figure 4) requires a 10-fold higher concentration of the virus peptide to induce significant T-cell proliferation. However, the cross-reactive response of this line is not restricted to the measles peptide, but is also observed with the parainfluenza type III peptide. This cross-reactivity is not seen in all

$MOG_{198-208}$ T cell lines. Moreover, the converse is also true in so far that only one T cell line selected using the measles peptide has been seen to cross-react with $MOG_{198-208}$ peptide.

This study is very much in its preliminary stages and the molecular basis of this cross-reactive response with respect to MHC class II restriction, identification of the critical amino acids required for MHC binding/T cell receptor interaction and T cell activation are at present being determined. It also remains to be established if there is any association between this cross-reactive response involving the L protein of the paramyxoviruses and MOG with MS, or post-measles virus encephalomyelitis. However, as far as we are aware this study provides the first indication that molecular mimicry can and does occur between measles and myelin components in man.

Summary

The myelin oligodendrocyte glycoprotein (MOG) is the first myelin antigen formally demonstrated to elict both a demyelinating autoantibody response and an encephalitogenic T cell response in EAE. The combination of this autoaggressive immune response produces in the animal model a pathology very similar to that seen in MS. The presence of both T and B cell responses to MOG in patients with MS is highly suggestive that synergy between these responses is involved in the immunopathogenesis of this disease. It is unknown if the autoimmune response to MOG in MS is a primary event in the induction of the disease. However amino acid sequence homologies with two environmental antigens, one dietary - butyrophilin, and the other, the paramyxoviruses, infectious provide multiple sites to trigger a primary autoimmune response by molecular mimicry. The presence of multiple cross-reactive epitopes on a single, immunodominant myelin autoantigen may explain the relatively poor association of MS suceptiblity with any singly MHC gene product.

Acknowledgements

We wish to thank all those who have supported the above studies, in particular Jean-Marie Matthieu and Minnetta Gardinier for sharing their MOG cDNA data, Klaus Dornmair for the preparation of the MOG-fusion proteins, Hans Lassmann and his co-workers for their continuing histopathological studies on MOG induced EAE and Hartmut Wekerle for his continued support and helpful discussions.

REFERENCES

1. Martin, R., McFarland, H.F. and McFarlin, D.E. Immunological aspects of demyelinating diseases. *Annu.Rev.Immunol.* 10:153-187, 1992.
2. Lyman, W.D., Roth, G.A., Chiu, F.-C., Brosnan, C.F., Bornstein, M.B. and Raine, C.S. Antigen-specific T cells can mediate demyelination in organotypic central nervous system cultures. *Cell.Immunol. 102:217-226, 1986.*
3. Brosnan, C.S., Selmaj, K. and Raine, C.S. Hypothesis: A role for tumor necrosis factor in immune-mediated demyelination and its relevance to multiple sclerosis. *J.Neuroimmunol.* 18:87-94, 1988.
4. Prineas, J.W. and Raine, C.S. Electron microscopy and immunoperoxidase studies of early multiple sclerosis lesions. *Neurology* 26:29-32, 1976.
5. Prineas, J.W. and Graham, J.S. Multiple sclerosis: Capping of surface immunoglobulin G on macrophages engaged in myelin break down. *Ann.Neurol.* 10:149-158, 1981.
6. Compston, D.A.S., Morgan, B.P., Oleesky, D., Fifield, R. and Campbell, A.K. Cerebrospinal fluid C9 in demyelinating disease. *Neurology* 36:1503-1506, 1986.
7. Scolding, N.J., Morgan, B.P., Houston, W.A.J., Linington, C., Campbell, A.K. and Compston, D.A.S. Vesicular removal by oligodendrocytes of membrane attack complexes formed by activated complement. *Nature* 339:620-622, 1989.

8. Compston, D.A.S., Morgan, B.P., Campbell, A.K., Wilkins, P., Cole, G., Thomas, N.D. and Jasani, B. Immunocytochemical Localization of the Terminal Complement Complex in Multiple Sclerosis *Neuropathol. Appl. Neurobiol.* 15:307-316, 1989.
9. Esiri, M.M. Multiple sclerosis: a quantitative and qualitative study of Ig-containing cells in the CNS. *Neuropathol. Appl. Neurobiol.* 6:9-21, 1980.
10. Newcombe, J.I.A., Gahan, S. and Cuzner, M.L. Serum antibodies against central nervous system proteins in human demyelinating disease. *Clin.Exp.Immunol.* 59:383-390, 1985.
11. Waksman, B.H., Porter, H., Lees, M.B., Adams, R.D. and Folch, J. A study of the chemical nature of components of bovine white matter effective in producing allergic encephalomyelitis in the rabbit. *J.Exp.Med.* 100:451-471, 1954.
12. Raine, C.S. Experimental allergic encephalomyelitis and experimental allergic neuritis. In: *Handbook of Clinical Neurology*, edited by Vinken, Bruyn, and Klawans, Elsevier Sc.Publ., 1985, p. 429-466.
13. Lassmann, H. *Comparative neuropathology of chronic experimental allergic encephalomyelitis und multiple sclerosis*, Berlin:Springer Verlag, 1983.
14. Williams, R.M., Lees, M.B., Cambi, F. and Macklin, W.B. Chronic experimental allergic encephalomyelitis induced in rabbits with bovine white matter proteolipid apoprotein. *J.Neuropathol.Exp.Neurol.* 41:508-521, 1982.
15. Lassmann, H., Brunner, C., Bradl, M. and Linington, C. Experimental allergic encephalomyelitis: The balance between encephalitogenic T lymphocytes and demyelinating antibodies determines size and structure of demyelinated lesions. *Acta Neuropathol.* 75:566-576, 1988.
16. Bornstein, M.B. and Raine, C.S. Multiple sclerosis and experimental allergic encephalomyelitis: Specific demyelination of CNS in culture. *NeuropatholAppl.Neurobiol.* 3:359-367, 1977.
17. Brosnan, C.F., Stoner, G.L., Bloom, B.R., Wisniewski, H.M. Studies on demyelination by activated lymphocytes in the rabbit eye II. Antibody-dependant cell-mediated demyelination. *J. Immunol.* 118:2103-2110, 1977.
18. Lassmann, H., Stemberger, H., Kitz, K. and Wisniewski, H.M. In vivo demyelinating activity of sera from animals with chronic experimental allergic encephalomyelitis. Antibody nature of the demyelinating factor and the role of complement. *J.Neurol.Sci.* 59:123-137, 1983.
19. Raine, C.S., Johnson, A.B., Marcus, D.M., Suzuki, A. and Bornstein, M.B. Demyelination in vitro. Absorption studies demonstrate that galactocerebroside is a major target. *J.Neurol.Sci.* 52:117-131, 1981.
20. Lebar, R. and Voisin, G.A. Studies on auto-immune encephalomyelitis in the guinea pig. I. Influence of different immunizations on antibodies specificity and biological properties, relation to the incidence of the disease. *Int.Arch.Allergy Appl.Immunol.* 46:82-103, 1974.
21. Lebar, R., Lubetzki, C., Vincent, C., Lombrail, P. and Poutry, J.-M. The M2 autoantigen of central nervous system myelin, a glycoprotein present in oligodendrocyte membranes. *Clin.Exp.Immunol.* 66:23-443, 1986.
22. Linington, C. and Lassmann, H. Antibody responses in chronic relapsing experimental allergic encephalomyelitis: Correlation of serum demyelinating activity with antibody titer to myelin/oligodendrocyte glycoprotein (MOG). *J.Neuroimmunol.* 17:61-70, 1987.
23. Linington, C., Webb, M. and Woodhams, P.L. A novel myelin-associated glycoprotein defined by a mouse monoclonal antibody. *J.Neuroimmunol.* 6:387-396, 1984.
24. Sun, J., Link, H., Olsson, H., Xiao, B., Andersson, G., Ekre, H.-P., Linington, C. and Diener, P. T and B cell responses to myelin-oligodendrocyte glycoprotein in multiple sclerosis. *J.Immunol.* 146:1490-1495, 1991.
25. Xiao, B., Linington, C. and Link, H. Antibodies to myelin-oligodendrocyte glycoprotein in cerebrospinal fluid from patients with multiple sclerosis and controls. *J.Neuroimmunol.* 31:91-96, 1991.
26. Linington, C., Bradl, M., Lassmann, H., Brunner, C. and Vass, K. Augmentation of demyelination in rat acute allergic encephalomyelitis by circulating mouse monoclonal antibodies directed against a myelin/oligodendrocyte glycoprotein. *Am.J.Pathol.* 130:443-454, 1988.
27. Brunner, C., H. Lassmann, T.V. Waehneldt, J.-M. Matthieu and C. Linington: Differential ultrastructural localisation of myelin basic protein, myelin/oligodendrocyte glycoprotein and 2',3'-cyclic nucleotide 3'-phosphodiesterase in the CNS of adult rats. J. Neurochem. 52, 296-304 (1989)
28. Amiguet, P., Gardinier, M.V., Zanetta, J.-P. and Matthieu, J.-M. Purification and partial structural and functional characterization of mouse myelin/oligodendrocyte glycoprotein. *J.Neurochem.* 58:1676-1682, 1992.
29. Gardinier, M.V., Amiguet, P., Linington, C. and Matthieu, J.-M. Myelin/oligodendrocyte glycoprotein is a unique member of the immunoglobulin superfamily. *J.Neurosci.Res.* 33:177-187, 1992.
30. Lassmann, H., Zimprich, F., Rossler, K. and Vass, K. Inflammation in the nervous system. Basic mechanisms and immunological concepts. *Rev.Neurol.* 147:763-781, 1991.

31. Lassmann, H. and C. Linington: The role of antibodies against myelin surface antigens in chronic EAE. In: A multidisciplinary approach to Myelin diseases. Ed: G. S. Crescenzi. NATO ASI Series Vol A142, Plenum Press New York, pp 219-226, 1987.
32. Linington, C., B.P. Morgan, N.J. Scolding, S. Piddlesden, P. Wilkins and D.A.S. Compston. The role of complement in the pathogenesis of experimental allergic encephalomyelitis. *Brain* 112:895-911, 1989.
33. Kerlero de Rosbo, N., Honegger, P., Lassmann, H. and Matthieu, J.-M. Demyelination induced in aggregating brain cell cultures by a monoclonal antibody against myelin/oligodendrocyte glycoprotein. *J.Neurochem.* 55:583-587, 1990.
34. Lebar, R. and Voisin, G.A. Studies on auto-immune encephalomyelitis in the guinea pig. I. Influence of different immunizations on antibodies specificity and biological properties, relation to the incidence of the disease. *Int.Arch.Allergy Appl.Immunol.* 46:82-103, 1974.
35. Schluesener, H.J., Sobel, R.A., Linington, C. and Weiner, H.L. A monoclonal antibody against a myelin oligodendrocyte glycoprotein induces relapses and demyelination in central nervous system autoimmune disease. J.Immunol. 139:4016-4021, 1987.
36. Linington, C., Lassmann, H., Morgan, B.P. and Compston, D.A.S. Immunohistochemical localization of terminal complement component C'9 in experimental allergic encephalomyelitis. *Acta Neuropathol.* 79:78-85, 1989.
37. Piddlesden, S., Lassmann, H., Laffafian, I., Morgan, B.P. and Linington, C. Antibody-mediated demyelination in experimental allergic encephalomyelitis is independent of complement membrane attack complex formation. *Clin.Exp.Immunol.* 83:245-250, 1991.
38. Linington, C., T. Berger, L. Perry, S. Weerth, D. Hinze-Selch, Y. Zhang, H.-C. Lu, H. Lassmann and Wekerle, H. T cells specific for the myelin oligodendrocyte glycoprotein (MOG) mediate an unsual autoimmune inflammatory response in the central nervous system. *Eur. J. Immunol.* accepted for publication.
39. Wekerle, H. Myelin specific, autoaggressive T cell clones in the normal immune repertoire: their nature and their regulation. *Intern. Rev. Immunol.* 9:231-241, 1992.
40. Dyrberg, T. Molecular mimicry in autoimmunity, *in*: Molecular Autoimmunity, Academic Press, New York (1991)..
41. Jahnke, U., Fischer, E.H. and Alvord, E.C. Sequence homology between certain viral proteins and proteins related to encephalomyelitis and neuritis. *Science* 229:282-284, 1985.
42. Fujinami, R.S. and Oldstone, M.B.A. Amino acid homology between the encephalitogenic site of myelin basic protein (MBP) and virus: Mechanism for autoimmunity. *Science* 230:1043-1046, 1985.
43. Jack, L.J.W. and Mather, I.H. Cloning and analysis of cDNA encoding bovine butyrophilin, an apical glycoprotein expressed in mammary tissue and secreted in association with the milk-fat globule membrane during lactation. *J. Biol. Chem.* 265:14481-14486, 1990..
44. Karjalainen, J., Martin, J.M., Knip, M., Ilonen, J. et al. A bovine albumin peptide as a possible trigger of insulin-dependent diabetes mellitus. *New. Eng. J. Med.* 327:302-307, 1992.
45. Longenecker, B.M. and Mossman, T.R. "Natural" antibodies to chicken MHC antigens are present in mice, rats, humans, alligators, and allogeneic chickens. *Immunogenetics* 11:293-302.
46. Blumberg, B.M., Crowley, J.C., Silverman, J.I., Menonna, J., Cook, S.D. and Dowling, P.C. Measles virus L protein evidences elements of ancestral RNA polymerase. Virology 164:487-497, 1988.

INTRATHECAL CYTOKINE SYNTHESIS IN INFLAMMATORY AND DEMYELINATING DISEASES OF THE CENTRAL NERVOUS SYSTEM

Bruno Tavolato and Paolo Gallo

Institute of Neurology
Second Neurologic Clinic
University of Padua
School of Medicine
Padua, Italy

INTRODUCTION

The immune response is strictly and continuously regulated by a highly complex network of soluble mediators (i.e., cytokines and growth factors) that usually act as autocrine and/or paracrine regulators of cell growth and function. Under pathological conditions, however, when these mediators are produced in larger amounts in tissues or by circulating cells, they can be detected in biological fluids, and may exert hormone-like effects. Thus, the analysis of body fluids (such as serum, cerebrospinal fluid, urine and synovial fluid) for the presence of cytokines may constitute a strategy to define the role of these factors in inflammatory reactions *in vivo*.

The cerebrospinal fluid (CSF) is contiguous with the extracellular spaces of brain tissue and, during the course of central nervous system (CNS) immunological and infectious diseases, its analysis often provides useful information. In fact, quantitative and qualitative abnormalities in immunoglobulin patterns, as well as modifications in the number and/or subset distribution of mononuclear cells can easily be demonstrated. Since cytokines and growth factors can be produced by both invading/activated mononuclear cells [T and B lymphocytes, macrophages, natural killer cells (NK)] and by reactive/proliferating glial cells [astrocytes and microglia] it seemed reasonable to look for intrathecally synthesized cytokines as an expression of immunopathological processes taking place within the brain.

Many recent reports have described the detection of cytokines in serum and CSF in CNS diseases, with special regard to multiple sclerosis (MS), viral and bacterial meningoencephalitis, and human immunodeficiency virus type 1 (HIV-1) infection. Here we summarize our data on this topic, review published findings, and address the relevance of intrathecally synthesized cytokine to the immunopathogenesis of infectious and demyelinating diseases of the CNS.

A Multidisciplinary Approach To Myelin Diseases II
Edited by S. Salvati, Plenum Press, New York, 1994

INTERLEUKIN 1 (IL-1)

IL-1 does not cross the blood-brain barrier (BBB). However, systemically produced IL-1ß clearly affects the periventricular structures of the CNS (sites were the BBB is interrupted) inducing fever, sleep and the release of a variety of neuropeptides. IL-1ß can be produced by activated astrocytes, microglial cells, and, under pathological conditions, by invading/infiltrating monocytes/macrophages as well. IL-1 receptors (IL-1R) were found to be distributed throughout the rat brain. The physiological and pathological effects of IL-1 arise from a complex network of signals induced by IL-1ß, IL-1α, IL-1R, sIL-1R, and IL-1R antagonist (Dinarello, 1987, 1989; Dinarello and Wolff, 1993).

Although the chromatographic separation of IL-1 from binding and/or inhibitory substances seems to be an important step in identifying serum or plasma IL-1 by radio- or enzyme-linked immunoassay, many reports described its detection in *native* biological fluids. Plasma IL-1ß concentrations in normal subjects are usually below the detection limit (20-40 pg/ml) of the available assays. Strenuous exercise and ovulation induce a rise in circulating IL-1. High IL-1 levels were detected in synovial fluid during acute exacerbation of rheumatoid arthritis, in patients with renal allograft rejection, alcoholic hepatitis, burns and sepsis (for review, see Dinarello, 1991).

IL-1 was demonstrated in the CSF of patients with closed head trauma, bacterial and viral meningitis, HIV-1 infection, brain tumors, degenerative disk disease, and cervical spondylosis (Gorczynski and Keystone, 1986; McClain et al., 1987; Ramilo et al., 1989, 1990). Bioactive IL-1 was detected in the CSF of guinea pigs with chronic relapsing experimental allergic encephalomyelitis (Symons, et al., 1987).

IL-1 was found to be frequently increased in MS CSF in only one study (Hauser et al., 1990) that included patients with non-inflammatory disorders, stroke and migraine; increased IL-1 levels were also found in patients with other neurological diseases (28%). At least five independent investigations, in which different methods were used, failed to detect IL-1 in MS CSF (Gallo et al., 1989, 1991; Maimone et al., 1991; Peter et al., 1991; Tsukada et al., 1991; Weller et al., 1991).

A more recent study of patients with optic neuritis did not find IL-1ß in the serum but detected low amounts in 6/20 CSF (Deckert-Schluter et al., 1992).

We found that high levels of biologically active IL-1 were intrathecally synthesized only in patients with bacterial meningitis/meningoencephalitis (Gallo et al., 1989a, 1989b). IL-1 was demonstrated in up to 50% of the patients with viral infections (such as HSV-1 and HIV-1 encephalitis) but at lower levels than in bacterial infections. IL-1ß could also be detected in the CSF and cystic fluid of brain tumor patients; low levels of IL-1ß were found in MS CSF only very rarely.

Since IL-1 can be produced by cells of monocyte/macrophage lineage (microglia included), as well as by reactive glial cells (astrocytes), fibroblasts and endothelial cells, the wide spectrum of CNS diseases associated with intrathecal IL-1 production is not surprising. The role of this cytokine in bacterial and viral-induced meningeal inflammation is discussed in the TNFα section (see below).

INTERLEUKIN 2 (IL-2)

IL-2 not only acts on cells of lymphoid lineage, but may also functionally and morphologically influence cells outside the immune system. With regard to the CNS, it was demonstrated that oligodendrocytes react with anti-IL-2 receptor antibody *in vitro*; however, the biological effect of IL-2 on these cells is unclear, since it seems to mediate both growth stmulation and inhibition.

Given that IL-2 is not produced by glial cells, its intrathecal cellular source is restricted to infiltrating activated IL-2-producing lymphoid cells, such as T lymphocytes, NK and LAK, both at the brain tissue level and in the CSF.

Several investigations detected IL-2 in both the CSF and serum of MS patients (Gallo et al., 1988, 1989, 1991; Adachi et al., 1989, 1990; Trotter et al., 1989, 1990; Peter et al., 1991; Sharief et al., 1991; Weller et al., 1991). However, the available data on this topic are not consistent and in part contradictory. It seems that both relapsing-remitting (RR-MS) and chronic-progressive (CP-MS) MS patients may have increased IL-2 levels in serum and, less frequently, in the CSF. Only 1 out of 12 reports found all the patients studied to be negative (Weller et al., 1991). The clinical utility of IL-2 measurement is also a matter of debate; indeed, the correlation found between IL-2 levels and disease activity/progression (Trotter et al., 1989, 1990) was not confirmed by two independent studies (Gallo et al., 1991; Peter et al., 1991).

From an immunopathological point of view, the detection of high IL-2 levels in MS serum is particularly interesting for the following reasons : 1) a major sequela of immunotherapy with human recombinant IL-2 (hrIL-2) is the development of the "vascular leak syndrome", an accumulation of extracellular fluid (Cotran et al., 1987) ; 2) hrIL-2 therapy is often complicated by transient focal neurologic deficits (Denicoff et al., 1987; Bernard et al., 1990); 3) acute fatal leucoencephalopathy (with clinicopathological features closely resembling acute perivenous leucoencephalomyelitis) was described after hrIL-2 therapy (Vecht et al., 1990); 4) infusion of both hrIL-2 and mouse recombinant IL-2 in rats alters BBB permeability, and induces alterations in incerebrovascular morphology, demyelination and occasional axonal degeneration (Ellison et al., 1990; Ellison and Merchant, 1991), and 5) systemic IL-2 administration in humans increases expression of adhesion molecules (i.e., ICAM and ELAM) and class I and II MHC antigens on endothelial cells (Cotran et al., 1987). Since IL-2 does not activate endothelial cells directly, but enhances the production of other cytokines, such as IL-1ß, TNFα and IFN (which were demonstrated to exert profound effects on endothelial cells *in vitro*), it was suggested that T-cell-derived IL-2 initiates a cytokine cascade that induces endothelium activation and, ultimately, perivascular infiltration of immunocompetent cells. Such a mechanism might be active in and relevant to MS immunopathology.

The frequent detection of increased levels of soluble IL-2 receptors (sIL-2R) in MS serum and/or CSF (Greenberg et al., 1988; Adachi et al., 1989, 1990; Gallo et al., 1989, 1991; Hartung et al., 1990; Kittur et al., 1990; Sharief et al., 1991) confirms the migration of systemically activated T cells into the CNS, and further stresses the role of the IL-2/sIL-2R/IL-2R(CD25) circuit in MS immunopathology. A crucial role for IL-2/IFN -producing Th1/CD4+CDw49d+ lymphocytes in the pathogenesis of acute experimental allergic encephalomyelitis was recently advanced (Baron et al., 1993). This seems to be in agreement with our previous observation that, unlike IL-2, T cell growth factor IL-4, which is produced by Th2 lymphocytes, was never detected in MS serum and CSF.

Oligodendrocytes were shown to express IL-2R, but, as mentioned above, the effect of IL-2 on these cells is disputed, and no definitive conclusions can be drawn on its possible role in demyelinating/remyelinating processes.

TUMOR NECROSIS FACTOR ALPHA (TNFα)

TNFα is intimately linked to the central pathophysiological processes of several acute and chronic human disease states, and is perhaps the most prominent mediator of metabolic changes at all organization levels - cell, tissue and organism. TNFα administration reproduces nearly every feature of sepsis in detail. In a primate model, passive pre-immunization with anti-TNFα antibodies prevented septic shock and death, despite ongoing untreated bacteremia (for review, see Manogue et al., 1991).

At the CNS level, TNFα can be produced by reactive glia (Frei et al., 1987; Robbins et al., 1987), and by infiltrating cells of monocyte/macrophage lineage. As TNFα at high concentrations was shown to exert demyelinating activity *in vitro*, (Selmaj and Raine, 1988), a role for this cytokine in immune-mediated demyelination was advanced (Brosnan et al., 1988).

TNFα has been closely associated with bacterial meningitis/meningo-encephalitis, where particularly high levels were detectable in the CSF, but not in the corresponding serum (McCracken et al., 1989; Mustafa et al., 1989a; Nadal et al., 1989; Waage et al., 1989). TNFα was also detected in the CSF of HIV-1-infected patients (Grimaldi et al., 1991), but several studies could not confirm this finding (Gallo et al., 1989b, 1989d; Shaskan et al., 1992).

The available data on TNFα presence in the CSF of MS patients appear somehow confusing. While at least five independent studies failed to detect this cytokine (Frei et al., 1988; Gallo et al., 1988; Franciotta et al., 1989; Peter et al., 1991; Weller et al., 1991), 4 others reported percentages of positivity ranging from 20 to 100% (Hauser et al., 1990; Maimone et al., 1991; Sharief et al., 1991; Tsukada et al., 1991). For instance, Hauser et al., also found TNFα in 12/18 cases of non inflammatory neurological disease, 5/5 stroke, 11/19 inflammatory disorders of the CNS, and 2/4 migraine; while Tsukada et al. (who found TNFα in all MS CSF studied) detected TNFα only in inflammatory (7/8 Guillain-Barré, 5/7 CIDP) and never in non inflammatory neurologic diseases.

Our data were obtained in a large series of patients with several neurological diseases, and show that intrathecal TNFα synthesis is detectable only in patients with bacterial meningitis; indeed, high levels can be demonstrated a few hours from the onset of disease. Since TNFα and related cytokines, such as IL-1ß and IL-6, may play a crucial role in meningitis outcome (Quagliarello and Scheld, 1992), our observation is particularly worthy of interest, and is further confirmed by several experimental findings: 1) direct inoculation of IL-1ß and/or TNFα in the CSF of experimental animals induces inflammation and impairs the BBB in a time- and dose-dependent manner (Ramilo et al., 1990; Quagliarello et al., 1991); 2) when these cytokines are administered with their respective neutralizing antibodies an almost complete suppression of meningeal and CSF inflammatory changes is observed (Mustafa et al., 1989b; Ramilo et al., 1990; Saukkonen et al, 1990); and 3) the CSF and meningeal inflammatory response to *H. influenzae* lipopolisaccharyde and *Listeria Monocytogenes* is largely prevented by the simultaneous inoculation of antibodies to TNFα, IL-1ß or both (Leist et al., 1988; Ramilo et al.,1990).

The recognition that macrophage-derived endotoxin-induced inflammatory cytokines are critical mediators of meningeal inflammation, and may negatively influence disease outcome, provides the rationale for the use of corticosteroids and/or non-steroidalanti-inflammary drugs in the treatment of bacterial meningitis. Moreover, these studies suggest other therapeutic strategies, such as new antibiotics with more bactericidal than bacteriolytic activity, cytokine antagonists, and monoclonal antibodies to cytokine-induced adhesion glycoproteins.

INTERLEUKIN 6 (IL-6)

IL-6 can be readily demonstrated in the serum of subjects with a variety of diseases, including inflammation and malignancy, and is thought to account for the polyclonal B-cell abnormalities as well as the autoantibody production seen in such conditions (Nijsten et al., 1987; Van Oers et al., 1988; Hirano, 1991). High serum IL-6 levels were described in patients with cardiac myxoma, Castleman's disease, rheumatoid arthritis (also in synovial fluid) (Helle et al., 1991), alcoholic liver cirrhosis, HIV infection, multiple myeloma, and plasma cell leukemias. IL-6 can be detected in urine samples from patients with mesangial proliferative glomerulonephritis, and it acts as an autocrine growth factor for mesangial cells (Hirano, 1991).

Monocyte/macrophages, astrocytes, microglial cells and endothelial cells produce IL-6 (Frei et al., 1989; Mantovani and Dejana, 1989; Benveniste et al., 1990; Wesselingh et al., 1990).Since IL-6 induces the terminal differentiation of activated B lymphocytes into immunoglobulin secreting cells, and an intrathecal B cell response with oligo/polyclonal IgG production is a peculiar feature of MS, it was suggested that locally produced IL-6 might account for intrathecally synthesized IgG. Thus, several groups investigated IL-6 presence in MS CSF using both biological and immunoenzymatic techniques. In this case as well the data are not concordant. The failure to detect this cytokine in MS CSF was described by several workers (Frei et al., 1988; Houssiau et al., 1988; Gallo et al., 1990; Hauser et al., 1990), but two independent groups demonstrated IL-6 in respectively 29% (Maimone et al., 1991) and 60% (Weller et al., 1991) of the MS CSF studied.

Very high IL-6 levels can be demonstrated in the CSF from patients with viral and bacterial meningitis and encephalitis (Frei et al., 1988; Houssian et al., 1988; Helfgott, et al., 1989; Waage et al., 1989; Gallo et al., 1990; Weller et al., 1991), and from patients with malignant and benign CNS neoplasias (Leppert et al., 1989; Van Meir et al., 1990). The detection of IL-6 in the CSF of patients with glioma is particularly interesting since IL-6 mRNA was found to be inducible in cultured astrocytoma and glioblastoma cell lines (Yasukawa et al., 1987), thus suggesting a role for IL-6 as autocrine growth factor for these tumors.

We found detectable levels of IL-6 in the CSF of HIV-1-infected patients with and without neurological complications (Gallo et al., 1989b, 1991b). The demonstration of IL-6 and IL-1ß in asymptomatic HIV seropositive subjects suggests that a subclinical immunopathological process taking place within the CNS may be one of the first manifestationsof HIV infection (Gallo et al., 1991a). Immunocytochemical analysis of IL-6 expression in brain sections from patients with HIV-1-related encephalitis/leukoencephalopathy disclosed that reactive astroglial cells were strongly stained by anti-IL-6 antibody, while cells of monocyte/macrophage lineage were only rarely and faintly stained. A similar staining pattern was obtained with anti-IL1ß antibody, thus suggesting that both cytokines were intrathecally produced by reactive astrocytes.

INTERFERON GAMMA (IFN)

The key function of IFN resides more in its activation of the immune reaction, than in its antiviral effects. Unlike members of the IFNα and IFNß families, which may be produced by any cell type, the IFN production is a specialized function of T lymphocytes. IL-1ß and IL-2 act synergistically to boost IFN production. IFN promotes immune responses (antibody production and development of cytotoxic T cells) by inducing/augmenting class II antigen expression and IL-1 production by many different accessory cells (antigen presenting cells), and thus stimulates their interaction with T cells. IFN -mediated class I and II MHC antigen expression can also be observed on T and B cells, and capillary endothelial cells (for review, see de Maeyer and de Maeyer-Guignard, 1991).

IFN produced by activated T lymphocytes that have colonized the CNS may induce class II MHC antigen expression on microglial cells and astrocytes, which may therefore function as antigen-presenting cells within the brain (Fontana et al., 1984).

Despite earlier promising findings (Hirsh et al., 1985; Abbot et al., 1987), we (Gallo et al., 1989c) and others (Lebon et al., 1987; Weller et al., 1991) failed to find CSF IFN in neuroimmunologic disorders, including MS, HIV-1 infection, post-infectious encephalomyelitis, and subacute sclerosing panencephalitis; CSF IFN was detected in herpes simplex virus type I encephalitis (Lebon et al., 1988) and, occasionally, in neoplastic leptomeningeal disease (Weller et al., 1991).

In mice infected with lymphocytic choriomeningitis virus (LCMV), intracerebral IFN production was demonstrated 6 to 7 days after infection (Frei et al., 1988). The CSF levels of IFN were up to 1,000-fold higher than the serum levels. In this experimental model, the sources of IFN are very likely LCMV-specific T cells which penetrate into the CNS in the course of LCMV disease; this is also suggested by the absence of IFN in CSF of LCMV infected athymic nu/nu mice which do not develop disease.

Taken all together, these data provide evidence that intrathecal IFN synthesis takes place during CNS viral infection. IFN may not only prompt electively antigen presenting cells (i.e., astrocytes, microglia) to fulfil their function, but may also play an important role in the regulation of intracerebral anti-viral humoral and cellular immune responses.

SUMMARY

During the course of demyelinating and inflammatory diseases of the CNS, as well as in neoplastic and autoimmune brain disorders, blood-derived mononuclear cells and reactive glial cells may produce cytokines and growth factors intracerebrally/intrathecally. The detection of such soluble immune-response mediators in the CSF may provide useful information on the immunopathological processes taking place within the CNS.

High IL-2 levels were demonstrated in serum and CSF of MS patients, thus suggesting a systemic T cell activation and a role for this cytokine in MS immunopathogenesis by activating endothelial cells at the BBB levels.

Intrathecal TNFα synthesis seems to be an exclusive feature of bacterial meningitis , since it was never found in other diseases, including MS, HIV-related encephalitis, brain tumors and viral meningoencephalitis. The outcome of patients with bacterial CNS infections may be strictly related to the levels of intrathecally produced TNFα and IL-1ß.

IL-1ß and IL-6 can be detected in the CSF of patients with viral infections (HIV included) and benign and malignant brain tumors (meningiomas and gliomas), and may reflect reactive gliosis and neoplastic glial transformation.

No definitive relationship between intrathecally synthesized cytokines and the oligodendrocyte damage observed in demylinating diseases emerges from the literature data. In addition, except for the established association of TNFα and IL-1ß with bacterial meningitis, the significance of, and the possible role(s) played by cytokines in most neuroimmunological diseases are questioned by contradictory findings. Finally, it has to be kept in mind that the failure in detecting cytokines in the CSF does not exclude their intracerebral production followed by complete absorbtion at the target cell level.

ACKNOWLEDGEMENTS

This work was supported by grants from Ministero della Sanità - Progetto AIDS 1992-1993, Italian Society for Multiple Sclerosis, and Regione Veneto.

REFERENCES

Abbott, R.J., Bolderson, I., Gruer, P.J.K., and Peatfield, R.C., 1987. Immunoreactive IFN-γ in CSF in neurological disorders. *J.Neurol.Neurosurg.Psychiat.* 50:882.

Adachi, A., Kumamoto., R., and Araki, S., 1989. Interleukin 2 receptor levels indicating relapse in multiple sclerosis. *Lancet* i:559.

Adachi, A., Kumamoto, R., and Arachi, S., 1990. Elevated soluble interleukin-2 receptor levels in patients with multiple sclerosis. *Ann.Neurol.* 28:687.

Baron, J.L., Madri, J.A., Ruddle, N.H., Hashim, G., Janeway, C.A.Jr., 1993. Surface expression of α4 integrin by CD4 T cells is required for their entry into brain parenchyma. *J.Exp.Med.* 177:57.

Benveniste, E.N., Sparacio, S.M., Norris, J.G., Grenett, H.E., and Fuller, G.M., 1990. Induction and regulation of interleukin-6 gene expression in rat astrocytes. *J.Neuroimmunol.* 30:201.

Bernard, J.T., Ameriso, S., Kempf, R.A., Rosen, P., Mitchell, M.S., Fisher, M., 1990. Transient focal neurologic deficits complicating interleukin-2 therapy. *Neurology* 40:154.

Brosnan, C.F., Selmaj, K.W., Raine, C.S., 1988. Hypothesis: a role for tumor necrosis factor in immune-mediated demyelination and its relevance to multiple sclerosis. *J.Neuroimmunol.* 18:87.

Cotran, R.S., Pober, J.S., Gimbrone, M.A., Springer, T.A.Jr., Wiebke, E.A., Gaspari, A.A., Rosemberg, S.A., and Lotze, M.T., 1987. Endothelial activation during interleukin 2 immunotherapy. A possible mechanism for the vascular leak syndrome. *J.Immunol.* 139:1883.

Deckert- Schlüter, M., Schlüter, D., and Schwendemann, G., 1992. Evaluation of IL-2, sIL-2R, IL-6, TNFα, and IL-1ß levels in serum and CSF of patients with optic neuritis. *J.Neurol.Sci.* 113:50.

de Maeyer, E., and de Maeyer-Guignard, J., 1991. Interferons, *in:*"The Cytokine Handbook", A. Thomson, ed., Academic Press Ltd., London.

Denicoff, K.D., Rubinoff, D.R., and Papa, M.Z., 1987. The neuropsychiatric effects of treatment with interleukin-2 and lymphokine-activated killer cells. *Ann.Intern.Med.* 107:293.

Dinarello, C.A., 1987. Interleukins, tumor necrosis factors (cachectin), and interferons as endogenous pyrogens and mediators of fever. *Limphokines* 14:1

Dinarello, C.A., 1989. Interleukin-1 and its biologically related cytokines, *Adv.Immunol.* 44:153.

Dinarello, C.A., 1991. Interleukin-1, *in* : "The Cytokine Handbook", A.Thomson, ed., Academic Press Ltd, London.

Dinarello, C.A., and Wolff S.M., 1993. The role of interleukin-1 in disease, *N.Engl.J.Med.* 328:106.

Ellison, M.D., Krieg, R.J., Povlishock, J.T., 1990. Differential central nervous system responses following single and multiple recombinant interleukin-2 infusions. *J.Neuroimmunol.* 28:249.

Ellison, M.D., and Merchant, R.E., 1991. Appearance of cytokine-associated central nervous system myelin damage coincides temporally with serum tumor necrosis factor induction after recombinant interleukin-2 infusion in rats. *J.Neuroimmunol.* 33:245.

Fontana, A., Fierz, W., and Wekerle, H., 1984. Astrocytes present myelin basic protein to encephalitogenic T-cell lines. *Nature* 307:273.

Franciotta, D.M., Grimaldi, L.M.E., Martino, G.V., Piccolo, G., Bergamaschi, R., Citterio, A., and Melzi d'Eril, G.V., 1989. Tumor necrosis factor in serum and cerebrospinal fluid of patients with multiple sclerosis. *Ann.Neurol.* 26:787.

Frei, K., Siepl, C., Groscurth, P., Bodmer, S., Schwerdel., C., and Fontana, A., 1987. Antigen presentation and tumor cytotoxicity by interferon-gamma treated microglial cells. *Eur.J.Immunol.* 17:1271.

Frei, K., Leist, T.P., Meager, A., Gallo, P., Leppert, D., Zinkernagel, R.M., and Fontana, A., 1988. Production of B cell stimulatory factor-2 and interferon gamma in the central nervous system during viral meningitis and encephalitis. Evaluation in a murine model infection and in patients. *J.Exp.Med.* 168:449.

Frei, K., Malipiero, U.V., Leist, T.P., Zinkernagel, R.M., Schwab, M.E., and Fontana, A., 1989. On the cellular source and function of interleukin 6 produced in the central nervous system in viral diseases. *Eur.J.Immunol.* 19:689.

Gallo, P., Piccinno, M.G., Pagni, S., and Tavolato, B., 1988. Interleukin-2 levels in serum and cerebrospinal fluid of multiple sclerosis patients. *Ann.Neurol.* 24:795.

Gallo, P., Argentiero, V., Piccinno, M.G., Giometto, B., Pagni, S., Bozza, F., and Tavolato, B., 1989a. Soluble mediators of the immune response in cerebrospinal fluid and serum of patients with multiple sclerosis, *in:* "Multiple Sclerosis Research", M.A. Battaglia, ed., Elsevier Science Publisher B.V. (Biomedical Division), Amsterdam.

Gallo, P., Frei, K., Rordorf, C., Lazdins, J., Tavolato, B., and Fontana, A., 1989b. Human immunodeficiency virus type 1 (HIV-1) infection of the central nervous system: an evaluation of cytokines in cerebrospinal fluid. *J.Neuroimmunol.* 23:109.

Gallo, P., Piccinno, M.G., Pagni, S., Argentiero, V., Giometto, B., Bozza, F., and Tavolato, B., 1989c. Immune activation in multiple sclerosis: study of IL-2, sIL-2R , and IFN gamma levels in serum and cerebrospinal fluid. *J.Neurol.Sci.* 92:9.

Gallo, P., Piccinno, M.G., Krzalic, L., and Tavolato, B., 1989d. Tumor necrosis factor alpha (TNFα) and neurological diseases. Failure in detecting TNFα in the cerebrospinal fluid from patients with multiple sclerosis, AIDS dementia complex, and brain tumors. *J.Neuroimmunol.* 23:41.

Gallo, P., Frei, K., Leppert, D., and Fontana, A., 1990. On the intrathecal synthesis of immunoglobulins: detection of BSF-2/IL-6 in CSF, *in*: "Trends in Neuroimmunology", M.G. Marrosu, C. Cianchetti, and B. Tavolato, eds., Plenum Press, New York.

Gallo, P., Giometto, B., and Tavolato, B., 1991a. Neuroimmunology of HIV-1 infection. *Arch. AIDS Res.* 5:53.

Gallo. P., Laverda, A.M., De Rossi, A., Pagni, S., Del Mistro, A., Cogo, P., Piccinno. M.G., Plebani, A., Tavolato, B., and Chieco-Bianchi, L., 1991b. Immunological markers in the cerebrospinal fluid of HIV-1-infected children. *Acta Paediatr.Scand.* 80:659.

Gallo, P., Piccinno, M.G., Tavolato, B., and Siden, A.,1991c. A longitudinal study on IL-2, sIL-2R, IL-4, and IFN gamma in multiple sclerosis CSF and serum. *J.Neurol.Sci.* 101:227.

Gorczynski, M., and Keystone, E.J., 1986. Interleukin-1 activity in human cerebrospinal fluid, *Immunol.Lett.* 13:231.

Greenberg, S.J., Marcon, L., Hurwitz, B.J., Waldmann, T.A., Nelson, D.L., 1988. Elevated levels of soluble interleukin-2 receptors in multiple sclerosis. *N.Engl.J.Med.* 319:1019.

Grimaldi, L.M.E., Marino, G.V., Franciotta, D.M., Brustia, R., Castagna, A., Pristerà, R., and Lazzarin, A., 1991. Elevated alpha-tumor necrosis factor levels in spinal fluid from HIV-1-infected patients with central nervous system involvement. *Ann.Neurol.* 29:21.

Hartung, H.-P., Hughes, R.A.C., Taylor, W.A., Heininger, K., Reiners, K., and Toyka, K.V., 1990. T cell activation in Guillain-Barré Syndrome and in MS : elevated serum levels of soluble IL-2 receptors. *Neurology* 40:215.

Hauser, S.L., Doolittle, T.H., Lincoln, R., Brown, R.H., and Dinarello, C.A., 1990. Cytokine accumulation in CSF of multiple sclerosis: frequent detection of interleukin-1 and tumor necrosis factor but not interleukin-6. *Neurology* 40:1735.

Helfgott, D.C., Tatter, S.B., Santhanam, U., Clarick, R.H., Bhardway, N., May, L.T., and Sehgal, P.B., 1989. Multiple forms of IFN-β_2/IL-6 in serum and body fluids during acute bacterial infection. *J.Immunol.* 142:948.

Hirsh, R.L., Panitch, H.S., and Johnson, K.P., 1985. Lymphocytes from multiple sclerosis patients produce elevated levels of gamma interferon in vitro. *J.Clin.Immunol.* 5:386.

Houssiau, F.A., Bukasa, K., Sindic, C.J.M., Van Damme, J.V., and Van Snick, J., 1988. Elevated levels of the 26K human hybridoma growth factor (interleukin 6) in cerebrospinal fluid of patients with acute infection of the central nervous system. *Clin.Exp-Immunol.*71:320.

Helle, M., Boeije, L., de Groot E., de Vos, A., and Aarden, L.A., 1991. Sensitive ELISA for interleukin-6.Detection of IL-6 in biological fluids: synovial fluids and sera. *J.Immunol.Meth.* 138:47.

Hirano, T., 1991. Interleukin-6, *in* :"The Cytokine Handbook", A.Thomson, ed., Academic Press, London.

Kittur, S.D., Kittur, D.S., Soncrant, T.T., Rapoport, S.I., Tourtellotte, W.W., Nagel, J.E., and Adler, W.H., 1990. Soluble interleukin-2 receptors in the cerebrospinal fluid from individuals with various neurological disorders. *Ann.Neurol.* 28:168.

Lebon, P., Schuller, E., Degos, J.-D., Lyon-Caen, O., and Ponsot, G., 1987. CSF alpha and gamma interferons in acute and subacute encephalitis and in multiple sclerosis: comparative study, *in*:"Cellular and Humoral Immunological Components of Cerebrospinal Fluid in Multiple Sclerosis", A.Lowenthal and J.Raus, eds., Plenum Press, New York.

Lebon, P., Boutin, B., Dulac, O., Ponsot, G., and Arthuis, M., 1988. Interferon in acute and subacute encephalitis. *Br.Med.J.* 296:9.

Leist, T.P., Frei, K., Kam-Hansen, S., Zinkernagel, R.M., and Fontana, A., 1988. Tumor necrosis factor α in cerebrospinal fluid during bacterial, but not viral meningitis. Evaluation in murine model infections and in patients. *J.Exp.Med.* 167:1743.

Leppert, D., Frei, K., Gallo, P., Yasargil, M.G., Hess, K., Baumgartner, G., and Fontana, A., 1989. Brain tumor: detection of B-cell stimulatory factor-2/interleukin-6 in the absence of oligoclonal bands of immunoglobulins. *J.Neuroimmunol.* 24:259.

Maimone, D., Gragory, S., Arnason, B.G.W., and Reder, A., 1991. Cytokine levels in the cerebrospinal fluid and serum of patients with multiple sclerosis. *J.Neuroimmunol.* 32:67.

Manogue, K.R., van Deventer, S.J.H., and Cerami, A., 1991. Tumor necrosis factor alpha or cachectin, *in*: "The Cytokine Handbook", A. Thomson, ed., Academic Press, London.

Mantovani,A., and Dejana, E., 1989. Cytokines as communication signals between leukocytes and endothelial cells. *Immunol.Today* 10:370.

McClain, J., Cohen, D., Ott, L., Dinarello, C.A., and Young, B., 1987. Ventricular fluid interleukin-1 activity in patients with head injury, *J.Lab.Clin.Med.* 110:48.

McCracken, G.H.Jr., Mustafa, M.M., Ramilo, O., Olsen, K.D., and Risser, R.C., 1989. Cerebrospinal fluid interleukin-1ß and tumor necrosis factor concentrations and outcome from neonatal gram-negative enteric bacillary meningitis. *Pediatr.Infect.Dis.J.* 8:155.

Mustafa, M.M., Lebel, M.H., Ramilo O., Olsen, K.D., Resch, J.S., Beutler, B., and McCracken, G.H.Jr., 1989a. Correlation of interleukin-1ß and cachectin concentration in cerebrospinal fluid and outcome from bacterial meningitis. *J.Paediatr.* 115:208.

Mustafa, M.M., Ramilo, O., Olsen, K.D., Franklin, P.S., Hansen, E.J., Beutler, B., and McCracken, G.H., 1989b. Tumor necrosis factor in mediating experimental Haemophilus Influenzae type b meningitis. *J.Clin.Invest.* 84:1253.

Nadal, D., Leppert, D., Frei, K., Gallo, P., Lamche, H., and Fontana, A., 1989. Tumor necrosis factor-α in infectious meningitis. *Arch.Dis.Child.* 64:1274.

Nijsten, M.W.N., de Groot, E.R., ten Duis, H.J., Klasen, H.J., Hack, C.E., and Aarden, L.A., 1987. Serum levels of interleukin-6 and acute phase responses. *Lancet* ii:921.

Peter, J.B., Boctor, F.N., Tourtellotte, W.W., 1991. Serum and CSF levels of IL-2, sIL-2R, TNFα, and IL-1ß in chronic progressive multiple sclerosis: expected lack of clinical utility. *Neurology* 41:121.

Quagliarello, V.J., Wispelwey, B., Long, W.J., and Scheld, W.M., 1991. Recombinant human interleukin-1 induces meningitis and blood-brain barrier injury in the rat : characterization and comparison with tumor necrosis factor. *J.Clin.Invest.* 87:1360.

Quagliarello, V.J., and Scheld, W.M., 1992. Bacterial meningitis: pathogenesis, pathophysiology, and progress. *N.Engl.J.Med.* 327:864.

Ramilo, O., Mustafa, M.M., Sáez-Llorens, X., Mertsola, J., Ohkawara, S., Yoshinaga, M., Hansen, E.J., and McCracken, G.H.Jr., 1989. Role of interleukin 1-beta in meningeal inflammation. *Ped.Infect.Dis.J.* 12:909.

Ramilo, O., Sáez-Llorens, X., Mertsola, J., Jafari, H., Olsen, K.D., Hansen, E.J., Yoshinaga, M., Ohkawara, S., Nariuchi, H., and McCracken, G.H.Jr., 1990. Tumor necrosis factor α/cachectin and interleukin 1ß initiate meningeal inflammation. *J.Exp.Med.* 172:497.

Robbins, D.S., Shirazi, Y., Drysdale, B.-E., Lieberman, A., Shin, H.S., and Shin, M.L., 1987. Production of cytotoxic factor for oligodendrocytes by stimulated astrocytes. *J.Immunol.* 139:2597.

Saukkonen, K., Sande, S., Cioffe, C., Wolpe, S., Sherry, B., Cerami, A., and Tuomanen, E., 1990. The role of cytokines in the generation of inflammation and tissue damage in experimental gram-positive meningitis. *J.Exp.Med.* 171:439.

Selmaj, K.W., and Raine, C.S., 1988. Tumor necrosis factor mediates myelin and oligodendrocyte damage in vitro. *Ann.Neurol.* 23:339.

Sharief, M.K., Hentges, R., and Thompson, E.J., 1991a. The relationship of interleukin-2 and soluble interleukin-2 receptors to intrathecal immunoglobulin synthesis in patients with multiple sclerosis. *J.Neuroimmunol.* 32:43.

Sharief, M.K., Phil, M., and Hentges, R., 1991b. Association between tumor necrosis factor-α and disease progression in patients with multiple sclerosis. *N.Engl.J.Med.* 325:467.

Shaskan, E.G., Thompson, R.M., and Price, R.W., 1992. Undetectable tumor nercrosis factor-alpha in spinal fluid from HIV-1-infected patients. *Ann.Neurol.* 31:687.

Symons, A., Bundick, R.V., Suckling, A.J., and Rumsby, M.G., 1987. Cerebrospinal fluid interleukin-1-like activity during chronic relapsing experimental allergic encephalomyelitis. *Clin.Exp.Immunol.* 68:648.

Tsukada, N., Miyagi, K., Matsuda, M., Yanagisawa, N., and Kone, K., 1991. Tumor necrosis factor and interleukin-1 in the CSF and sera of patients with multiple sclerosis. *J.Neurol.Sci.* 102:230.

Van Meir, E., Sawamura, Y., Diserens, A.-C., Hamou, M.F., and de Tribolet, N., 1990. Human glioblastoma cells release interleukin 6 in vivo and in vitro. *Cancer Res.* 50:6683.

Van Oers,M.H.J., Van der Heyden, A.A.P.A.M., and Aarden, L.A., 1988. Interleukin 6 (IL-6) in serum and urine of renal transplant recipients. *Clin.Exp.Immunol.* 71:314.

Vecht, C.J., Keohane, C., Menon, R.S., Henzen-Logmans, S.C., Punt., C.J.A., Stoter, G., 1990. Acute fatal leukoencephalopathy after interleukin-2 therapy. *N.Engl.J.Med.* 323:1146.

Waage, A., Brandtzaeg, P., Halstensen, A., Kierulf, P., and Espevik, T., 1989a. The complex pattern of cytokines in serum from patients with meningococcal septic shock. Association between interleukin 6, interleukin 1, and fatal outcome. *J.Exp.Med.* 169:333.

Waage, A., Halstensen, A., Shalaby, R., Brandtzaeg, Kierulf, P., and Espevik, T., 1989b. Local production of tumor necrosis factor α, interleukin 1ß, and interleukin 6 in meningococcal meningitis. *J.Exp.Med.* 170:1859.

Weller, M., Stevens, A., Sommer, N., Melms, A., Dichgans, J., and Wiethölter, H., 1991. Comparative analysis of cytokine patterns in immunological, infectious, and oncological neurological disorders. *J.Neurol.Sci.* 104:215.

Wesselingn, S.L., Gough, N.M., Finlay-Jones, J.J., and McDonald, P.J., 1990. Detection of cytokine mRNA in astrocyte cultures using the polymerase chain reaction. *Lymphok.Res.* 9:177.

Yasukawa, K., Hirano, T., Watanabe, Y., Muratani, K., Matsuda, K., Nakai, S., and Kishimoto, T., 1987. Structure and expression of human B cell stimulatory factor-2 (BSF-2/IL-6) gene. *EMBO J.* 6:2939.

LINEAGES, CELL FATE DETERMINATION, AND MIGRATION IN CNS GLIOGENESIS

James E. Goldman

Department of Pathology and
The Center for Neurobiology and Behavior
Columbia University
New York, NY 10032

The differentiation of oligodendrocytes from immature neuroectodermal cells is relevant to an understanding of multiple sclerosis and other demyelinating disorders. In this talk, I shall focus on several aspects of oligodendrocyte development *in vivo*. What are the sources of oligodendrocytes in the mammalian CNS and what is the nature of these progenitors? Do oligodendrocytes or their precursors migrate through the brain, and if so, how is this migration controlled? Are there immature glia in the adult CNS, and can these cells differentiate into oligodendrocytes and myelinate axons?

1. Sources of oligodendrocytes

The genesis of oligodendrocytes in the mammalian CNS is a late gestational and early post-natal event. Oligodendrocytes develop from immature neuroectodermal cells deep within the CNS. Patterns of development and sources of progenitors vary somewhat from area to area. In the forebrain, oligodendrocytes arise from an immature, proliferating population of cells in the subventricular zone (SVZ), next to the lateral ventricles [1-5]. Cerebellar oligodendrocytes arise from immature progenitors at the base of the cerebellum in the roof of the IVth ventricle [4,6]. Oligodendrocytes of the optic nerve are thought to arise from a germinative zone at the base of the IIIrd ventricle, just above the chiasm, although this source has not been directly demonstrated [7]. The spinal cord does not contain a well-formed SVZ when oligodendrocytes are generated, but progenitors may lie in the ventral aspect of the central cord [8].

2. Nature of oligodendrocyte progenitors

Oligodendrocyte progenitors have been only partially characterized. While morphological examination of the developing CNS can identify oligodendrocytes and astrocytes once they have differentiated, identifying glial progenitors has been difficult,

A Multidisciplinary Approach To Myelin Diseases II
Edited by S. Salvati, Plenum Press, New York, 1994

since immature cells tend not to have distinguishing features. Antibody binding studies have somewhat the same problems. Trying to identify a glial progenitor before it expresses a mature glial antigen like GFAP (an intermediate filament of astrocytes) or the myelin proteins, MAG, PLP, or MBP, has been problematic. Nevertheless, several antigens are expressed by immature glia. These include GD3 ganglioside, a simple ganglioside present in large amounts in the immature CNS; a variety of complex gangliosides recognized by the monoclonal antibody A2B5; nestin, an intermediate filament protein expressed in immature neuroectodermal cells; the intermediate filament protein vimentin; and the embryonic form of N-CAM (E-NCAM) [9-11]. These antigens are not specific to glial progenitors, however, and therefore their expression must be interpreted in an appropriate context. Immature populations are heterogeneous with respect to the expression of these antigens, however, and we do not yet know if all are expressed by all immature glia, possibly in some temporally-regulated sequence, or whether different markers define different lineages. Early stages of oligodendrocyte development are additionally marked by the expression of sulfated glycolipids bound by the monoclonal antibody 04, and by the NG2 proteoglycan [12-14].

3. Tracing oligodendrocyte development and migrational pathways *in vivo*

A large number of studies over the past dozen years have described oligodendrocyte differentiation in primary cultures from young mammalian CNS. These studies have been invaluable in delineating stages of development, based upon morphology, antigen expression, and responsiveness to growth factors, and in defining a set of growth factors that can regulate proliferation, differentiation, and survival of oligodendrocytes and their progenitors, and have been considered in recent reviews [15-18].

A number of approaches have been used to examine oligodendrocyte development *in vivo*. Purely morphological analyses define glial cells that have matured, but have been less successful at defining immature stages. Lineage relationships have been difficult to determine, since distinguishing between an uncommitted glial progenitor and a committed oligodendrocyte or astrocyte progenitor (if such exist) in their early stages is not possible.

Approaches using antigen expression have suggested that oligodendrocytes develop from germinal zones and migrate therefrom into gray and white matter [4,5,6,19]. These studies have examined antigens known to be expressed at different stages of early oligodendrocyte development in culture, such as GD3 ganglioside, the gangliosides bound by A2B5, the 04 antigen, and galactocerebroside, as well as molecules long known associated with myelin, such as carbonic anhydrase II and proteolipid protein. Development and migration were inferred from observing cells labeled with early markers near germinal zones in neonates and cells with more complex morphologies and later markers at later times in white matter. All of these studies are consistent with the idea that oligodendrocytes first arise in germinal zones, migrate from them through developing white and gray matter, acquire further antigens and more complex morphologies, continue to divide as they migrate (and even after they migrate), and eventually stop migrating. The sequence of expression of the various oligodendrocyte antigens is known in general terms, although systematic comparisons among all antigens have not been performed. A2B5 and GD3 are expressed early on, then carbonic anhydrase, 04, and CNPase are expressed at intermediate stages, then GC, and finally MAG, PLP, and MBP. Immunocytochemical studies allow us to visualize large numbers of glia in one section and to define at what stages and in what areas of the brain certain antigens are expressed. Such studies are not useful for lineage analysis, however, and will not definitively define migrational pathways.

Morphological analysis and immunocytochemical studies, combined with mitotic labeling, has been useful in determining that immature glia continue to divide while they migrate [1,2,3,20,21] Retroviral studies indicate that some glia continue to divide even after they stop migrating (see below). Oligodendrocytes apparently continue to divide up to and including the early stages of expression of sulfatide and perhaps galactocerebroside [19,20,21], but slow or stop when later myelin antigens are expressed, such as MBP.

Our laboratory has been tracing oligodendrocyte development *in vivo* using replication-deficient retroviral vectors. We chose this strategy because the vectors introduce reporter genes as heritable markers [22,23,24]. Thus, in principle, the progeny of a single, infected progenitor can be identified. Our method has been to label selectively the immature cells of the forebrain SVZ of neonatal and young rats by stereotactic injection [25]. Since these viral vectors cannot replicate and since they do not remain infectious for more than a few hours and infections are limited to the SVZ, we can follow the migration and developmental fates of the infected cells. Thus, with increasing times after injection, labeled cells are observed farther and farther away from the SVZ. Developmental fates were assessed by morphological analyses of the infected cells and by immunostaining with a variety of antibodies to glial and neuronal antigens. In neonatal rats, SVZ cells migrate into both gray and white matter of the hemisphere. In gray matter, they differentiate into both oligodendrocytes and astrocytes. The astrocytes appear to be largely of the classical, bushy, "protoplasmic" variety. Several types of oligodendrocytes developed, including perineuronal satellite cells, perivascular cells, and myelinating oligodendrocytes. By contrast, in white matter, the large majority of SVZ cells differentiated into oligodendrocytes, both myelinating and non-myelinating. In the striatum, a mixture of astrocytes and oligodendrocytes emerge.

Glial cells tended to cluster together in tightly-knit groups. An analysis using simultaneous injection of two different retroviruses, one expressing the beta-galactosidase gene, the other expressing the alkaline phosphatase gene, indicated that cells within each cluster were related to each other. Most clusters were homogeneous, containing either astrocytes or oligodendrocytes, but some contained both glial types, and the occasional cluster contained cells with a neuronal morphology in addition to glia. Thus, although many glial progenitors tend to give rise to offspring of similar types, some can maintain developmental plasticity even after migration, and can generate a variety of cell types during their last mitotic divisions.

Developmental plasticity can also be demonstrated by infecting SVZ cells *in vivo* and then removing the SVZ and placing the cells in culture (Levison, SW, Chuang, C, Abramson, B and Goldman, JE, The migrational patterns and developmental fates of glial precursors in the rat subventricular zone are temporally regulated. Development, in press). Many of the clones generated *in vitro* by individual SVZ cells are heterogeneous, containing both astrocytes and oligodendrocytes. Interestingly, most of the astrocytes are of the "type 1" variety. This observation suggests that the SVZ cells represent an earlier glial progenitor than the "O-2A" progenitors of the optic nerve (see below).

The large majority of SVZ cells from the neonatal CNS that migrate into and remain in white matter develop into oligodendrocytes. More strikingly, SVZ cells labeled at P14 come to reside almost exclusively in white matter, and develop only into oligodendrocytes (Levison, SW, Chuang, C, Abramson, B and Goldman, JE, The migrational patterns and developmental fates of glial precursors in the rat subventricular zone are temporally regulated. Development, in press). Yet, when these SVZ cells are placed in culture, they can differentiate into astrocytes. We interpret these findings as evidence not only of developmental plasticity of SVZ cells, but also of important developmental cues provided by the environment. In particular, these observations are consistent with a model in which immature white matter promotes oligodendrocyte differentiation and/or inhibits astrocyte differentiation.

This model appears to reconcile two disparate schemes for gliogenesis in the optic nerve, perhaps the most extensively studied zone of gliogenesis in the mammalian CNS. In one scheme, based upon morphological observations and thymidine incorporation *in vivo* [20], a prenatal wave of astrocyte genesis precedes a post-natal wave of oligodendrocyte genesis. In the other, based upon studies of gliogenesis *in vitro*, a prenatal generation of astrocytes ("type 1") is followed by a post-natal generation of a bipotential progenitor ("O-2A") that gives rise to both oligodendrocytes and another form of astrocyte ("type2") [9,15]. The astrocytes that arise early are likely to be derivatives of radial glial systems established even earlier (see [26] for discussion). Radial glia transform into astrocytes *in vivo* [27,28], and in culture give rise to "type 1" astrocytes [29]. Beginning around birth, the so-called "O-2A" progenitors migrate into the optic nerve. If these cells come from a germinal zone at the base of the IIIrd ventricle, and if they are similar or identical to the migrating SVZ cells of the hemispheres, as seems likely, then we would predict first, that they would indeed be bipotential, but second, that they would differentiate preferentially into oligodendrocytes in the white matter environment of the optic nerve. Indeed, visualization of O-2A progenitors in the developing optic nerve using quisqualate-induced uptake of cobalt delineates cells with the morphologies of developing (and eventually mature) oligodendrocytes but not astrocytes [30], supporting the premise that these bipotential progenitors differentiate preferentially into oligodendrocytes in nascent white matter.

The glial progenitors of the anterior forebrain SVZ migrate extensively into neocortex, subcortical white matter and striatum (Levison, SW, Chuang, C, Abramson, B and Goldman, JE, The migrational patterns and developmental fates of glial precursors in the rat subventricular zone are temporally regulated. Development, in press). Most of the migration occurs in a coronal plane, however, with little anterior-posterior travel. This finding appears to differ from the more extensive migration of transplanted immature glia. It may be that progenitors that begin their travels in the SVZ follow from the beginning a set of pathways that restrict their migratory behavior, while transplanted cells may not find such pathways and therefore such constraints. Further investigation of the migration of transplanted glia is in order, with particular attention to the developmental stage of both the transplanted cells and the host.

We have observed notable migrational changes during early post-natal gliogenesis, allowing SVZ cells initially to colonize gray and white matter, but restricting them to white matter at a later time (Levison, SW, Chuang, C, Abramson, B and Goldman, JE, The migrational patterns and developmental fates of glial precursors in the rat subventricular zone are temporally regulated. Development, in press). This phenomenon could result from either a lineage restriction of the cells themselves or from migrational restriction due to environmental changes. Lineage restriction within the SVZ seem improbable, since SVZ cells labeled with retrovirus at either P2 or P14 are capable of developing into both astrocytes and oligodendrocytes. Lineage restriction during migration is probable, since those cells that remain in white matter differentiate into oligodendrocytes. Migrational restriction due to non-lineage related factors is likely, although specific substrates for migration are not known. One possible substrate is the scaffolding of radial glial processes, used at an earlier developmental point to guide neuroblast migration. These are still present during the first post-natal week of rat CNS development, but collapse during the second week as radial glia transform into astrocytes [31]. By P14, few are left, thus removing this possible guidance mechanism for directing glial progenitors into the neocortex. How progenitors migrate along white matter, and what substrate(s) they move along therein is not known. O-2A cells in culture are migratory, and slow as they differentiate into oligodendrocytes. The same phenomenon may occur in vivo. Signals that control when a progenitor begins to migrate, continues to migrate, or stops will be important to determine.

4. Glial progenitor differentiation in the adult mammalian CNS

A number of studies indicate that the adult mammalian CNS contains immature glia. Furthermore, there is evidence that at least some of these cells actually develop into oligodendrocytes and astrocytes. A series of studies combining thymidine incorporation with morphological analyses (summarized in [32] show a low rate of mitotic activity in the adult rodent CNS. The majority of cells that are labeled at short times after thymidine injection appear to be immature, whereas oligodendrocytes and astrocytes are labeled at longer times after injection, indicating differentiation of the immature cells. Thus, the adult CNS is able to support the differentiation of glia, although its capacity may be limited. The source of glial progenitors is not known, however, and cannot be determined from thymidine studies because of the mitotic dilution of the label. There may be two sources. One is the SVZ, which persists into adulthood as a small residue of dividing cells. Although a recent report concluded that adult SVZ cells do not migrate, and in fact, die [33], this matter should be re-examined. Direct labeling of SVZ cells with dyes or retroviral vectors should be able to answer the question. The second source is the pool of extra-SVZ immature glia themselves. If this were to be the case, then divisions of progenitors must at some point be asymmetrical, producing one dividing progenitor and one cell that will differentiate into a mature glial cell, a characteristric of stem cells.

Indeed, the most detailed characterization of an immature oligodendrocyte that has properties of a glial stem cell in the adult CNS is the extensive work by Noble and colleagues, who have defined a small, unipolar, slowly cycling, A2B5+/04+ (but Gal C-) cell in the adult rodent optic nerve (summarized in [18]. These cells can be isolated from adult optic nerve, they cycle slowly (like stem cells in epithelia), and they appear to undergo asymmetric divisions, producing offspring like themselves as well as oligodendrocytes. Further consistent with the idea that the adult CNS contains immature oligodendrocytes is the isolation of 04+/GalC- cells from the adult human CNS [34].

Can these immature glia in the adult brain be stimulated to proliferate more extensively then they normally do, to migrate, and to remyelinate areas of demyelination? The answer is not known. However, optic nerve O-2A adult progenitors *in vitro* can be stimulated to become a more rapidly proliferating population when exposed to PDGF and basic FGF [35], although this stimulation has limits.

A major challenge will be to understand the capacity of the adult CNS to support increased oligodendrocyte progenitor proliferation, differentiation, and myelination, either by endogenous glia or by transplanted glia.

Acknowledgments

The author thanks Steve Levison, Pierre Vaysse, Bernetta Abramson, and Cathy Chuang for their major efforts in our research. This work was supported by NIH Grant NS17125.

References

1. J Altman and GP Das, Autoradiographic and histological studies of postnatal neurogenesis. I. A longitudinal investigation of the kinetics, migration, and transformation of cells incorporating tritiated thymidine in neonatal rat, with special reference to postnatal neurogenesis in some brain regions. J Comp Neurol 126: 337-390, 1966.
2. JA Paterson, A Privat, EA Ling, and CP Leblond, Investigation of glial cells in semithin sections III Transformation of subependymal cells into glial cells as shown by radioautography after 3H-thymidine injection into the lateral ventricle of the brain of young rats. J. Comp. Neurol. 149: 83-102, 1973.
3. K Imamoto, JA Paterson, and CP Leblond, Radioautographic investigation of gliogenesis in the corpus callosum of young rats. I. Sequential changes in oligodendrocytes. J. Comp. Neurol. 180: 115-138, 1978.

4. SM LeVine and JE Goldman, Spatial and temporal patterns of oligodendrocyte differentiation in rat cerebrum and cerebellum. J.Comp.Neurol. 277, 441-455: 1988.
5. R Hardy and R Reynolds, Proliferation and differentiation potential of rat forebrain oligodendrocyte progenitors both in vitro and in vivo. Development 111: 1061-1080, 1991.
6. R Curtis, J Cohen, J Fok-Seang, MR Hanley, NA Gregson,R Reynolds, and GP Wilkin, Development of macroglial cells in rat cerebellum. I. Use of antibodies to follow early in vivo development and migration of oligodendrocytes. J.Neurocytol. 17: 43-54, 1988.
7. R Small, P Riddle, and M Noble, Evidence for migration of oligodendrocyte-type-2-astrocyte progenitor cells into the developing rat optic nerve. Nature 328:155-157.
8. BC Warf, J Fok-Seang, and RH Miller, Evidence for the ventral origin of oligodendrocyte precursors in the rat spinal cord. J .Neurosci. 11:2477-2488, 1991.
9. MC Raff, RH Miller, and M Noble, A glial progenitor cell that develops in vitro into an astrocyte or an oligodendrocyte depending on culture medium. Nature 303: 390-396, 1983.
10. G Levi, F Aloisi, and G Wilkin, Differentiation of cerebellar bipotential glial precursors into oligodendrocytes in primary cultures: developmental profile of surface antigens and mitotic activity. J. Neurosci. Res. 18: 407-417, 1987.
11. JE Goldman, SS Geier, and M Hirano, Differentiation of astrocytes and oligodendrocytes from germinal matrix cells in primary culture. J. Neurosci. 6: 512-60, 1986.
12. WBStallcup and L Beasly, Bipotential glial precursor cells of the optic nerve express the NG2 proteoglycan.J. Neurosci. 7:2737-2744, 1987.
13. I Sommer, and M Schachner, Monoclonal antibodies (01 to 04) to oligodendrocyte cell surfaces: an immunocytochemical study in the central nervous system. Dev. Biol. 83: 311-327, 1981.
14. R Bansal, AE Warrington, AL Gard,and SE Pfeiffer,Multiple and novel specificities of monoclonal antibodies O1, O4, and R-mAb used in the analysis of oligodendrocyte development. J. Neurosci. Res. 24: 548-557, 1989.
15. RH Miller, C ffrench-Constant, and MC Raff, The macroglial cells of the rat optic nerve. Annu. Rev. Neurosci. 12: 517-534, 1989.
16. WD Richardson, MC Raff, and M Noble, The oligodendrocyte-type-2 astrocyte lineage. Semin Neurosci 2:444-454, 1990.
17. JE Goldman, Regulation of oligodendrocyte differentiation. TINS 15:359-362, 1992.
18. M Noble, D Wren, and G Wolswijk, The O-2A adult progenitor cell: a glial stem cell of the adult central nervous system. Sem in Cell Biology 3:413-422, 1992.
19. AE Warrington and SE Pfeiffer, Proliferation and differentiation of 04+ olgiodendrocytes in postnatal rat cerebellum: analysis in unfixed tissue slices using anti-glycolipid antibodies. J Neurosci Res 33:338-353, 1992.
20. RP Skoff and PE Knapp, Division of astroblasts and oligodendroblasts in postnatal rodent brain: evidence for separate astrocyte and oligodendrocyte lineages. Glia 4: 165-174, 1991.
21. R Reynolds and GP Wilkin, Oligodendroglial progenitor cells, but not oligodendroglia, divide during normal development of the rat cerebellum. J Neurocytol 20:216-224, 1991.
22. D Turner and C Cepko, Cell lineage in the rat retina: a common progenitor for neurons and glia persists late in development. Nature, Lond. 328: 131-136, 1987.
23. JR Sanes, Analysing cell lineage with a recombinant retrovirus. TINS 12: 21-28, 1989.
24. DS Galileo, GCGray, GC Owens, J Majors, and JR Sanes, Neurons and glia arise from a common progenitor in chicken optic tectum: demonstration with two retroviruses and cell-type-specific antibodies. Proc. Natl. Acad. Sci. USA 87: 458-462, 1990.
25. SW Levison and JE Goldman, Both oligodendrocytes and astrocytes develop from progenitors in the subventricular zone of postnatal rat forebrain. Neuron, 10: 201-212, 1993.
26. SW Levison and JE Goldman, (1993). Astrocyte origins, in "Pharmacology of Astrocytes," S. Murphy, ed., Academic Press, New York, NY pp. 1-22.
27. J-P Misson, T Takahashi, and VS Caviness, Jr., Ontogeny of radial and other astroglial cells in murine cerebral cortex. Glia 4: 138-148, 1991.
28. T Voigt, Development of glial cells in the cerebral wall of ferrets: Direct tracing of their transformation from radial glia into astrocytes. J.Comp.Neurol. 289: 74-88, 1989.
29. SM Culican, NL Baumrind, M Yamamoto, and AL Pearlman, Cortical radial glia: Identification in tissue culture and evidence for their transformation to astrocytes. J.Neurosci. 10:684-692, 1990.
30. B Fulton, JF Burne, and M Raff, Visualization of O-2A progenitor cells in developing and adult rat optic nerve by quisqualate-stimulated cobalt uptake. J. Neurosci. 12:4816-4833, 1992.
31. SM LeVine and JE Goldman, Embryonic divergence of oligodendrocyte and astrocyte lineages in developing rat cerebrum. J.Neurosci. 8: 3992-4006, 1988.
32. GF McCarthy and CP Leblond, Radioautographic evidence for slow astrocyte turnover and modest oligodendrocyte production in the corpus callosum of adult mice infused with 3H-thymidine. J. Comp. Neurol. 271: 589-603, 1988.

33. CM Morshead and D van der Kooy, Postmitotic death is the fate of constitutively proliferating cells in the subependymal layer of the adult mouse brain. J Neurosci, 12:249-256, 1992.
34. RC Armstrong, HH Dorn, CV Kufta, E Friedman, and ME Dubois-Dalcq, Pre-oligodendrocytes from adult human CNS. J. Neurosci. 12: 1538-1547, 1992.
35. G Wolswijk and M Noble, Cooperation between PDGF and FGF converts slowly dividing O-2A adult progenitor cells to rapidly dividing cells with the characteristics of their perinatal counterparts. J Cell Biol 118: 889-900.

AREAS OF DEMYELINATION REPAIRED BY MOUSE OLIGODENDROCYTES AND RAT TYPE-1 ASTROCYTES CREATE A NEW MODEL OF IMMUNE MEDIATED DEMYELINATION

W.F. Blakemore and A.J. Crang

MRC Cambridge Centre for Brain Repair
Department of Clinical Veterinary Medicine
University of Cambridge
Madingley Road
Cambridge CB3 OES
United Kingdom

INTRODUCTION

Inflammatory demyelination and failure of remyelination are features of multiple sclerosis. A number of in vitro studies have demonstrated that cultured oligodendrocytes are damaged and even killed by serum factors such as complement, inflammatory cytokines and products of activated macrophages[1,2]. These mediators of inflammation can gain access to the CNS when there is breakdown of the blood-brain barrier. This has led to the suggestion that breakdown of the blood-brain barrier and/or inflammation alone may be sufficient to bring about CNS demyelination, and some have even suggested that such events may be crucially involved in the pathogenesis of multiple sclerosis.

Transplantation of glial cells into demyelinating lesions in CNS offers an experimental approach which allows investigation of the complex interactions that occur between CNS glia, Schwann cells and axons during remyelination and repair. Such studies have shown that type-1 astrocytes are central to the exclusion of Schwann cells from areas of glia-free demyelination[3], however for type-1 astrocytes to be established in a manner which prevents Schwann cell remyelination of CNS axons, cells of the O-2A lineage are also required[4,5]. Here we investigate the possibility of repopulating glia-free areas of demyelination with a mixture of isogeneic type-1 astrocytes and mouse O-2A cells in order to see if it is possible to establish mouse oligodendrocytes within a rat astrocytic environment. This

A Multidisciplinary Approach To Myelin Diseases II
Edited by S. Salvati, Plenum Press, New York, 1994

would then allow us to examine if the inflammation associated with rejection of the mouse cells would have any effect on oligodendrocyte remyelination.

THE ESTABLISHMENT OF MOUSE OLIGODENDROCYTES IN A RAT ASTROCYTE ENVIRONMENT - THE CHIMERIC GLIAL ENVIRONMENT

Injection of ethidium bromide into the white matter tracts of the spinal cord results in death of astrocytes and oligodendrocytes. This leaves demyelinated axons clumped together in a glia-free environment. Such lesions are normally remyelinated by Schwann cells with central remyelination restricted to the edge of the lesion where the demyelinated axons abut normal tissue[6]. When mixed glial cell cultures are injected into such lesions the nature of repair can be changed from predominantly Schwann cell to predominantly oligodendrocyte remyelination[3]. However, when cultures that are depleted of either cells of the O-2A lineage[7] or type-1 astrocytes[3], are injected into the lesion, the majority of axons are still remyelinated by Schwann cells. These results indicate that when faced with a strong inherent Schwann cell response neither cells of the O-2A lineage alone, nor type-1 astrocytes alone, possess the capacity to prevent Schwann cell remyelination of demyelinated axons.

Using the information gained by these experiments, together with the knowledge that mouse oligodendrocytes will myelinate rat axons[8], it was possible to conceive an experiment in which an attempt could be made to establish an area of mouse myelination in a rat which, on subsequent rejection of the mouse cells, would not be followed by Schwann cell remyelination. The rationale behind this experiments can be summarised thus:

(i). In the face of inherent Schwann cell remyelination of ethidium bromide lesions extensive oligodendrocyte remyelination can only be achieved if mixed glial cell cultures are injected into the area of demyelination.

(ii). The injection of a chimeric culture composed of mouse O-2A lineage cells and isogeneic rat type-1 astrocytes into immunosuppressed rats would create a situation where mouse O-2A cells myelinate the demyelinated axons, and at the same time facilitate the establishment of co-transplanted rat isogeneic rat astrocytes in a manner which would limit Schwann cell invasion of the lesion.

(iii). Once this had been achieved the area would then be demyelinated by permitting the rejection of the mouse cells. As the lesion will contain non-rejectable isogeneic astrocytes, this second wave of demyelination would not be followed by Schwann cell invasion. The axons demyelinated following the rejection of mouse cells will then be available for remyelination by host derived oligodendrocytes.

VALIDATION OF THE STRATEGY

To test this strategy, ethidium bromide lesions were made in the spinal cord of 3 month-old rats and cell preparations composed of isogeneic rat type-1 astrocytes depleted of cells of the O-2A lineage by immunocytolysis using a cocktail of antibodies directed against phenotypic markers of that lineage (O4, A2B5 and tetanus toxin receptor)[7], and mouse O-2A lineage cells, prepared by shaking off the top-dwelling cells from a mixed glial culture, were injected into the demyelinating area three days after the ethidium bromide injections. The mouse cells were prepared from male animals in order to enable detection with a mouse Y-chromosome-specific probe.

In transplanted animals given 15 mg/kg of cyclosporin daily the demyelinated axons in the lesions were mainly remyelinated by oligodendrocytes at 2 or 4 weeks after transplantation (Group A, Table 1). The extent of oligodendrocyte remyelination was significantly different from lesions injected with the type-1 astrocyte preparation alone, and from lesions receiving no injection (Figure 1). This result shows that a culture of rat astrocytes and mouse oligodendrocytes is as effective as an isogeneic mixed glial cell culture in controlling Schwann cell remyelination (Figures 1 and 2). The presence of mouse cells within the area of demyelination was confirmed by in-situ hybridisation using a probe directed against repeat sequences present on the mouse Y-chromosome[9] (Figure 2).

Table 1 Summary of rat type-1 astrocyte/mouse O-2A lineage cell transplantation experiments

Group	Immunosuppression following transplantation (weeks)	Local X-irradiation at termination of immuno-suppression	Survival following termination of immunosuppression (weeks)	Survival post transplantation (weeks)
A	2 or 4	No	0	2 or 4
B	2	No	4	6
C	2	No	2	4
D	2	Yes	4	6

In a second group of animals (Group B, Table 1) immunosuppression was maintained for the first 2 weeks and then discontinued to allow rejection of the xenogeneic cells. On examination of these animals 4 weeks after removal of cyclosporin, remyelination was again found to have been carried out almost completely by oligodendrocytes (Figures 1 and 2). In our proposed scenario, these oligodendrocytes should be of host-origin. That this was the case was demonstrated by the failure of the mouse Y-chromosome probe to detect any mouse cells in the regions of remyelination (Figure 2).

Confirmation that rejection of mouse oligodendrocytes produced demyelination which was subsequently remyelinated by host oligodendrocytes was obtained from the results of a further two groups of transplanted animals.

In a group of animals killed 2 weeks after cessation of cyclosporin treatment, evidence of active demyelination associated with extensive

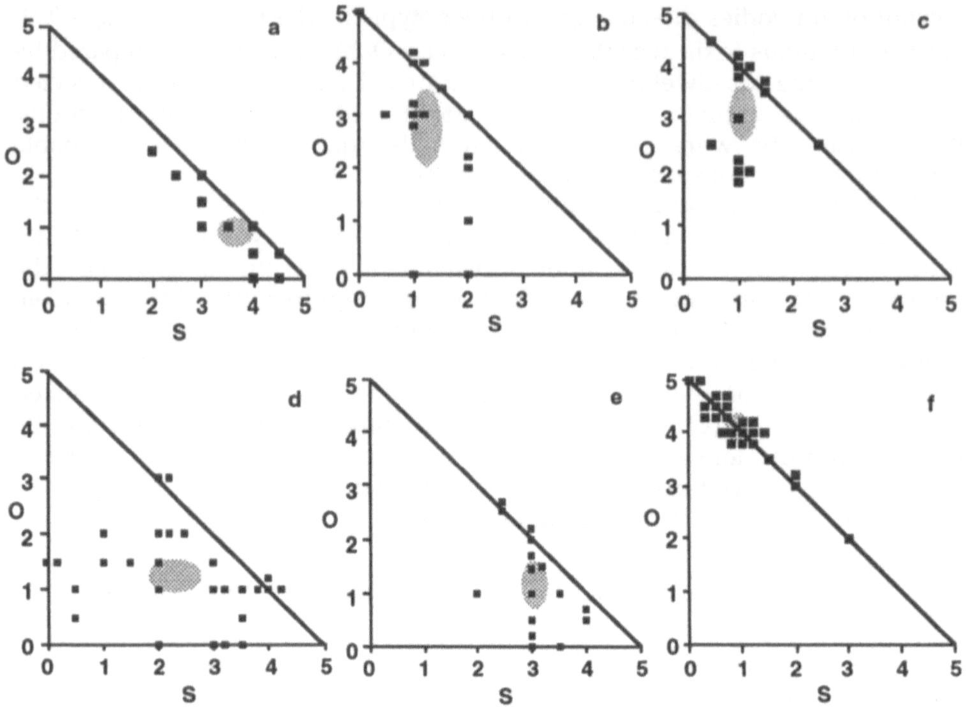

Figure 1. The nature and extent of the repair of spinal cord white matter ethidium bromide lesions can be documented quantitatively by using plots of oligodendrocyte (O) versus Schwann cell (S) remyelination. The proportion of axons remyelinated by oligodendrocytes and Schwann cells, and the proportion of axons which remain demyelinated are estimated for one micron plastic sections cut from each lesion-containing block. Thus, a section in which all the available axons are remyelinated by oligodendrocytes would have an oligodendrocyte remyelination score of 5, a Schwann cell remyelination score of 0, and a demyelinated axons score of 0. The mean oligodendrocyte and Schwann cell remyelination scores are indicated ± 2 SEM and the shaded domains enclosed by these limits represent an average repair for the group of animals. Non-overlapping repair domains in these representations therefore indicate significantly different results; the direction of their displacement from each other indicates which parameters contribute to the difference.

Following the injection of ethidium bromide into spinal cord white matter, the demyelinated axons are remyelinated mainly by Schwann cells (a). When isogeneic mixed glial cell cultures are injected into the lesion (b) or transplants comprising isogeneic type-1 astrocytes and mouse O-2A cells are introduced to lesions in immunosuppressed rats (c), remyelination is mainly carried out by oligodendrocytes. Transplantation of isogeneic O-2A cells alone (d) or type-1 astrocytes alone (e) into lesions results in only a small increase in the extent of oligodendrocye remyelination compared with non-transplanted animals. At four weeks following the rejection of mouse cells from a mixed species transplant all the axons are remyelinated, the vast majority by oligodendrocytes (f).

infiltration of inflammatory cells was observed in some lesions (Group C, Table 1; Figure 3). In these areas of inflammation evidence of oligodendrocyte remyelination was found close to axons undergoing macrophage stripping of remyelinated myelin sheaths (Figure 3). In no lesions were large numbers of demyelinated axons present. These observations show that without cyclosporin the myelin sheaths made by the mouse oligodendrocytes are removed, and following their removal, the demyelinated axons are rapidly remyelination by rat oligodendrocytes.

It is possible to inhibit remyelination by exposing tissue to 40 Grays of X-irradiation prior to demyelination[10]. Therefore, the final group of animals received local X-irradiation to the spinal cord at the time the animals were removed from cyclosporin treatment (Group D, Table 1). The effect of this treatment is to inhibit any remyelination that takes place following rejection of the mouse cells on withdrawal of immunosuppression. Thus, it is not only possible to identify the axons initially remyelinated by mouse cells, but also those that were remyelinated by rat cells during this first round of remyelination. When these animals were examined 4 weeks after cessation of cyclosporin treatment and exposure to X-irradiation, large areas of demyelination were found and only a small number of axons were remyelinated (Figure 4). The demyelinated axons were surrounded by astrocyte process such that the lesions resembled the demyelinated plaques of multiple sclerosis (Figure 4). This experiment shows that the majority of axons within the lesion are being remyelinated by rat oligodendrocytes following rejection of the mouse oligodendrocytes. Since the astrocyte cultures had been depleted of cells of the O-2A lineage by immunocytolysis[7], the result also indicates that the cells remyelinating the non-X-irradiated lesions post rejection are of host origin. However, in a further series of experiments we have demonstrated that although our O-2A depleted cultures do not generate O-2A lineage cells in vitro, they do contain progenitor cells capable of generating oligodendrocytes in vivo[11]. Therefore, although we consider it unlikely, we cannot exclude the possibility that transplant-derived rat cells are contributing to the generation of the oligodendrocytes which remyelinate the axons following the immune-mediated demyelination.

DISCUSSION

These experiments demonstrate that by using chimeric cultures composed of mouse O-2A lineage cells and isogeneic rat type-1 astrocytes it is possible to establish mouse oligodendrocytes in a rat astrocytic environment in such a manner that the normal Schwann cell remyelination of ethidium bromide lesions in the rat spinal cord is inhibited. The reconstitution of the lesion-area with rat astrocytes ensures that the demyelination associated with rejection of the mouse cells will not be followed by Schwann cell remyelination. This means that the effect of immune-mediated inflammation on oligodendrocyte remyelination can be examined. As the lesions remyelinate rapidly, indeed demyelination and remyelination can be observed in the same local area, our results indicate that an inflammatory response of the type associated with the rejection of the mouse oligodendrocytes does not inhibit remyelination of demyelinated axons by oligodendrocytes. on the contrary, the rapid remyelination observed in the lesions suggests that it may in fact stimulate remyelination as has been suggested by a number of authors[12,13,14].

Although still at a preliminary stage, these mixed species transplant experiments show that it is possible to manipulate the repair of areas of

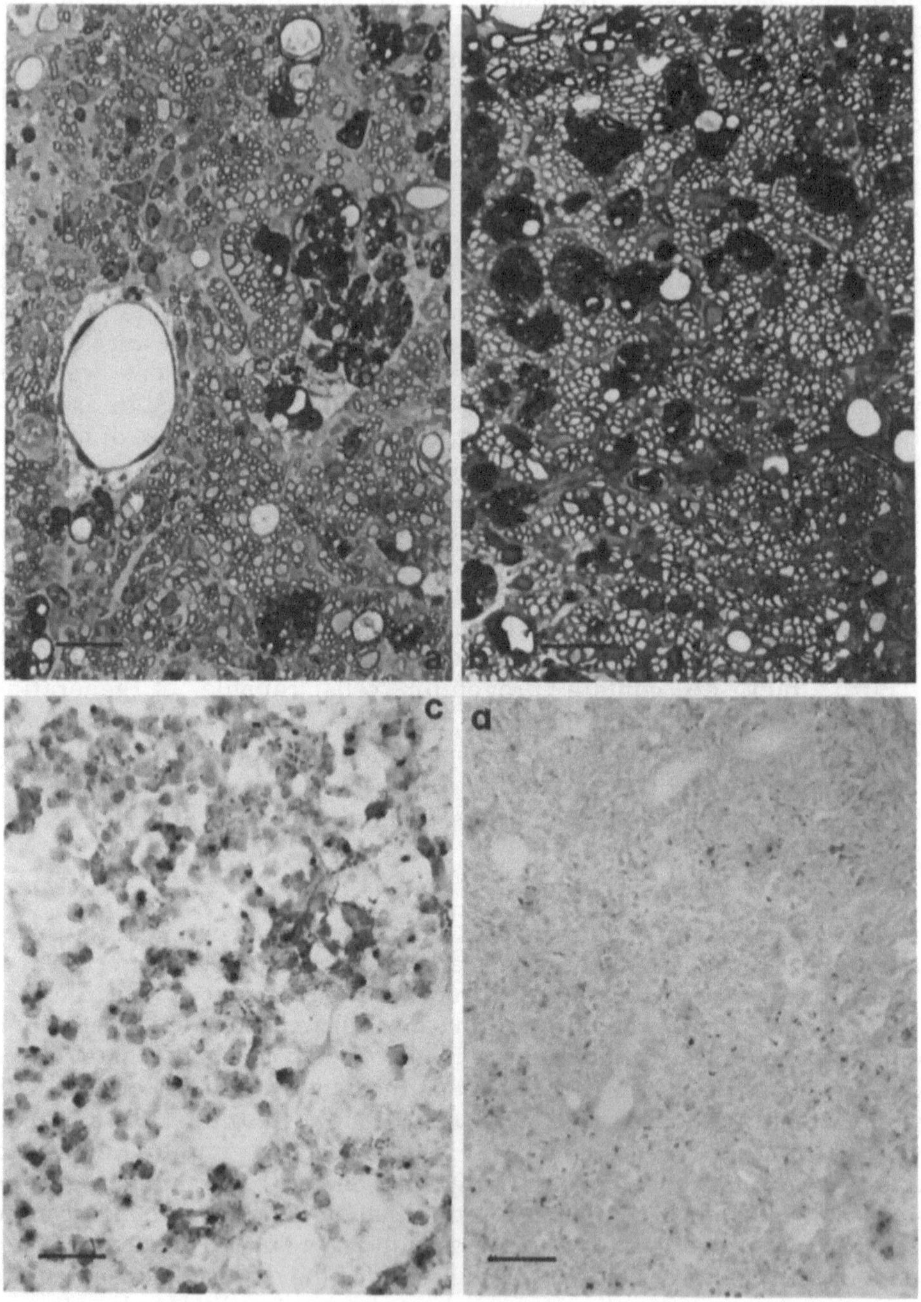

Figure 2. (a) In lesions transplanted with chimeric cultures and maintained on cyclosporin, demyelinated axons are remyelinated by mouse oligodendrocytes. (b) In lesions transplanted with chimeric cultures examined 4 weeks after removal of cyclosporin, the demyelinated axons are remyelinated by rat oligodendrocytes. (c) The presence of mouse cells within lesions injected with chimeric cultures and maintained on cyclosporin can be demonstrated by in situ hybridisation using a probe to repeat sequences on the mouse Y-chromosome. (d) No mouse cells can be detected in remyelinated lesions 4 weeks after removal of cyclosporin. Scale bars 20 μm.

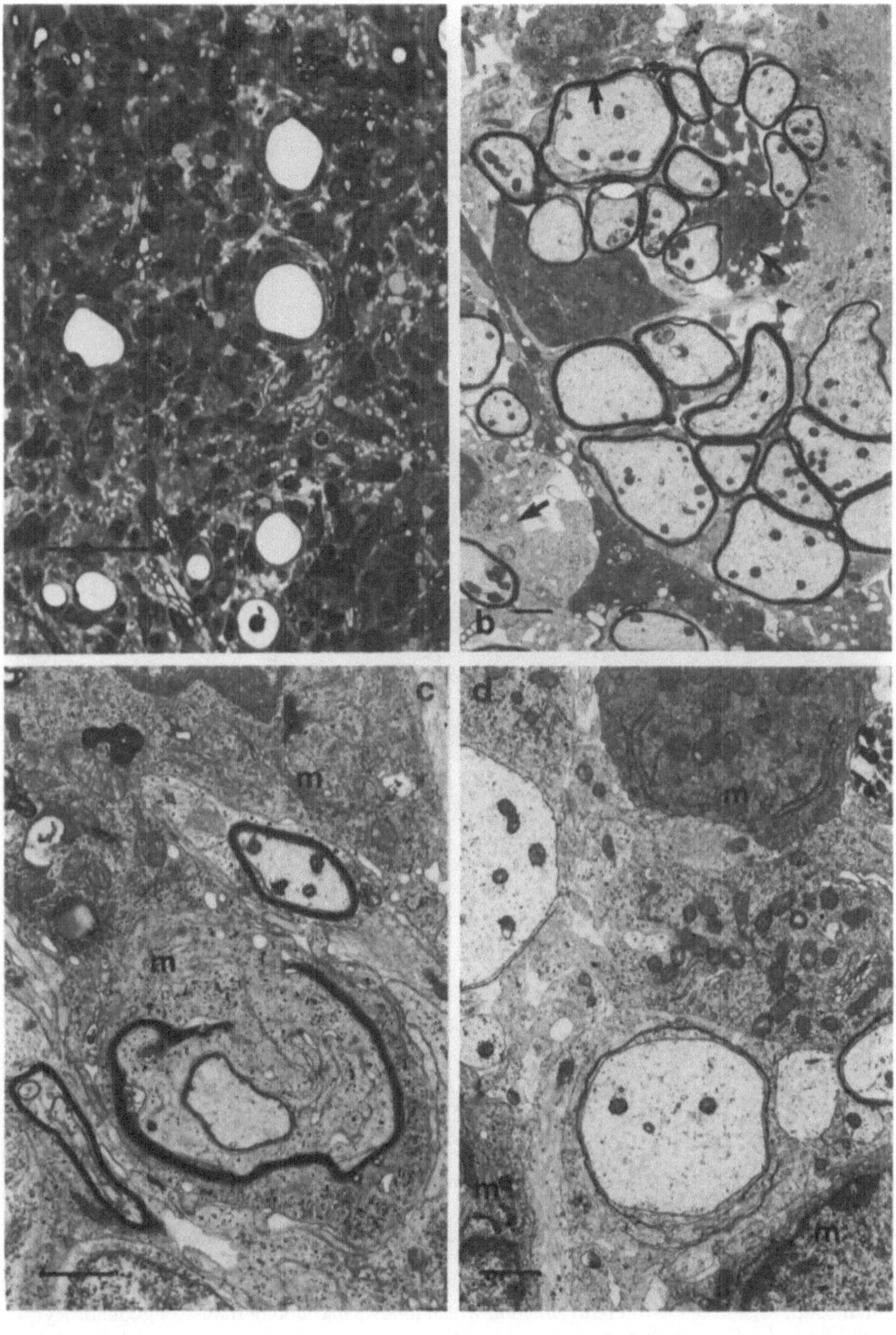

Figure 3. (a) Lesions examined 2 weeks after removal of immunosuppression are infiltrated by inflammatory cells - scale bar 20 μm. (b) Inflammatory cells (arrows) infiltrating areas of remyelination 2 weeks after removal of immunosuppression - scale bar 1 μm. (c) Macrophage (m) stripping of a myelin sheath - scale bar 1 μm. (d) The presence of axons surrounded by thin cytoplasm-rich myelin sheaths provides evidence of oligodendrocyte remyelination in an area which still contains macrophages (m) - scale bar 1 μm.

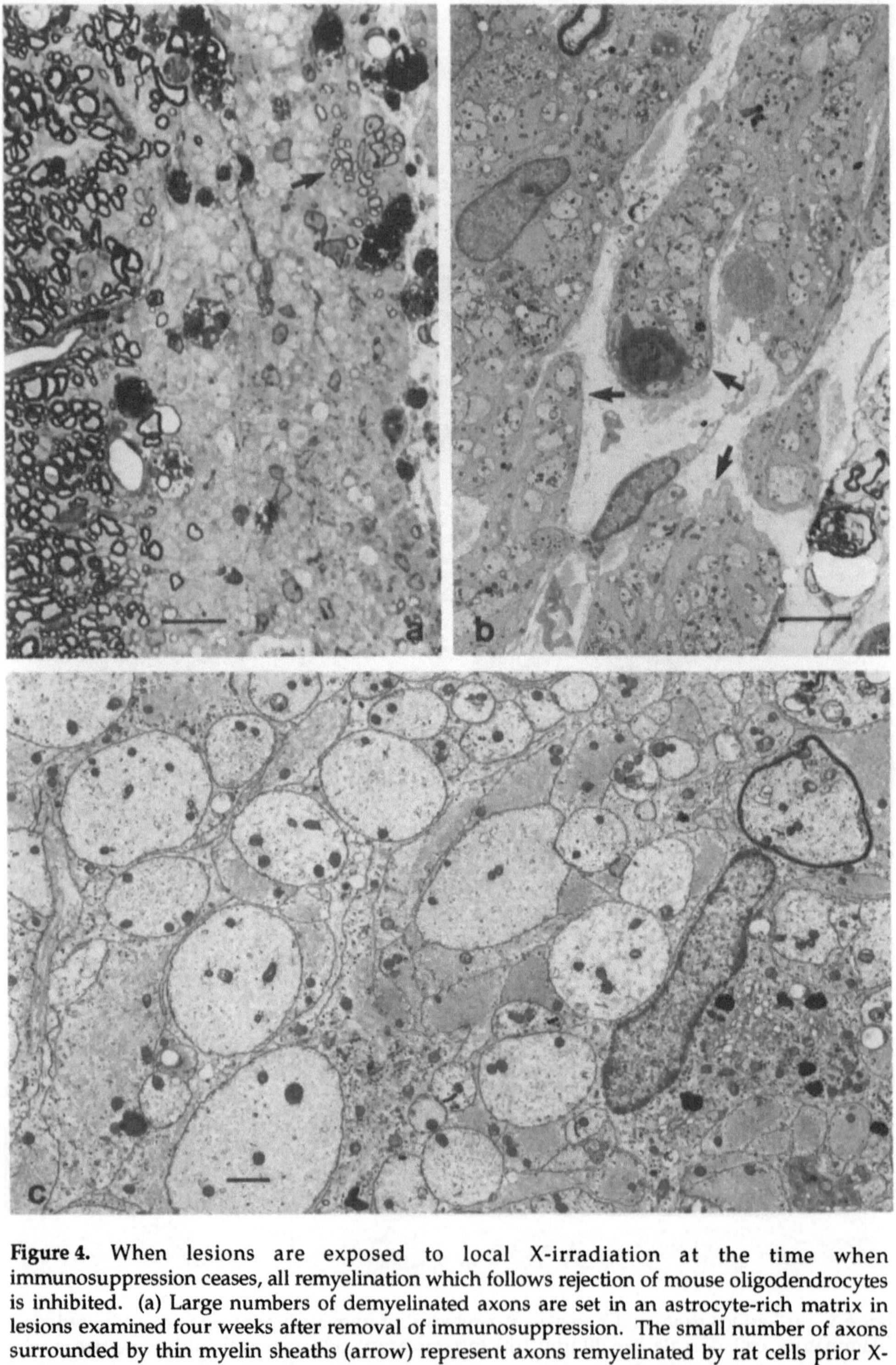

Figure 4. When lesions are exposed to local X-irradiation at the time when immunosuppression ceases, all remyelination which follows rejection of mouse oligodendrocytes is inhibited. (a) Large numbers of demyelinated axons are set in an astrocyte-rich matrix in lesions examined four weeks after removal of immunosuppression. The small number of axons surrounded by thin myelin sheaths (arrow) represent axons remyelinated by rat cells prior X-irradiation - scale bar 20 μm. (b) Demyelinated axons contained within an astrocyte environment are delineated by a glia limitans (arrows) - scale bar 4 μm. (c) Following rejection of mouse oligodendrocytes and suppression of remyelination, axons remain demyelinated and are separated by astrocyte processes - scale bar 1 μm.

demyelination in a manner which facilitates remyelination of demyelinated axons by oligodendrocytes. Of particular interest is the observation that an immune response, which is self limiting and lacks specificity for oligodendrocyte specific epitopes, does not inhibit remyelination. This observation calls into question the concept that products of inflammation *per se* can induce demyelination or inhibit remyelination.

ACKNOWLEDGEMENTS

We would like to thank Jennifer Gilson and Mike Stocker for their excellent technical contributions to the work and the Multiple Sclerosis Society of Great Britain for financial support. The in situ hybridisation was carried out be S. Ryder who holds an ARRC Veterinary Schools Scholarship. The probe (pY353/B) was made available by Dr C.E. Bishop.

REFERENCES

1. K.Selmaj, C.S. Raine, M. Farooq, W.T. Norton and C.S. Brosnan, Cytokine cytotoxicity against oligodendrocytes.Apoptosis induced by lymphotoxin, *J. Immunol.* 147:1522 (1991).
2. A.Compston, N. Scolding, D. Wren amd M. Noble, The pathogenesis of demyelinating disease: insights from cell biology, *TINS*, 14:175 (1991).
3. W.F. Blakemore and A.J. Crang, The relationship between type-1 astrocytes, Schwann cells and oligodendrocytes following transplanation of glial cell cultures into demyelinating lesions in the adult rat spinal cord. *J. Neurocytol.*, 18: 519 (1989).
4. A.J. Crang and W.F. Blakemore, The effect of the number of oligodendrocytes transplanted into x-irradiated, glial-free lesions on the extent of oligodendrocyte remyelination. *Neurosci. Lett.*, 193: 269 (1989).
5. R.J.M. Franklin, A.J. Crang and W.F. Blakemore, Type-1 astrocytes fail to inhibit Schwann cell remyelination of CNS axons in the absence of cells of the O-2A lineage. *Devel. Neurosci.*, 14:85 (1992) .
6. D.L. Graca and W.F. Blakemore, Delayed remyelination in rat spinal cord following ethidium bromide injection. *Neuropathol. Appl. Neurobiol.*, 12:593 (1986).
7. R.J.M. Franklin, A.J. Crang and W.F. Blakemore, Transplanted type-1 astrocytes facilitate repair of demyelinating lesions by host oligodendrocytes in adult rat spinal cord. *J. Neurocytol*, 20:420 (1991) .
8. A.J. Crang and W.F. Blakemore, Remyelination of demyelinated rat axons by transplanted mouse oligodendrocytes. *Glia*, 4: 305 (1991).
9. A.R. Harvey, Y. Fan, M.W. Beilharz and M.D. Grounds, Survival and migration of transplanted male glia in adult female mouse brains monitored by a Y-chromosome-specific probe. *Mol.Brain.Res.*, 12:339 (1992) .
10. W.F. Blakemore and A.J. Crang, The use of cultured autologous Schwann cells to remyelinate areas of persistent demyelination in the central nervous system, *J. Neurol. Sci.*, 70: 207 (1985).
11. R.J.M. Franklin, A.J. Crang and W.F. Blakemore, The reconstruction of an astrocytic environment in glia-deficent areas of white matter, *J. Neurocytol.*, in press (1993).
12. S.K. Ludwin, Chronic demyelination inhibits remyelination in the central nervous system, *Lab. Invest.*, 43:382 (1980).
13. C.S. Raine, Multiple sclerosis: immunolopathologic mechanisms in the progression and resolution of inflammatory demyelination, In *Mechanisms in Neurologic and Psychriatiric Disease* Ed. B.H. Waksman Raven press New York, p 37 (1990).
14. E.N. Benveniste and J.A. Merrill, Stimulation of oligodendroglial proliferation and maturation by interleukin-2, *Nature*, 321:610 (1986).

TRANSPLANTATION OF MYELINATING CELLS INTO THE CENTRAL NERVOUS SYSTEM

Ian D. Duncan, and David R. Archer

Department of Medical Sciences
University of Wisconsin School of Veterinary Medicine
2015 Linden Drive
Madison, WI 53706, USA

INTRODUCTION

Transplantation of myelinating cells into the central nervous system (CNS) has two major applications, firstly as an experimental approach to study the development and cellular interactions of these cells following their transplantation, and secondly to test their ability to myelinate axons in the recipient animal. As Bjorklund[1] stated, neural transplantation is an experimental tool with clinical possibilities. Using transplantation strategies, it is possible to test the abilities of different glial cell preparations to migrate, divide, mature, and finally myelinate axons. Such strategies are also being used to test the ability of transplanted oligodendrocytes or Schwann cells to repair focal or multi-focal areas of loss of myelin, or achieve global repair in dysmyelinating states. Current data documenting extensive myelin formation following transplantation of glial cells,[2-5] suggests that such an approach may have an important future therapeutic role in human myelin disorders. The recent functional successes with transplantation of human fetal dopaminergic neurons into the brains of patients with spontaneous and drug-induced Parkinson's disease,[6,7] and similar rodent tissue into experimental models of Parkinson's,[8] augurs well for the potential use of glia, although the scale of repair required in multi-focal or diffuse myelin disease, provides additional challenges.

Glial cell transplantation has been studied in two model systems, firstly using the myelin mutants as recipients, or secondly transplanting cells into persistently demyelinated, focal areas of the spinal cord.[9] While these approaches can be used to ask similar questions, each has certain advantages. The identification of remyelinating cells in animals with chemically induced, focal areas of demyelination, requires rigorous testing of the source of cells responsible for the remyelination, because of the propensity for host repair. Persistent demyelination can usually only be achieved by preventing the normal host remyelinating response by prior irradiation[9] of the demyelinated area. Irradiation however, can complicate the pathological changes seen in the demyelinated zones. In contrast, in the myelin mutants, host myelination is usually minimal and proof of identification of myelinating cells is straightforward. This chapter will discuss transplantation into the myelin mutants and the studies on each of the mutants used as recipients of glial cell transplants will be reviewed then a summary of these results presented.

IDENTIFICATION OF TRANSPLANTED CELLS

A central issue in all transplantation studies is the ability to unequivocally identify the implanted cells and differentiate them from host tissue. Perhaps the easiest way to accomplish this is to use a mutant that has a known absence of a specific glial cell or myelin protein, and to subsequently identify this protein in the transplanted CNS. This approach has been used successfully in two myelin mutants, most notably in the shiverer (*shi*) mouse which lacks myelin basic protein (MBP) and also in the myelin deficient (*md*) rat, which lacks proteolipid protein (PLP). These studies will be described below.

It is also possible to pre-label the cells to be transplanted. Such labelling methods include, 1) the use of fluorescent dyes such as fast blue,[10] Hoechst[11] and carbocyanine DiI,[12] 2) pre-labelling of dividing glia with tritiated thymidine, and 3) the use of molecular markers such as the expression of the lacZ gene. Another approach, which is however cumbersome, is the specific rejection of the transplanted cells[13] (see quaking mouse below).

1) QUAKING MOUSE

This was the first myelin mutant used as a recipient for myelinating cells.[13] As axons in the quaking (*qk*) mouse spinal cord are either hypomyelinated or ensheathed by oligodendrocyte processes but not myelinated, focal areas of demyelination were created prior to implantation of myelin-forming cells. The protocol involved focal injections of lysolecithin followed shortly thereafter by the implantation of cultured fetal rat Schwann cells on a collagen bed. The mice were immunosuppressed with daily injections of anti-lymphocytic serum. Three to four weeks after surgery, immunosuppression was stopped and an immune cell transfer given[13] in an attempt to identify whether the Schwann cell myelin seen at the site of transplantation had been made by host or rat cells. At the site of the implant, numerous Schwann cell myelinated axons were seen, most with a normal thickness myelin sheaths in contrast to the adjacent *qk* myelinated fibres. In these areas there was notable evidence of macrophage invasion, suggesting rejection of these presumptive rat Schwann cells. In those mice which immunosuppression was continued, no cellular infiltrate was seen.

This study provided the first evidence that cultured cells, transplanted as a xenograft could myelinate host axons of a myelin mutant. There was no spread of the transplanted cells away from the site of implantation however, probably because most *qk* axons are ensheathed and there was no stimulus for transplanted cells to migrate. The *qk* mouse therefore has limitations in its use in transplant studies where questions regarding migration and large-scale repair by transplanted glial cells are being asked.

2) SHIVERER MOUSE

The shiverer (*shi*) mutant mouse was first described in 1973 (see Gumpel[14] for review). This mutation, which is inherited as an autosomal recessive trait, results in a major deletion of the gene for MBP and consequent hypomyelination of the CNS. Affected mice develop a tremor at around 12 days of age and later develop seizures leading to death between 100-150 days. They have a notable lack of myelin in the brain and optic nerves (about 25% of normal in the optic nerve[15]), but the myelin deficiency is not as severe in the spinal cord. Shiverer myelin is negative for MBP when stained immunocytochemically with antibodies to this protein. The major dense line, the major site of localization of MBP, is not seen in the majority of myelinated fibres in the CNS.

The lack of MBP in the *shi* mouse, led Gumpel and Lachapelle[16,17] and their colleagues to use this mutant as a recipient of normal glial cells, assaying the success of transplantation by the presence of MBP positive myelin in the CNS and the presence of a normal major dense line. Since their first report, they have used *shi* for extensive studies on oligodendrocyte and Schwann cell transplantation. The basic technique used by Gumpel's group has been the transplantation of fragments of the CNS from embryonic or neonatal normal mouse brain, into the brain of newborn *shi* mice, or the spinal cord of adult *shi* mice, using charcoal as a marker of the site of transplantation. They have also transplanted aggregates of purified adult oligodendrocytes into neonatal *shi* brain.[18] These studies have clearly shown that normal oligodendrocytes make MBP positive myelin when transplanted into the *shi* CNS. They also showed that oligodendrocytes (or their precursors) could migrate as far as the brainstem, cerebellum and cervical spinal cord from the site of transplantation in the forebrain.[19] The success of transplantation is dependent upon the age of the donor tissue and the age of the recipient. Using older donor tissue or transplanting tissue into older mice resulted in less myelin and shorter migration of the transplanted cells. The first studies of Gumpel and Lachapelle were initially confirmed by two other groups who transplanted fetal or neonatal brain tissue,[20] or bulk-separated oligodendrocytes[21] into *shi* brain. More recently, transplantation of oligodendrocyte progenitors isolated by immunopanning, has also proven the ability of immature cells to mature and to myelinate *shi* axons.[22] In contrast to these immature cells, cells isolated as the long-term surviving oligodendrocytes or adult progenitors in the axotomized optic nerve of the rat, have also been found to myelinate *shi* axons.[23] Fragments of optic nerve from rats which had been enucleated up to two years previously, were implanted into newborn *shi* brain.[23] MBP positive myelin was seen within and adjacent to the area of the implant, although it was not known whether this myelin had been made by persistent adult progenitors[24] within the optic nerves or surviving, but de-differentiated oligodendrocytes.

One of the most remarkable features of the early studies of Gumpel et al. was the observation of "long-stage migration" of cells from the site of transplantation.[19] Whether the migrating cells (which later made MBP positive myelin) were progenitors or mature oligodendrocytes was not known, although it was thought most likely that they were immature cells. However, they later showed that preparations of adult oligodendrocytes isolated from 4-6 week old rat brain (95% purity as identified by immunolabelling), could also migrate long distances in the brain.[18] The most likely pathway for migration of these cells was suggested to be along major axonal tracts.[19] It has been suggested, however, that the migration of the transplanted cells reported in these studies could also have been the result of their deposition in the ventricular system[22] but the ability of cells to migrate and invade the parenchyma from the ventricular system needs to be shown.

In a recent analysis of glial cell migration, Gansmuller et al.[25] labelled cells in fragments from telencephalic periventricular zone of newborn mice with Hoechst 333432 dye. By double labelling with antibodies identifying immature and mature oligodendrocytes (04 and GalC, respectively), they followed the migration and development of the transplanted cells. Long distance migration along axonal pathways occurred (although the actual distance was not stated), when the cells were 04 positive and GalC negative, but migration across grey matter only occurred when cells were GalC positive. Migration of transplanted oligodendrocytes and Schwann cells towards focal areas of demyelination in the spinal cord has also been shown.[11,26,27] In these experiments, focal areas of demyelination were made by injection of lysolecithin, and cells injected some distance, caudal or rostral to the lesion. To confirm the origin of the cells, Schwann cells labelled with Hoechst dye were found to be able to migrate up to 8 mm towards the demyelinated lesion; migration of oligodendrocytes was confirmed by MBP immunocytochemistry.

The *shi* mouse has also been used as a recipient of transplanted glial cells to answer other questions. Firstly, it has shown that xenografts (human,[28] ovine,[29] and rabbit[30]) can myelinate *shi* axons. Secondly, it has been frequently noted that there is never any sign of rejection of allografts in the transplanted *shi* CNS. This is likely to be due to the fact that transplantation is carried out when the host *shi* is usually only 2-3 days of age, a time of likely immuno-tolerance. However, no note has been made of rejection of cells transplanted into adult *shi* spinal cord. Finally, the longevity of Schwann cells transplanted into *shi* brain has also been examined in detail.[31] Although labelled Schwann cells were found up to 40 days after transplantation, by 120 days, none were detectable. These cells showed modest evidence of migratory ability in the brain, usually in association with axons. Their disappearance in the long-term was thought to be a result of an astrocytic response which eventually results in their expulsion from the brain parenchyma.[31]

3) MYELIN DEFICIENT RAT

This X-linked mutant rat was reported in 1979 by Csiza and de Lahunta[32] in a Wistar rat colony. A tremor is first noted when the rats begin to move at 10-12 days of age, seizures begin at 18-20 days and they die by 23-25 days. A longer-lived "strain" of the *md* rat has been reported[33,34] which is currently being used as a transplant recipient. In the CNS of the *md* rat there is an almost total absence of myelin as a result of a point mutation in the proteolipid protein (PLP) gene.[35] The occasional myelinated fibers found in the CNS are MBP positive but PLP negative (Fig. 1).[36]

The first studies of transplantation into the *md* rat used a dissociated spinal cord preparation from 0-1 day old female littermates.[2] Injection of these cells as an allograft into the spinal cord of their male littermates ($\approx$ 50% of whom are hemizygous for the mutant gene) resulted in large areas of PLP positive myelin in the dorsal or ventral columns (Fig. 1[a-d]). Since this report, we have used such a preparation in many *md* rats (Fig. 2) with almost 100% success, i.e., this experiment is very repeatable. In order to determine whether glial cells derived from adult animals or purified adult oligodendrocytes could myelinate *md* axons, two experiments were performed.[3] In the first, a dissociated cell preparation, made from 2 month old rat spinal cord and which contained at least 50-60% of adult oligodendrocytes (01 positive) was injected into the *md* rat. The rest of the preparation contained 30-40% microglia and 2% with a progenitor phenotype - 04 positive/01 negative. This resulted in large areas of myelin at the site of injection, similar in amount to that produced by transplanted neonatal cells. In the second part of this experiment, adult oligodendrocytes as identified by the 01 antibody, were sorted by fluorescent activated cell sorting, resulting in a 93-95% purity for the adult cell. These cells also made myelinated axons in the *md* spinal cord (Figs. 3 and 4), although less than with the dissociated cell preparation.

Others have also used the *md* rat as a recipient of transplanted glia although with some differences in results from our studies.[4] In these studies, transplantation of tissue from normal rat donors older than 9 days of age produced no myelin. In contrast, dissociated preparations from 15-18 old fetal rat spinal cord produced copious amounts of myelin when transplanted into the *md* rat spinal cord.[4]

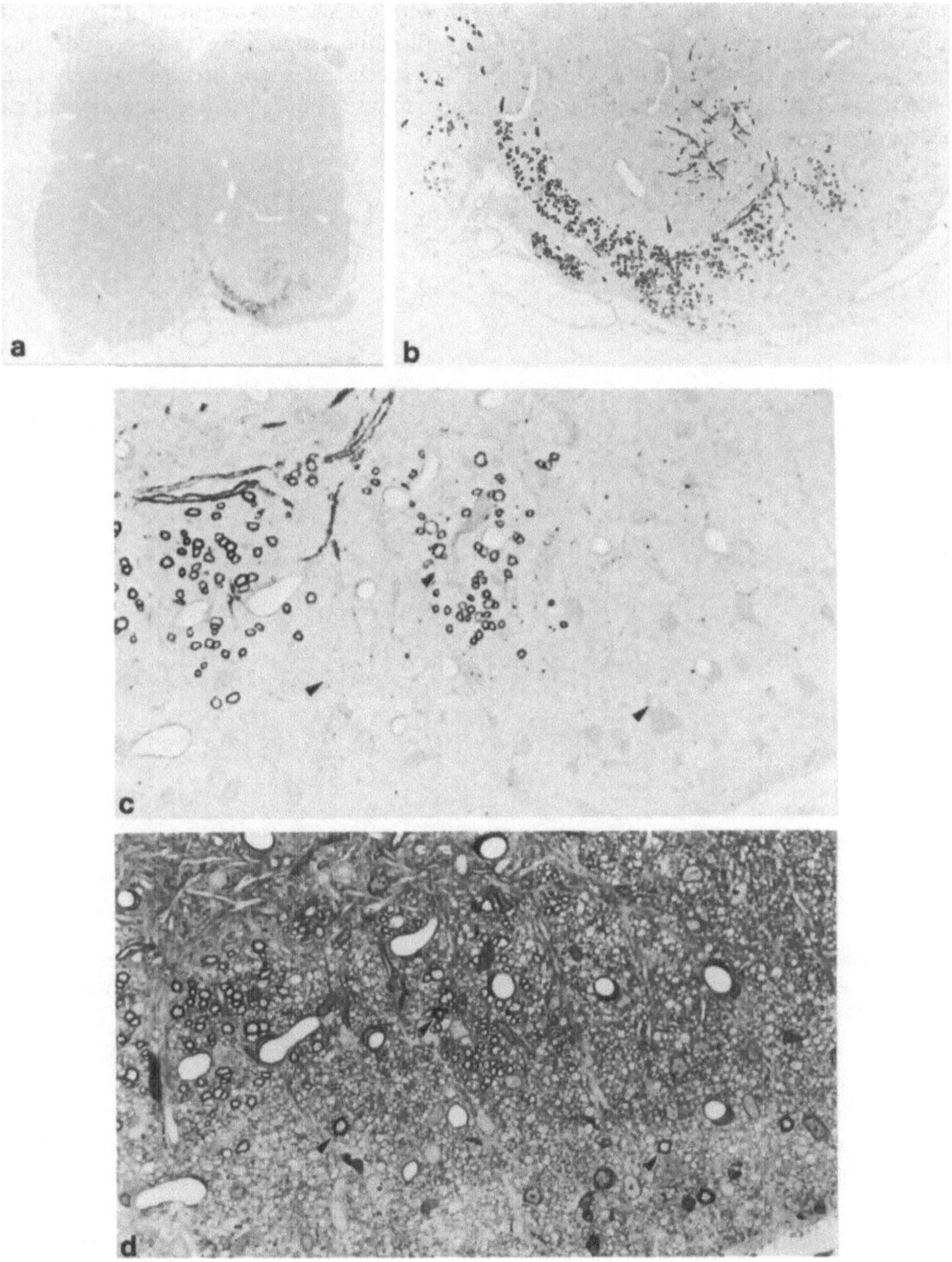

Figure 1. a) Spinal cord of a 22 day old *md* rat transplanted 19 days earlier with a dissociated cell suspension from a female littermate. A small area of PLP positive myelin is seen in the right ventral columns. b) On higher power of this area, individual PLP positive fibres can be seen, some of which are in the grey matter. c,d) Adjacent areas from the site shown in b, stained for PLP (c) and with toluidine blue (d). Outside the PLP positive area of the graft, four large diameter fibers are seen in the toluidine blue stained section (d) which are negative for PLP (c).

We have also examined the ability of cryopreserved glia, either transplanted as xenografts (canine origin) or allografts, to produce myelin in the *md* rat (Archer and Duncan, in preparation). Cryopreserved allografts were very successful in myelinating *md* axons but xenografts produced much less myelin with evidence of a cellular response and early rejection of the grafted tissue. This cellular infiltration, could however be considerably reduced by the daily injection of cyclosporin A. The success of xenografts in our study is in contrast to that reported by Rosenbluth et al.[4] where mouse glia transplanted into the *md* rat failed to make myelin.

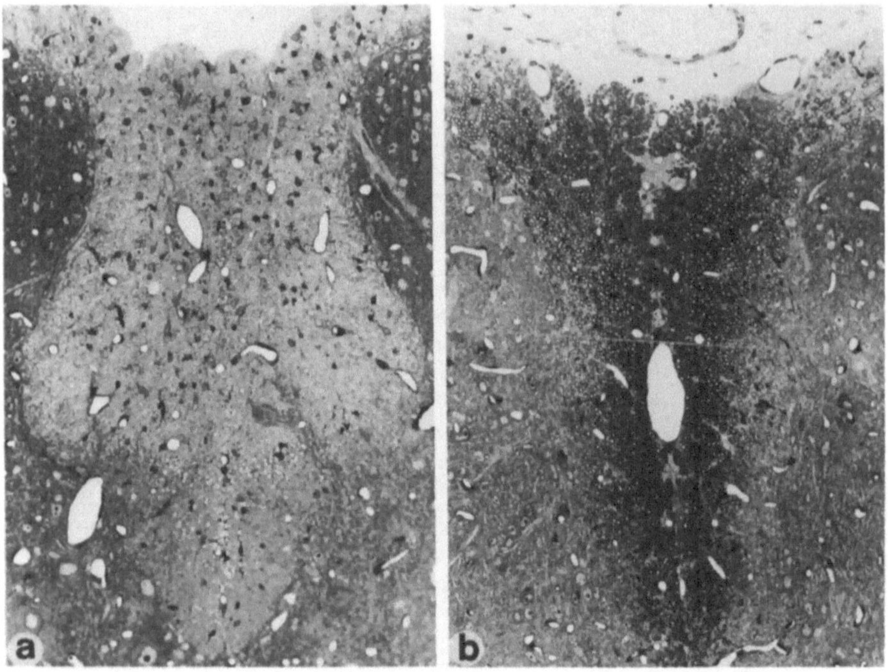

Figure 2. Dorsal columns of an uninjected (a) and injected *md* rat (b), which had received a dissociated cell preparation 14 days previously. In the injected rat, almost the entire dorsal columns have been myelinated by the transplant

Transplantation of glia into the *md* rat brain has also been carried out to study aspects of glial cell development *in vivo*. Fast blue labelled oligodendrocytes have been found to migrate from the site of transplantation in the *md* brain.[10] In an extension of this study, transplantation of a fast blue labelled cell preparation, enriched for the 0-2A progenitor was studied post-operatively, double labelling for GalC or glial fibrillary acidic protein.[37] The authors suggested that the data implies that only mature oligodendrocytes developed *in vivo*, with no evidence of the transplanted cells becoming Type II astrocytes.

The functional implications of glial cell transplantation are starting to be explored in the *md* rat. In collaboration with J. Kocsis and colleagues at Yale, experiments have been carried out which study the physiological properties of axons myelinated by transplanted cells. These studies have shown a return of nerve conduction velocity near to normal in the areas of myelination, suggesting that glial cell transplantation could be of great therapeutic value.[38]

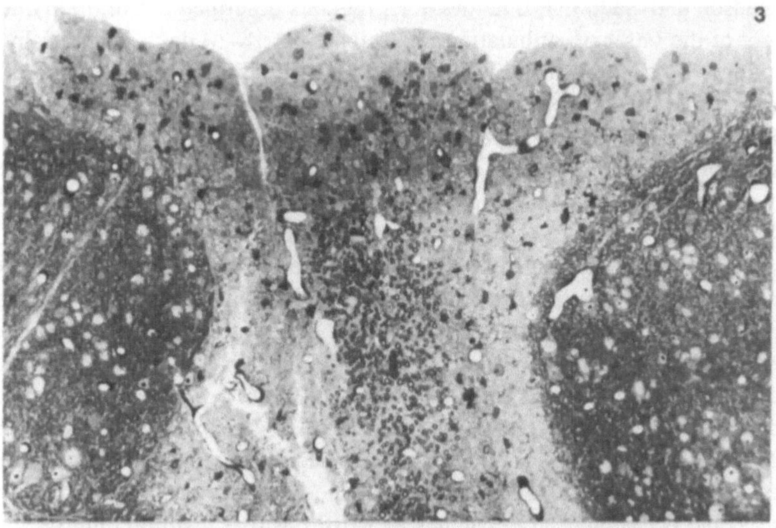

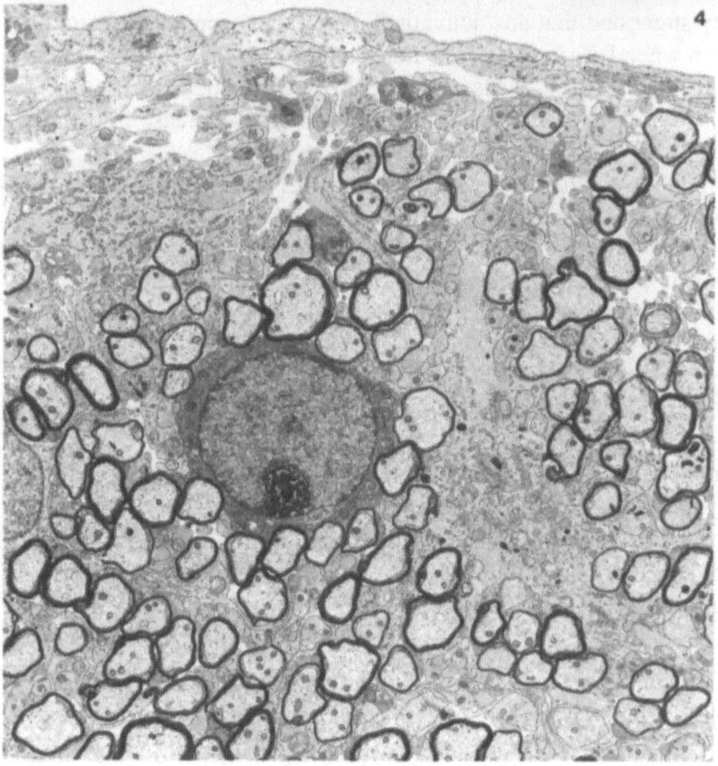

Figure 3. Mature (01 positive) cells are also capable of forming myelin the *md* rat as is seen in the dorsal columns of this rat. The amount of myelin not as great as that produced by neonatal cells.

Figure 4. An electron micrograph from a rat which had received a mature oligodendrocyte transplantation. In the center of the field a healthy looking oligodendrocyte, unlike that seen in the uninjected *md* rat, is seen.

4) JIMPY MOUSE

The jimpy (*jp*) mouse has been the most investigated of all the X-linked mutants, with a known splice site mutation in the PLP gene, leading to a deletion of exon 5 (for review see Hudson and Nadon[39]). Like the other mutants described above, jimpy develops a marked tremor at the onset of ambulation and seizures at 18-20 days followed by death at 25-30 days of age. There is a generalized paucity of myelin throughout the CNS although in the spinal cord, more myelin is seen than in the *md* rat.[46] The myelin present in *jp* spinal cord is MBP positive and PLP negative.[39]

Because of the short life-span of *jp*, there have been few transplant studies in this mutant. However, three studies have reported the results of cross-transplantation of tissue between *jp, shi* and normal mice.[40-42] In the first study, transplantation of newborn *jp* brain tissue into *shi* produced rare small patches of MBP positive myelin.[40] Transplantation of normal tissue into *jp* brain produced patches of PLP positive myelin as did transplants of *shi* tissue in *jp*. Thus, the *jp* environment did not have a negative influence on the myelinating capacity of normal or *shi* oligodendrocytes. Lachapelle et al.[42] compared the myelinating capacity of fragments of corpus callosum from P13 *shi, jp* and normal mice when transplanted in similar cross-transplantation fashion as[40,41] their previous experiments to the results of transplantation of tissue from newborn animals. While they confirmed their earlier observations that *jp* oligodendrocytes myelinated poorly on transplantation compared to normal or *shi* cells, occasional large patches of MBP positive myelin could be seen in *shi* recipients. They suggested that an "activating" factor, not secreted by astrocytes (as similar results were seen when P13 *jp* tissue was transplanted to newborn *jp*) stimulated the mitotic potential of grafted cells.[42] This however, remains speculative, as there was no autoradiographic evidence or Brdu labelling to confirm an increase in cell division in such patches.

Two important points can be emphasized from these combined studies. Firstly, that despite *in vitro* evidence that *jp* oligodendrocytes produce PLP in culture when normal conditioned media is added,[43] the myelin produced by *jp* tissue in *shi* is apparently not PLP positive. Secondly, there does not appear to be an environmental, inhibiting or noxious factor in *jp* which prevents the myelinating function of transplanted, normal or *shi* oligodendrocytes.

5) SHAKING PUP

This canine mutant seen in Spaniel dogs, was first described by Griffiths et al.[44] The neurological symptoms are similar to those seen in the rodent myelin mutants but with a later onset of intermittent seizures. If hand-reared, these dogs can live indefinitely. The myelin defect is severe in the brain but less so in the spinal cord where there are similarities with the amount of myelin seen in *shi.* There is an increasing amount of myelin found with time however. A point mutation has been identified in the shaking (sh) pup PLP gene[39] but *sh* pup myelin is weakly PLP positive[45] and so PLP immunolabelling is not as useful a marker of transplantation as in the *md* rat.

Preliminary transplantation studies have been carried out in the *sh* pup with promising results.[5] In summary, these results demonstrate that injection of dissociated cell preparations from normal dog spinal cord taken at embryonic or neonatal time-points, can myelinate large areas of the dog spinal cord at the site of injection (Fig. 5). Fetal tissue which, as expected, contains a large number of oligodendrocyte progenitor cells, would appear to be the best source of myelinating cells with considerable abilities to migrate. The myelin formed by the transplanted cells was normal in thickness and persisted in one dog with no evidence of rejection (and no immunosuppression) for 27 weeks. Ongoing studies are aimed at determining the extent to which larger scale repair can be achieved, both in the spinal cord and brain, and whether functional recovery following glial cell transplantation, can be achieved.

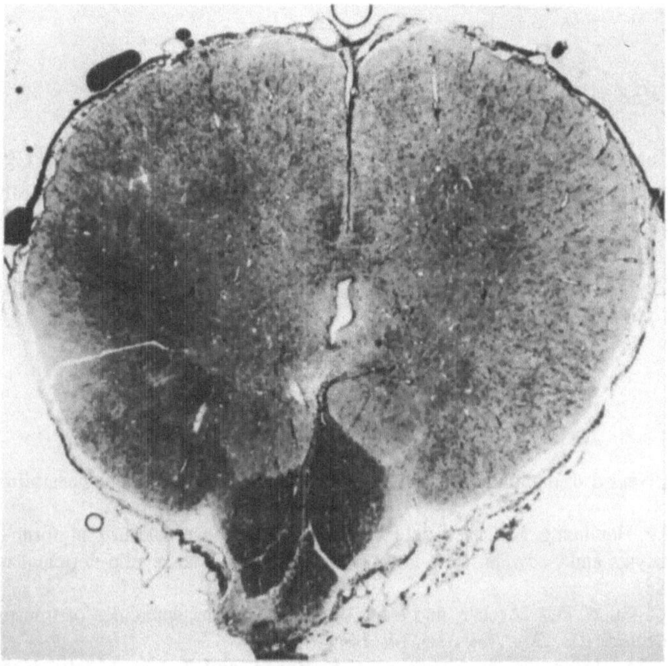

Figure 5. A thoracic spinal cord segment from a 10 week old shaking pup which had received a transplant of cells from the brain of a normal E45 dog, 8 weeks previously. The dorsal columns appear normally myelinated and there is a spread of myelin through the right lateral and ventral white matter.

SUMMARY

There is clearly a growing interest in glial cell transplantation as a means to ask both basic and applied questions about glial cell biology. The myelin mutants discussed above have provided numerous experimental opportunities to answer such questions. Each mutant has its own relative advantages and disadvantages. In those with the least myelin (*jp* mouse and *md* rat), the results of transplantation are usually easily identified. However, these mutants die by 3-4 weeks of age, and so the effects of transplantation can only be studied in the short-term. An exception to this is the longer-lived strain of *md* rat which lives for three times as long. *Shi* lives longer and its lack of MBP provides a good marker for transplanted cells. However, there are more myelinated axons in the spinal cord than *jp* or *md* and so identification of transplanted cells in the cord, unless they are labelled, is problematic. As the *sh* pup lives for a long time and as its neuraxis is so much closer in size to that of man, it allows for opportunities to perform long-term transplantation studies aimed at global repair, which could be of significance to man. To date, distinguishing myelin made by transplanted versus host oligodendrocytes in the *sh* pup has not been difficult because of the normal thickness of the myelin made by the transplanted cells. However, a marker of transplanted cells would be useful.

Future issues which should be addressed include the determination of the optimal cell to be transplanted and its route of access to the CNS. In addition, a combination of the techniques used to label cells and identify their myelinating capacity will be important, i.e., fluorescent dye labelling methods and analysis of tissue for these cells and myelin they produce is not compatible with high resolution (1 μm, toluidine blue stained) sections. These are just some of the technical advances that will help accelerate this research towards therapeutic trials.

ACKNOWLEDGEMENTS

Our studies on the *md* rat and *sh* pup have been supported by NIH grant NS23124, The Myelin Project and Elisabeth Elser Doolittle Charitable Trust. We would like to thank our collaborators, Drs. P. M. Wood, and J. Kocsis for their collaborations with parts of this work. We would also like to acknowledge the technical assistance of R. Hoffman, K. Hoffman and Z. Ghuo. Figure 1 is reproduced by permission of Chapman and Hall (Duncan et al., 1988) and Figures 2b, 3 and 4 by permission of S. Karger AG (Duncan et al., 1992).

REFERENCES

1. A. Bjorklund, Neural transplantation - an experimental tool with clinical possibilities, *TINS* 14:319 (1991).
2. I.D. Duncan, J.P. Hammang, K.F. Jackson, P.M. Wood, R.P. Bunge, and L. Langford, Transplantation of oligodendrocytes and Schwann cells into the spinal cord of the myelin-deficient rat, *J. Neurocytol.* 17:351 (1988).
3. I.D. Duncan, C. Paino, D.R. Archer, and P.M. Wood, Functional capacities of transplanted cell-sorted adult oligodendrocytes, *Dev. Neurosci.* 14:114 (1992).
4. J. Rosenbluth, M. Hasegawa, N. Shirasaki, C.L. Rosen, and Z. Liu, Myelin formation following transplantation of normal fetal glia into myelin-deficient rat spinal cord, *J. Neurocytol.* 19:718 (1990).
5. D.R. Archer, P.A. Cuddon, and I.D. Duncan, myelination by glial cells following transplantation into the CNS of the shaking pup, *Restor. Neurol. Neurosci.* 4:12.P1 (Abstract) (1992).
6. H. Widner, J. Tetrud, S. Rehncrona, B. Snow, P. Brundin, B. Gustavii, A. Bjorklund, O. Lindvall, and J.W. Langston, Bilateral fetal mesencephalic grafting in patients with severe MPTP-induced parkinsonism: results after 2 years, *Restor. Neurol. Neurosci.* 4:21.S9 (Abstract) (1992).
7. O. Lindvall, P. Brundin, H. Widner, S. Rehncrona, B. Gustavii, R. Frackowiak, K.L. Leenders, G. Sawle, J.C. Rothwell, C.D. Marsden, and A. Bjorklund, Grafts of fetal dopamine neurons survive and improve motor function in Parkinson's disease, *Science* 247:574 (1990).
8. H. Nishino, T. Hashitani, M. Kumazaki, H. Sato, F. Furuyama, Y. Isobe, N. Watari, M. Kanai, and S. Shiosaka, Long-term survival of grafted cells, dopamine synthesis/release, synaptic connections, and functional recovery after transplantation of fetal nigral cells in rats with unilateral 6-OHDA lesions in the nigrostriatal dopamine pathway, *Brain Res.* 534:83 (1990).
9. W.F. Blakemore and R.J.M. Franklin, Transplantation of glial cells into the CNS, *Trends Neurosci.* 14:323 (1991).
10. A. Espinosa de los Monteros, M.S. Zhang, M. Gordon, M. Aymie, and J.de Vellis, Transplantation of cultured premyelinating oligodendrocytes into normal and myelin-deficient rat brain, *Dev. Neurosci.* 14:98 (1992).
11. A. Baron-Van Evercooren, A. Gansmuller, E. Clerin, and M. Gumpel, Hoechst 33342; a suitable fluorescent marker for Schwann cells after transplantation in the mouse spinal cord, *Neurosci Letters* 131:241 (1991).
12. A. Gansmuller, F. Kruger, M. Gumpel, and A. Baron-Van Evercooren, Photoconverted carbocyanine DiI allows direct visualization of transplanted glial cells at the ultrastructural level, *Neurosci Letters* 147:151 (1992).
13. I.D. Duncan, A.J. Aguayo, R.P. Bunge, and P.M. Wood, Transplantation of in vitro cultures of rat Schwann cells into the mouse spinal cord, *J. Neurol. Sci.* 41:241 (1981).
14. M. Gumpel, Shiverer and other marker models used in intracerebral transplantations of glial cells, *Methods in Neuro. Sci.* 7:510 (1991).

15. S. Billings-Gagliardi, M.K. Wolf, D.A. Kirschner, and A.-L. Kerner, Shiverer jimpy double mutant mice.II. Morphological evidence supports reciprocal intergenic suppression, *Brain Res.* 374:54 (1986).
16. M. Gumpel, O. Gout, and A. Gansmuller, Spontaneous remyelination and intracerebral grafting of myelinating cells in mammals, in: "Neuronal grafting and Alzheimer disease," F. Gage A. Privat, and Y. Christen., eds., Springer, Berlin 43 (1989).
17. F. Lachapelle, M. Gumpel, M. Baulac, C. Jacque, P. Duc, and N. Baumann, Transplantation of CNS fragments into the brain of shiverer mutant mice: Extensive myelination by implanted oligodendrocytes, *Dev. Neurosci.* 6:325 (1983).
18. C. Lubetzki, A. Gansmuller, F. Lachapelle, P. Lombrail, and M. Gumpel, Myelination by oligodendrocytes isolated from 4-6-week old rat central nervous system and transplanted into newborn shiverer brain, *J. Neurol. Sci.* 88:161 (1988).
19. M. Baulac, F. Lachapelle, O. Gout, B. Berger, N. Baumann, and M. Gumpel, Transplantation of oligodendrocytes in the newborn mouse brain: extension of myelination by transplanted cells. Anatomical study, *Brain Res.* 420:39 (1987).
20. E. Friedman, G. Nilaver, P.W. Carmel, M. Perlow, L. Spatz, and N. Latov, Myelination by transplanted fetal and neonatal oligodendrocytes in a dysmyelinating mutant, *Brain Res.* 378:142 (1986).
21. S. Kohsaka, K. Yoshida, and Y. Inoue, Transplantation of bulk-separated oligodendrocytes into the brain of shiverer mutant mice: Immunohistochemical and electron microscopic studies on the myelination, *Brain Res.* 372:137 (1986).
22. A.E. Warrington, E. Barbarese, and S.E. Pfeiffer, Differential myelinogenic capacity of specific developmental stages of the oligodendrocyte lineage upon transplantation into hypomyelinating hosts, *J. Neurosci. Res.* 34:1 (1993).
23. S.K. Ludwin, Oligodendrocytes from optic nerves subjected to long term wallerian degeneration retain the capacity to myelinate, *Acta Neuropath.* 84:530 (1992).
24. G. Wolswijk and M. Noble, Identification of an adult-specific progenitor cell, *Dev.* 105:387 (1989).
25. A. Gansmuller, E. Clerin, F. KrHger, M. Gumpel, and F. Lachapelle, Tracing transplanted oligodendrocytes during migration and maturation in the shiverer mouse brain, GLIA 4:580 (1991).
26. O. Gout, A. Gansmuller, and M. Gumpel, Remyelination of a chemically induced demyelinated lesion in the spinal cord of the adult shiverer mouse by transplanted oligodendrocytes, *NATO ASI Series* H 43:185 (1990).
27. Baron-Van Evercooren,A., A. Gansmuller, E. Duhamel, F. Pascal, and M. Gumpel, Repair of a myelin lesion by Schwann cells transplanted in the adult mouse spinal cord, *J. Neuroimmunol.* 40:235 (1992).
28. M. Gumpel, F. Lachapelle, A. Gansmuller, M. Baulac, A. Baron-Van Evercooren, and N. Baumann, Transplantation of human embryonic oligodendrocytes into shiverer brain, Ann. *N. Y. Acad. Sci.* 495:71 (1988).
29. S.K. Ludwin and S. Szuchet, Myelination by mature ovine oligodendrocytes in vivo and in vitro: evidence that different steps in the myelination process are independently controlled, *GLIA* (1992). (In Press)
30. C. Jacque, J. Quinonero, P.V. Collins, H. Villarroya, and I. Suard, Comparative migration and development of astroglial and oligodendroglial cell populations from a brain xenograft, *J. Neurosci.* 12:3098 (1992).
31. A. Baron-Van Evercooren, E. Clerin-Duhamel, P. Lapie, A. Gansmuller, F. Lachapelle, and M. Gumpel, The fate of Schwann cells transplanted in the brain during development, *Dev. Neurosci.* 14:73 (1992).
32. C.K. Csiza and A. de Lahunta, Myelin deficiency (md): A neurologic mutant in the Wistar rat, *Am. J. Path.* 95:215 (1979).
33. K.F. Jackson, I.D. Duncan, M.R. Wells, and S.F. Worth, Observations on the CNS of longer lived myelin deficient rats, *Soc. Neurosci.* 14:829 (Abstract) (1988).
34. N.L. Nadon, I.D. Duncan, D.R. Archer, K. Hoffman, C. Czisa, and M. Wells, Oligodendrocyte survival and myelin gene expression in the older strain of the myelin deficient rat, *Soc. Neurosci.* 18:272 (Abstract) (1992).
35. I.D. Duncan, Dissection of the phenotype and genotype of the X-linked myelin mutants, *Ann. NY Acad. Sci.* 605:110 (1990).
36. I.D. Duncan, J.P. Hammang, and B.D. Trapp, Abnormal compact myelin in the myelin-deficient rat: Absence of proteolipid protein correlates with a defect in the intraperiod line, Proc. *Natl. Acad. Sci. USA* 84:547 (1987a).
37. Espinosa de los Monteros, M. Zhang, and J. de Vellis, O2A progenitor cells transplanted into the neonatal rat brain develop into oligodendrocytes but not astrocytes, *Proc. Natl. Acad. Sci. USA* 90:50 (1993).
38. I.D. Duncan, D.A. Utzschneider, D.R. Archer, J.D. Kocsis, and S.G. Waxman, Myelination of myelin deficient rat axons by transplanted glia restores conduction, *J. Neuropath. Exp. Neurol. Abstr. 52:287* (1993).
39. L.D. Hudson and N.L. Nadon, Amino acid substitutions in proteolipid protein that cause dysmyelination,

in: "Myelin: biology and chemistry," R.E. Martenson., ed., CRC Press, Boca Raton 677 (1992).
40. F. Lachapelle, P. Lapie, J.L. Nussbaum, and M. Gumpel, Immunohistochemical studies on cross-transplantations between jimpy, shiverer, and normal newborn mice, *J. Neurosci. Res.* 27:324 (1990).
41. F. Lachapelle, P. Lapie, A.T. Campagnoni, and M. Gumpel, Oligodendrocytes of the jimpy phenotype can be partially restored by environmental factors in vivo, *J. Neurosci. Res.* 29:235 (1991).
42. F. Lachapelle, P. Lapie, and M. Gumpel, Oligodendrocytes from jimpy and normal mature tissue can be 'activated' when transplanted in a newborn environment, *Dev. Neurosci.* 14:105 (1992).
43. W.P. Bartlett, P.E. Knapp, and R.P. Skoff, Glial conditioned medium enables jimpy oligodendrocytes to express properties of normal oligodendrocytes: Production of myelin antigens and membranes, *GLIA* 1:253 (1988).
44. I.R. Griffiths and I.D. Duncan, Shaking Pups: A disorder of central myelination in the Spaniel dog, J. Neurol. Sci. 50:423 (1981).
45. K. Yanagisawa, J.R. Moller, I.D. Duncan, and R.H. Quarles, Disproportional expression of proteolipid protein and DM-20 in the X-linked, dysmyelinating shaking pup mutant, *J. Neurochem.* 49:1912 (1987).
46. I. D. Duncan, J. P. Hammang, S. Goda, and R. H. Quarles, Myelination in the jimpy mouse in the absence of proteolipid protein, GLIA 2: 148 (1989).

GENETIC DISORDERS: CLINICAL AND THERAPEUTICAL ASPECTS

Marco Cappa[1], Enrico Bertini[2], Paola Cambiaso[1], Patrizia del Balzo[1], Patrizia Bardelli[3], Graziella Uziel[3], Antonella Di Biase[4], and Serafina Salvati[4]

[1]Divisione di Endocrinologia, [2]Neurologia Ospedale Pediatrico Bambino Gesu', IRCCS, Rome, Italy
[3]Istituto Neurologico C. Besta, IRCCS, Milan, Italy
[4]Istituto Superiore di Sanita', Rome, Italy

INTRODUCTION

Leukodystrophies are a group of degenerative diseases which involve the white matter. Enzymatic defects in myelin lipid metabolism lead to excessive tissue deposition of a normal component of myelin lipids or of breakdown products of myelin. This paper will focus on X-linked adrenoleukodystrophy that constitute part of genetic leukodystrophies.

X-linked Adrenoleukodystrophy (ALD) is an inherited disorder affecting 1/20,000 males, characterized by progressive demyelination of the Central Nervous System (CNS) and hypoadrenalism. The biochemical defect of ALD provokes impairment in the degradation of very long chain fatty acids (VLCFA), particularly C24:0 and C26:0, resulting in an increase of them in the plasma and an accumulation in the tissues. The enzymatic defect of ALD has been identified as a single peroxisomal beta-oxidation defect, namely as a defect of the lygnoceroil-CoA ligase enzyme activity[1]. The ALD gene has recently, been cloned; its product shares homology with a peroxisomal membrane protein involved in peroxisome biogenesis and belongs to the "ATP-binding cassette" superfamily of transporters[2]. The ALD locus was previously mapped to Xq28[3].

VLCFA accumulate in the CNS both as cholesterol esters or in the gangliosidic fraction, determining myelin damage by direct toxic effect. Progressive demyelination observed in this disorder is not fully explained by this mechanism; there is no direct correlation between VLCFA levels and the clinical phenotype[4]. The presence of perivascular lynphomonocyte infiltration in the demyelinated areas suggests that autoimmune factors may play an important role in the demyelination process[5].

At the adrenal gland level, VLCFA accumulation seems to be the only cause of

A Multidisciplinary Approach To Myelin Diseases II
Edited by S. Salvati, Plenum Press, New York, 1994

hypoadrenalism. Actually, deposition of VLCFA determines both an increase in the adrenocortical cell membrane microviscosity with subsequent alterations in the ACTH activity[6] and an abnormal increase of VLCFA-esterefied colesterol[7].

ALD is clinically heterogeneous even in the same sibship. We propose a classification in different clinical phenotypes (Table 1). Infantile ALD is a rapidly progressive disorder leading to a vegetative state in a few years; adolescent ALD has a slower progression. Neurological symptoms in adult cerebral ALD and AMN start after the age of 20 years. AMN generally spares cortical functions although 50% of these patients may present mental degradation during the follow-up with NMR and signs of cerebral white matter demyelination. Patients with Addison-only never show neurological symptoms and are probably an important percentage in the group of primary adrenal deficiency in children and adolescents[8]. Presymptomatic subjects simply show increased plasma VLCFA and are detected by screening males of a sibship with affected subjects. It is still being discussed whether children with adrenal insufficiency and increased VLCFA should be defined as Addison-only or presymptomatic; we define children who are neurologically normal with or without adrenal insufficiency as presymptomatic, and patients who have a normal neurological examination with isolated adrenal deficiency in adult age as Addison-only. Follow-up studies of presymptomatic patients are important because these patients can remain asymptomatic or show neurological symptoms at different ages. At the moment we have no predictive factors for the evolution of presymptomatic patients,therefore it is important to follow them serially by clinical and neuroimaging examinations.

Table 1. Clinical Definition of ALD phenotype

Phenotype	Definition
Childhood ALD	**Onset of neurological symptoms before puberty**
Adolescent ALD	**Onset of neurological symptoms after puberty but before 20 years**
Adult ALD Cerebral	**Central neurological symptoms after 20 years**
Adrenomyeloneuropathy	**Spinal cord and peripheral nerve involved mainly**
Addison only	**Adult with hypoadrenalism without any neurological alterations**
Presymtomatic	**Subject with ALD gene mutation free of neurological symptoms**

Different therapies for severe neurological symptoms were attempted including plasmapheresis, high doses of steroids, intravenous IgG, and immunosuppression without any clear improvement. First dietary therapeutic approaches by simply reducing food intake of VLCFA to less than 3 mg/die did not induce a substantial reduction of abnormal plasma VLCFA[9]. VLCFA have both an endogenous and exogenous origin so reduced food intake provokes an increase of endogenous biosynthesis. Rizzo et al.[10] and Moser et al.[11] proposed trials with a restrictive VLCFA-diet adding oleic acid (C18:1) in the triglyceric form. *In vitro* studies showed that, in the presence of oleic acid, synthesis of VLCFA from ALD patients' fibroblasts were reduced. This approach brought about a reduction of about 70% of VLCFA plasma levels but not to normal levels. Later, dietary therapy improved adding to trioleic acid other monoinsaturated very long chain fatty acids. Lorenzo's oil

(GTOE), a mixture of 74% of oleic acid with 24% of erucic acid and 2% of linoleic acid as well as other monoinsatured very long chain fatty acids, was then proposed[12]. Administration of GTOE with a VLCFA-restricted diet to ALD patients was able to normalize plasma VLCFA for the first time after only 2 months[13,14,15].

Intravenous IgG improved visual symptoms in an ALD patient during a period of 18 months[16]. Positive results were also obtained using intravenous IgG in animals with an acquired demyelination disorder[17]. However, our experience and also Aubourgs' (personal comunication) do not confirm these findings (see results).

The first experiments with bone marrow transplantation (BMT) were unsuccessful probably because they were attempted in patients with severe neurological conditions[18]. The first positive result was reported by Aubourg et al., who obtained clinical improvement and reduction of demyelinated areas in a patient with infantile ALD who was submitted to BMT in the very early stages of the disease[19].

MATERIALS AND METHODS

We observed, 54 patients with X-linked ALD and 11 patients with X-linked AMN at the Bambino Gesu' Children's Hospital and the Neurological Institute C. Besta (Table 2).

Diagnosis of male patients and female carriers was performed evaluating VLCFA in plasma and considering the absolute amount of C26:0 and C26/C22 ratio. After diagnostic confirmation, 49 patients were submitted to a VLCFA restrictive diet with the addition of GTOE, after the informed consent of patients or their parents. The VLCFA-resticted diet (below 3 mg daily) was obtained by avoiding food with a high content of VLCFA[15], (a free Mediterranean diet contains about 20 mg/day). Fatty acid intake was partially covered by GTOE, at the dose of 40-50 gr/day, using it cooked or raw (14,15). Diet salt content did not change during the trial. The follow-up of our patients submitted to this treatment ranged between 6 months to 4 and a half years.

We administered intravenous Immunoglobulin (Venoglobulin, Merieux, France) at dose of 1 g/kg every 2 weeks in 6 patients with severe infantile ALD for a period of 3 months and then every month. Patients with hypoadrenalism were treated with a substitutive therapy of Hydrocortison (10 mg/m^2/twice or thrice daily). Only one patient with adolescent ALD and adrenal deficiency since the age of 6 years needed 9-alpha-fluorocortisol 0.1 mg/day.

Plasma VLCFA levels were measured by previously described gaschromatographic methods[14]. ACTH test was performed by intravenous administration of 0.25 mg ACTH 1-24 (Synacthen,Ciba); plasma cortisol levels were measured by previously described methods at basal and 60 minutes after the introduction of ACTH (20).

Neurological evaluation was performed by the same neurologists (EB and GU), using the EDSS Kurtzke score[21]. Neuropsychological evaluation (NE) was used for the follow-up of patients paying particular attention to mood disorders, concentration difficulties and planning disabilities. NE consisted of the following tests: Wechler Intelligence Scale for Children-revised, Raven Progressive Matrix PM 47 or PM 38, Rey Complex Figures, Digitspan and Supraspan, Corsi Test span and supraspan, Free Drawing, Token test, Fluent verbal tests.

RESULTS

We studied sixty-three patients (Table 2), they belonged to 48 families, 17 families having more than one affected member. Forty-nine patients were submitted to the dietary trial and 3 of them underwent subsequent bone marrow transplantation.

Table 2. Kindreds with ALD studied

Phenotype	N. Patients	Onset of Neurological Symtoms (age: years[1])
Childhood ALD	28	7.4±1.6§
Adolescent ALD	7	11.9±4.8
Adult ALD Cerebral	3	28.6±2.3
Adrenomyeloneuropathy	11	28.2±6.8
Presymptomatic	16	
Heterozygotes	52	
Families	49	

[1] mean±SD

Table 3. Initial clinical neurological manifestations

Symptom	ALD	AMN
Behavior disturbances	16/38	0/11
Gait problems	9/38	11/11
Impaired vision	6/38	0/11
Impaired hearing	2/38	0/11
Seizures	4/38	0/11
Other	2/38	0/11

In 16 patients, the first signs of the disease were psychiatric disorders, particularly behavioral abnormalities, such as apathy and fatuity, and reduced attention ability (Table 3).

In most cases, changes in behavior together with poor performances at school, lead the families to seek psychological evaluation. Visual disturbances, such as strabismus or impaired visual capacities were early-onset symptoms in 6 patients. In one patient intracranial hypertension was the symptom at onset.

Hypoadrenalism was present in 44 subjects; only few of them presented with hyperpigmentation and, in 3 cases, adrenal crisis was the first manifestation of the disease. Presymptomatic subjects were recognized because of the presence of an affected member in the family. Interestingly, 11 out of 16 presymptomatic patients (68%) had hypoadrenalism. At diagnosis, hyperpigmentation, thin hair and malaise were present in 54% of all patients.

Plasma VLCFA levels returned to the normal range values after 2 months of C:26-restricted diet + GTOE administration (Fig.1); VLCFA levels remained normal throughout the study period in all subjects (Fig.2).

revealed by NMR, after 6 months and 2 years of GTOE administration, respectively. One of them underwent bone marrow transplantation with success.

b) Childhood ALD: all 24 patients deteriorated clinically during the trial. Five of them died; the remaining precipitated into a vegetative state during the period of one to 2 years. Intravenous immunoglobulins were given in 4 patients without improvement. Two patients were submitted to bone marrow transplantation, but neurological symptoms deteriorated shortly after immunosuppressive therapy.

c) Adolescent ALD: all 4 patients showed worsening of neurological symptoms and/or widening of demyelinated areas at NMR. Intravenous immunoglobulins were given to one patient. Two patients died after 2 and 3 years of treatment respectively.

d) AMN and cerebral ALD: AMN patients showed no progression of the disease after a period of 8 months - 4 years of GTOE administration; 2 of them showed improvement of urinary continence. Of the 2 patients with cerebral ALD, one showed a very slow progression and the other a rapid deterioration of the neurological condition. Intravenous immunoglobulins, administered to the last patient were unsuccessful.

Table 4 shows Kurtzke score values in the different clinical groups studied.

Table 4. Clinical evaluation of the patients submitted to dietary therapy, before and at the last observation. Data expressed in mean±SD. Kurtzke score (0=neurologically normal; 10= dead).

Phenotype	Months of treatment	Kurtzke score Before	Kurtzke score last observed
Childhood ALD	25.6±18.6	5.0±2.4	8.3±2.2
Adolesc. ALD	40.5±22.4	5.0±3.5	7.4±3.3
AMN	25.8±13.2	5.2±1.7	5.0±1.6
Presymtomatic[1]	24.6±18.3	0	0.3±0.9
Cerebral ALD[2]	8&46	6&4.5	6.5&9.5

[1]Two out 11 presymptomatic patients developed neurological symptoms during the therapeutical trial

[2]The data of the two patients shown separately.

According to the recently published data[22], GTOE administration resulted in a mild thrombocytopenia in 50% of patients. In 4 patients, total bilirubin levels were mildly but persistently elevated. In one patient, AST and ALT serum levels were elevated at the 6th month of the trial and remained above the normal ranges after 3 years. A mild and transitory elevation of AST was also observed in 2 more patients.

No toxic cardiac effect of erucic acid was observed. The cardiological exam, ECG and the echocardiographic exam performed each year in all patients were normal.

Hypoadrenalism was diagnosed in the presence of low basal plasma cortisol levels or of insufficient increase of cortisol following ACTH testing.

The neuropsychological evaluation was not performed in all subjects because we did not obtain enough collaboration and attention during the tests. Some patients with their families refused neuropsychological examination because they were worried about monitoring the progression of the disease. No specific cortical deficits were observed in our patients by NE. We observed abnormalities in attentive capacities with disturbances of cortical function organization. NE was particularly important in presymptomatic patients in the perspective of BMT, before the onset of clear neurological symptoms[19].

DISCUSSION

Therapeutical approaches are particularly difficult in ALD due to the still unknown mechanisms leading to CNS damage; moreover the great variability of clinical forms (i.e. different age of presentation, forms with slow progression of symptoms, and asymptomatic subjects) makes it difficult to verify in the short term the efficacy of treatment.

Our study confirms that C:26-restricted diet with GTOE supplementation is able to normalize plasma VLCFA levels in ALD patients[13,14]. This result, however, is not followed by a regression or stabilization of CNS demyelination, suggesting that other mechanisms, such as immune factors, might have a role in the pathogenesis of ALD. Therapy is not effective in the most rapid progressive form of ALD, such as childhood ALD, but even in the adolescent and cerebral forms, no substantial beneficial effect was noted. Moreover, the finding that 2 presymptomatic subjects on a GTOE + VLCFA-restricted diet manifested symptoms of the disease during the follow-up, excludes the fact that GTOE might have a role in preventing CNS demyelination in presymptomatic subjects.

This result appears particularly discouraging considering that early diagnosis among the family members of an index case, as well as prenatal diagnosis, are now readily available in specialized centers.

On the other hand, AMN patients taking GTOE showed, together with arrest of neurological deterioration, a beneficial effect on subjective symptomatology; in 2 AMN patients urinary continence improved.

Differently from CNS involvement, adrenal function appears strictly related to plasma VLCFA levels. Our data show that normalization of plasma VLCFA levels is followed by a tendency of PRA and aldosterone to return to normal range values.

In conclusion, the results of our dietary trial in ALD patients may be summarized as following:

a) No treatment is actually available for ALD patients presenting with advanced neurological damage. GTOE administration results in normalization of plasma VLCFA levels but does not change the clinical course of the disease;

b) Patients with initial neurological involvement and presymptomatic subjects might benefit from GTOE administration, slowing the progression of symptoms; however GTOE does not seem to have the ability to arrest the disease, as shown by our 2 presymptomatic subjects who manifested initial signs of disease during the trial;

c) In some selected cases, bone marrow transplantation has been shown to arrest the evolution of ALD, due to "metabolic enzyme replacement" (acquired capacity of VLCFA oxidation *in vitro*);

d) Intravenous immunoglobulin administration in ALD patients on a VLCFA-restricted diet + GTOE was unsuccessful in our experience;

e) Recent advances in genetics of ALD, will be, hopefully, followed by more successful and definitive therapeutic solutions for ALD patients.

REFERENCES

1. Tager JM,Brul S,Wiemer EAC et al.Peroxisomal Disorders: An updating. "Adrenoleukodystrophy and Other Peroxisomal Disorders". pp 3-15. Eds Uziel G, Wanders RJA & Cappa M. Amsterdam:Excerpta Medica.Elsevier Science Publisher (1990)
2. Mosser J, Douar AM, Sarde CO, K P, Fell R, Moser H, Poustka AM, Mandel JL and Aubourg P. Putative X-linked adrenoleukodystrophy gene shares unexpectd homology with ABC transpor ters. Nature 361: 726-730(1993).
3. Aubourg P, Feil R, Rocchiccioli & Mandel L.Genetics of Adrenoleukodystrophy. "Adrenoleukodystrophy

and Other Peroxisomal Disorders" pp 61-63. Eds Uziel G, Wanders RJA,& Cappa M. Amsterdam: Excerpta Medica Elsevier Science Publisher(1990).

4. Moser HW, Moser AB, Naidu S, Segal AH et al. X-linked Adrenoleukodystrophy: Epidemiology, Pathogenesis and Therapy. "Adrenoleukodystrophy and Other Peroxisomal Disorders" pp127-148.Uziel G, Wanders RJA & Cappa M. Amsterdam: Excerpta Medica Elsevier Science Publisher (1990).
5. Moser HW, Naidu S, Kumar AJ, Rosenbaum AE, The Adrenoleukodystrophies. "CRC Critical Reviews in Neurobiology" 3 29-88 (1987).
6. Withcomb RW, Linehan WM & Knazek RA, Effect of Long Chain saturated Fatty Acids on Membrane Microviscosity and Adrenocorticotropin Responsiveness of Human Adrenocortical Cells *in Vitro*. J Clin Invest 81 185-188, (1988).
7. Ogino T, Suzuki K Specifities of Human and Rat Brain Enzymes of Cholesterol Ester Metabolism Toward Very Long Chain Fatty Acids: Implication of Biochemical Pathogenesis of Adrenoleukodystrophy. J Neurochem 36:776-779(1981).
8. Sadeghi-Nejad A & Senior B, Adrenomyeloneuropathy Presenting as Addison Disease in Childhood N Engl J Med 322 13-16(1990).
9. Van Duyn MA, Moser AB, Brown FR, Sacktor N, Liv A, Moser HW, The Design of a Diet Restricted in Saturated Very Long Chain Fatty Acids: Therapeutic Application in Adrenoleukodystrophy Am J Clin Nutr 40: 277-282(1984).
10. Rizzo WB, Phillips MW, Dammann AL et al. Adrenoleukodystrophy: Dietary Oleic Acid Lowers Hexacosanoate Levels. Ann Neurol 21:232-239 (1987).
11. Moser AB, Borel J, Odone A et al. A new dietary therapy for Adrenoleukodystrophy Ann Neurol 21: 240-249 (1987).
12. Rizzo WB, Watkins PA, Phillips MW et al, Adrenoleukodystrophy: Oleic Acid Lowers Fibroblast-saturated C22:26 Fatty Acids. Neurology 36: 357-361 (1986).
13. Rizzo WB, Leshner RT, Odone A et al Dietary Erucic Therapy for X-linked Adrenoleukodystrophy. Neurology 39: 1415-1422 (1989).
14. Uziel G, Bertini E, Rimoldi M, Gambetti M: Italian Multicentric Dietary Therapeutical Trial "Adrenoleukodystrophy. Adrenoleukodystrophy and Other Peroxisomal Disorders" pp163-180 Eds Uziel G, Wanders RJA & Cappa M. Amsterdam Excerpta Medica. Elsevier Science Publisher B.V. (1990).
15. Gambarara M, del Balzo P, Cambiaso P, Bernabei S, Colombo AM, Sabetta G & Cappa M, Therapeutical Approaches of ALD: Problems with Diet in Italy. "Adrenoleukodystrophy and Other Peroxisomal Disorders pp 181-193 Eds Uziel G, Wanders RJA & Cappa M. Amsterdam: Excerpta Medica Elsevier Science Publisher (1990).
16. Miike T, Taku K, Tamura T, Ohta J, Ozaki M, Yamamoto C, Sakai T, Antoku Y, Yadomi C, Clinical Improvement of Adrenoleukodystrophy Following Intravenous Gammaglobulin Therapy. Brain Dev 11: 134-137(1989).
17. Rodriguez M and Lennon VA Immunoglobulin Promote Remyelination in The Central Nervous System. Ann Neurol 27: 12-17(1990).
18. Moser HW, Tutschka PJ, Brown FR III et al, Bone Marrow Transplant in Adrenoleukodystrophy. Neurol 34:1410-1417(1984).
19. Aubourg P, Blanche S, Jambaque I et al, Reversal of Early Neurologic and Neuroradiologic Manifestations of X-linked Adrenoleukodystrophy by Bone Marrow Transplantation. N Eng J Med 322:1860-1866 (1990).
20. Del Balzo P, Borrelli P, Cambiaso P, Danielli E, Cappa M, Adrenal Steroidogenenic Defects in Children with Precocious Pubarche, Horm Res 37: 180-184 (1992).
21. Kurtzke JF Rating Neurologic Impairement in Multiple Sclerosis: an Expanded Disability Status Scale (EDDS). Neurology (Cleveland)33: 1444-1452 (1983).
22. Zinckham WH, Klicker T, Borel J, Moser HW, Lorenzo's Oil and Thrombocytopenia in Patients with Adrenoleukodystrophy N Engl J Med 328:1126-1127 (1993).

ROUND TABLE: CLINICAL ASPECTS OF NEW DISCOVERIES GENETIC MARKERS AND THEIR CLINICAL AND THERAPEUTIC IMPLICATIONS

Michel Clanet,[1] Marie Paule Roth,[2] Eric Champagne,[2]
Anne Cambon-Thomsen[2]

[1]Service de Neurologie - CHU Toulouse Purpan - 31059 Toulouse Cedex
[2]C.R.P.G. - CNRS - UPR 8291 - CHU Toulouse Purpan - 31059 Toulouse Cedex

In the last decade many reports have confirmed the old assumption of a "genetic susceptibility to multiple sclerosis". These studies have been investigated in three orientations:

→ genetic analysis of pedigrees and study of concordance for MS in twins: the main results are a weak genetic susceptibility with a probable multigenic influence.

→ association studies with candidate genes and linkage analysis in multiplex families: among these genetic markers, the HLA genes are assumed to be of the greatest importance, particularly in Caucasoid patients, the alleles DRB*1501, DQA1* 0102 and DQB1*0602.

→ by analogy with Experimental Allergic Encephalomyelitis (EAE) molecular genetic studies were undertaken in an attempt to define the functional role of these markers in the pathophysiology of the disease more specifically the immunogenic response of T lymphocytes against Myelin Basic Protein (MBP).

The evolution of the concepts in understanding the role of genetic factors in MS is similar in some others auto-immune diseases like rheumatoid arthritis, insulin dependent diabetes, disseminated lupus etc... These new concepts have led to new therapeutic strategies in experimental diseases and, in a few cases, in human diseases. The purpose of this issue is to discuss the incidence of these new developments on the clinical aspects of MS and on tentative treatment.

GENETIC STUDIES IN MULTIPLEX FAMILIES AND TWINS

The analysis of the concordance for a disease between monozygotic and dizygotic twins allows an evaluation to be made of the relative importance of genetic and environmental factors in the etiology of a disease. The two last major reports in twin studies are controversial. In the Canadian sample of 70 twin pairs, the concordance was 25.9 % in

MZ pairs compared to 2.3 % in DZ pairs (1). The sample of French twins was 54: the concordance was 5.9 % in MZ pairs, not different to the 2.7 % of DZ pairs. This rate of concordance was not significantly different when taking account of MRI abnormalities in clinically unaffected twins (2). These discrepancies may suggest methodological differences in the recruitment of MS twins, an incorrect estimation of the concordance rate or a geographic influence on the genetic susceptibility. A bias in the recruitment of French twin pairs in the MS population would have overestimated the concordant sets. The small number of twin pairs in all the twin studies may result in an incorrect estimation. However, most of the previous studies, performed in areas of higher prevalence than in France, found a higher concordance rate in MZ twins. Moreover, in the French study, the concordance rate did not differ in unaffected MZ and DZ pairs with MRI abnormalities. It appears that the genetic influence is stronger in areas of high prevalence of MS. This is in agreement with other observations, like a stronger association of MS and HLA-DR2 in high prevalence areas, or more multiplex families in these zones. So, environmental factors seem to be essential to allow the expression of the genetic susceptibility with different genetic loci (3). The clinical aspects of MS in familial and sporadic cases are not different. However the clinical manifestations are not correlated among relatives (4). This observation implies that the prognosis may be different in affected patients in multiplex families. Moreover, a correlation in the clinical manifestations in multiplex families must be considered as a feature of misdiagnosis. Recent experience of clinicians collecting familial MS cases has underlined the need for more accurate diagnosis criteria than those of Poser (5). The usefulness of these criteria is demonstrated daily in the neurological assessment of MS patients.

GENETIC SUSCEPTIBILITY AND HLA MARKERS

The association between MS and HLA markers was described more than twenty years ago in Caucasoid populations, HLA-A3 and HLA-B7 antigens were found to be more frequent in patients than in controls.

The most recent data have confirmed that the susceptibility is confered by the haplotype HLA-DRw15 - DQw6- Dw2, in the genomic nomenclature HLA-DRB1* 1501 - DQA1* 0102 - DQB1* 0602 (6). It is not possible to settle the issue as to whether MS is more closely associated with either HLA-DR or HLA-DQ genes (7). There is no association with HLA-DP genes (8). The HLA class II alleles conferring the susceptibility are identical to those commonly expressed (9). Heterogeneity of remitting relapsing and primary progressive MS was suspected from the clinical and MRI background. An immunogenetic heterogeneity has also been found: in some populations the DR2-DQW6-DW2 haplotype was strongly associated with both forms but the haplotype DRW17(3)-DQW2 was more common in relapsing-remitting MS patients than in primary progressive MS (10). Any clinical trial involving patients in chronic progressive phase must clearly differentiate primary and secondary progressive MS.

The significance of confirming a weak association between MS and the HLA-DR-HLA-DQ allelles is always speculative:

- the hypothesis that HLA-DRw15 - HLA-DQw6 are only genetic markers cannot be completely excluded: there are many new non-class I and II genes within the MHC. RING genes that have been described recently: some of them, like RING 4 gene lies within the class II region (11). The study of their polymorphism may be of interest in MS.

- HLA markers may be involved by their functional role in the immune response by their ability to bind and present antigens to the T helper cell. In a Norwegian study of 61 MS patients, a comparison of the DQ ß chains linked with the HLA-DR antigens associated with MS showed shared A.A. sequences in the membrane distal domain of the molecule (12). Other reports did not confirm these previous findings (6). These shared sequences could trigger an aggressive immune reaction by presenting the same putative antigen (autoantigen?). Another possibility is that polymorphisms in gene regulatory sequences may favour class II overexpression in the central nervous system leading to an organ-specific autoimmune disease. Transgenic mice that express class I H-2K MHC molecules under the control of the myelin basic promoter specifically in the oligodendrocytes, developed hypomyelination in the CNS without involvement of the immune system. The question remained whether this demyelination resulted from the overexpression of the class I MHC antigens or interference with the production of MBP following the integration of the transgene (13).

In all these scenarios the development of strategies blocking the MHC class II molecules might be useful in the treatment of the disease. Experimentally, this blockade has been achieved with antibodies or with binding peptides preventing T cell activation (14).

- HLA molecules can act by shaping the repertoire of T cell receptor specificities that occur during thymic maturation of T cells. This assumption could be hypothesized if the susceptibility was associated more with HLA-DQ than with HLA-DR because it has been suggested that thymic repertoire selection may be the major role of HLA-DQ molecules (15).

HLA class II molecules can also influence the repertoire via non-inherited maternal HLA class II antigens. In rheumatoid arthritis it has been recently found that non-inherited maternal HLA-DR4 was more frequent in the mothers of DR4 negative patients than in controls (16). The mechanism by which an inherited maternal antigen may exert its influence is only speculative. No such study in MS has yet been published.

GENETIC SUSCEPTIBILITY AND TCR GENES

There is no conclusive hypothesis as to the role of TC receptor genes in inherited disease susceptibility.

In seeking to establish an association between TCR α chains and MS, some reports implicated specific alleles or haplotypes in MS risk, some did not (17,18,19). No linkage between the TCR α locus and risk of MS was found in multiplex families, either in US familial cases (20) or Canadian familial cases (19).

In the first instances, TCR ß genes were better candidates because association and linkage studies showed some polymorphisms at risk for MS, and inheritance of TCR ß haplotypes segregated in a non-random fashion (21,22). Further studies, however, failed to confirm these results (23,24).

These controversial findings cannot completely rule out that TCR genes are linked to MS. Discrepancies can be explained by technical differences in RFLP analysis of genomic DNA, methodological approaches used in studying the familial data, or complex biological problems inherent in these genes, like a high rate of crossing-over (25).

The development of new therapeutic strategies using TCR peptide therapy has become an exciting challenge arising from the analogy between EAE and MS rather than from a putative role of these TCR genes in conferring susceptibility. The following arguments may be mentioned (26)

- the number of MBP specific T cells in the blood and CSF of MS patients was found higher than in healthy individuals or controls.

- the MPB specificity of T cells was mainly directed against an immunodominant epitope in the middle of the molecule.

- human MBP-specific T cells were HLA restricted, HLA-DR2 being able to present several MBP epitopes, but TCR usage by MBP specific T cells derived from the blood of MS patients was more or less heterogeneous in the different studies.

- a restricted TCR ß chain usage was found in the brain of some MS patients: Vß 5.2, and to a lesser extent Vß6.1 genes in association with an ARG-GLY CDR3 motif were derived from MS CNS plaques. These same Vß genes were found to be overexpressed in the peripheral MBP response. These findings have led to phase 1 therapy of 11 MS patients with Vß5.2-39-59 and Vß6.1-39-59 (27).

However, even if the mechanisms of EAE are operative in MS, a variety of immune responses directed against different antigens are more likely than a single response against a dominant epitope. In humans, the immune reactions are considerably more diverse than in inbred rodents. Even in inbred mice, both MBP and PLP are encephalitogenic, each protein containing two or more encephalitogenic epitopes. In individual patients, only one epitope, different among MS cases, could induce the disease. This autoimmune reaction could also be triggered by multiple epitopes. In this case, experimental vaccination procedures with T cell receptor peptide would be less relevant.

GENETIC SUSCEPTIBILITY AND OTHER CANDIDATE GENES

Although polygenic inheritance of MS risk can be assumed to occur, other genetic loci are also involved. Some of them have been investigated: allotypic markers of immunoglobulins, TNF α and MBP genes.

Some associations were observed between MS and Gm allotypes of immunoglobulin heavy chains (28,29). Genomic RFLP analysis showed discordant results perhaps depending on the geographical distribution of the markers (30,31).

A putative role for TNF in the pathogenic mechanisms of MS was suggested by the findings of high levels of this cytokine in blood and CSF of active patients (32,33). TNFs were also observed in the active lesions in MS brains (34,35). TNFs secretion was found to be related to the HLA phenotype: monocytes from HLA DR2 individuals synthesized less TNF than cells from HLA DR3 or DR4 individuals (36). TNF A and TNF B genes are located between class I and class III genes and are polymorphic (37). One TNF haplotype was in linkage disequilibrium with HLA DR2 but did not seem to be an independent risk factor (38).

A multiallelic tetranucleotide repeat polymorphism identified as being 5' to the MBP gene on chromosome 18 is associated with MS (39). A linkage analysis performed recently in Finnish multiplex families suggested that a predisposition to MS was linked to the MBP gene (40). However these results were significant only when symptom-free subjects with abnormal MRI were included as affected. Confirmation of these data may imply that an isoform of MBP may confer a great vulnerability to the development of the disease or its expression in some specific cells by dynamic mutations.

CONCLUSIONS

Genetic susceptibility to Multiple Sclerosis is certain but weak, environmental factors playing a major role in inducing the disease. The susceptibility is dependent on multigenic loci, among which only few have been identified with certainty. Because of their functional role, the genes modulating the immune responses have been excessively studied: HLA genes only are firm disease markers and this relationship is not clearly understood. None of the speculations about the implication of an aggressive autoimmune reaction mediated by T cells are derived from any association between MS and T cell receptor polymorphisms but from the optimistic hypothesis that Multiple Sclerosis is a human Experimental Allergic Encephalomyelitis.

A large scale analysis of the genetic markers in Multiple Sclerosis will contribute to define putative subgroups of patients with different associations to a variety of candidate genes leading to the analysis of their interactions. A strategy to achieve this goal may be to look for new candidate genes from experimental diseases or fundamental findings, to determine a possible association with MS in a large population of sporadic patients and confirm their role in linkage analysis in well defined multiplex families (41). This identification will lead to a better understanding of the pathogenic mechanisms in the hope of new therapeutic interventions.

REFERENCES

1. G.C.Ebers, D.E.Bulman, A.D.Sadovnick et al, A population-based study of multiple sclerosis in twins, N Engl J Med.315: 1638 (1986)
2. French Research Group on Multiple Sclerosis, Multiple Sclerosis in 54 twinships: concordance rate is independant of zygosity, Ann Neurol.32:724 (1992)
3. H.F. McFarland, Twin studies in studies in multiple sclerosis, Ann. Neurol. 32: 722 (1992).
4. A.D.Sadovnick, L.L. Hashimoto, and S.A. Hashimoto, Heterogeneity in multiple sclerosis: comparison of clinical manifestations in relatives, Can. J. Neurol. 17: 387 (1990).
5. D.E. Goodkin, T.H. Doolittle, S.S. Hauser et al., Diagnostic criteria for multiple sclerosis research involving multiple affected families, Arch. Neurol. 48: 815 (1991).
6. O. Olerup, J. Hillert, HLA class II associated genetic susceptibility in multiple sclerosis : A critical evaluation, Tissues Antigens 38: 1 (1991).
7. A. Spurkland, K. Ronningen, B. Vandvik, E. Thorsby, F. Vardtal, HLA-DQA1 and HLA-DQB1 genes may jointly determine susceptibility to multiple sclerosis, Hum. Immunol. 30: 69 (1991).
8. M.P. Roth, H. Coppin, P. Descoins, J.B. Ruidavets, A. Cambon-Thomsen and M. Clanet, HLA-DPB1 gene polymorphism and MS : a large case control study in the northwest of France, J. Neuroimmunol. 34 : 215 (1991).
9. E.P. Cowan, M.L. Pierce, F.H. McFarland and D.E. McFarlin, HLA-DR and DQ allelic sequences in MS patients are identical to those found in the general population. Hum. Immunol.32: 203 (1991).
10. J. Hillert, M. Gronning, H. Nyland, H. Link, O. Olerup, An immunogenetic heterogeneity in multiple sclerosis, J. Neurol. Neurosurg. Psychiatry 55: 887 (1992).
11. A. Kelly, S.H. Powis, R. Glynne, E. Radley, S. Beck and J. Trowsdale, Second proteasome related gene in the human MHC class II region, Nature 353: 667 (1991).
12. F. Vartdal, L. Sollid, B. Vandvik, G. Markussen, E. Thorsby, Patients with MS carry DQ -B1 genes which encode shared polymorphic aminoacid sequences, Hum. Immunol. 25: 103 (1989).
13. A.M. Turnley, G. McRahan, H. Okano, O. Bernard, M. Katsuhiko et al. Dysmyelination in transgenic mice resulting from expression of class I histocompatibility molecules in oligodendrocytes, Nature 353: 566 (1991).
14. L. Adorini, J.C. Guery and S. Trembleau, Approaches toward peptide based immunotherapy of autoimmune diseases, Springer Semin. Immunopathol. 14 : 187 (1992).
15. D.M. Altmann, D. Sansom and S.G. Marsh, What is the basis for HLA-DQ associations with autoimmune disease ? Immunol. Today 12: 267 (1991).
16. S. Ten Wolde, F.C. Breedveld, R.R. P. Devries, J. D'Amaro, P. Rubinstein et al., Influence of non inherited material HLA antigens on occurence of rheumatoid arthritis, Lancet 341: 200 (1993).
17. J.R. Oksenberg, M. Sherrit, A.B. Begovich, H. Erlich, C.C. Bernard et al., T cell receptor Vα and Cα alleles associated with MS and myasthenia gravis, Proc. Nat. Acad. Sci USA 86: 988 (1989).
18. J. Hillert, C. Leng, O. Olerup, T cell receptor α chain germline gene polymorphisms in multiple sclerosis, Neurology 42: 80 (1992).
19. L.L. Hashimoto, T.W. Mak and G.C. Ebers, T cell receptor α chain polymorphisms in multiple sclerosis, J. Neuroimmunol 40 : 41 (1992).

20. S.G. Lynch, J.W. Rose, J.H. Petajan, M. Leppert, Discordance of the T cell receptor alpha chain in familial multiple sclerosis, Neurology 42: 839 (1992).
21. S.S. Beall, P. Concannon, P. Charmley, H. McFarland, R. Gatti et al., The germline repertoire of T cell receptor ß chain genes in patients with chronic progressive multiple sclerosis, J. Neuroimmunol 21: 59 (1989).
22. E. Seboun, M. Robinson, A. Doulille, T.A. Ciulla, T. Kindt, S.L. Hauser, A susceptibility locus for multiple sclerosis is linked to the T cell receptor ß chain complex, Cell 57: 1095 (1989).
23. J. Hillert, C. Leng and O. Olerup, No association with germline T cell receptor ß chain gene alleles or haplotypes in Swedish patients with multiple sclerosis, J. Neuroimmunol. 31: 141 (1991).
24. S.G. Lynch, J.W. Rose, J.H. Petajan, D. Stauffer, C. Kamerath, M. Leppert, Discordance of T cell receptor ß chain genes in familial multiple sclerosis, Ann. Neurol. 30: 402 (1991).
25. R.M. Ransohoff, T cell receptor germline genes and multiple sclerosis susceptibility, Neurology 42: 714 (1992).
26. R. Martin, H.F. McFarland, D.E. McFarlin, Immunological aspects of demyelinating diseases, Ann. Rev. Immunol. 10: 153 (1992).
27. H. Offner, G.A. Hashim, A.A. Vandenbark, T cell receptor peptide therapy in EAE and MS (1993).
28. M. Blanc, M. Clanet, C. Berr, J.M. Dugoujon, J.B. Ruidavets and A. Alperovitch, Immunoglobulin allotypes and susceptibility to MS, J. Neurol. Sci. 75: 1 (1986).
29. J.P. Salier, R. Sesboue, C. Martin Mondière, M. Daveau, P. Cesaro et al., Combined influences of Gm and HLA phenotypes upon multiple sclerosis susceptibility and severity, J. Clin. Invest. 78: 533 (1986).
30. J.S. Yu, J.P. Pandey, L. Massaceci, R. Lincoln, K. Osuku, E. Seboun, S. Hauser, Segregation of immunoglobulin heavy chain constant region genes in multiple sclerosis sibling pairs, J. Neuroimmunol. 42: 113 (1993).
31. M.A. Walter, W.T. Gibson, G.C. Ebers and D.W. Cox, Susceptibility to multiple sclerosis is associated with the proximal immunoglobulin heavy chain variable region, J. Clin. Invest. 87: 1266 (1991).
32. S.L. Hauser, T.H. Doolittle, R Lincoln, R.H. Brown, C.A. Dinarello, Cytokine accumulation in CSF of multiple sclerosis patients : frequent detection of IL1 and TNF but not IL6, Neurology 40: 1735 (1990).
33. M.K. Sharief, M. Phil and R. Hentges, Association between TNFα and disease progression in patients with MS, N. Engl. J. Med 325: 467 (1991).
34. F.M. Hofman, D.R. Hinton, K. Johnson and J.E. Merrill, TNF indentified in Multiple undentified in multiple sclerosis brain, J. Exp. Med 170: 607 (1989).
35. K. Selmaj, C.S. Raine, B. Canella and CF Brosnan, Identification of lymphotoxin and TNF in multiple sclerosis lesions, J. Clin. Invest. 87: 942 (1991).
36. O. Chaim, Z. Fronek, G.D. Lewis, M. Koo, J.A. Hansen and M.O. McDevitt, Heritable major histocompatibility complex class II associated differences in production of TNF α: relevance to genetic predisposition to systemic lupus erythematosis, Proc. Nat. Acad. Sci. USA 87: 1233 (1990).
37. C.V. Jongeneel, L. Briant, I.A. Udalova, A. Sevin, S.A. Nedospasov, and A. Cambon-Thomsen, Extensive genetic polymorphism in the human TNF region and relation to extended HLA haplotypes, Proc. Nat. Acad. Sci. USA 88: 9717 (1991).
38. L. Nogueira, M.P. Roth, C. Demangel, M. Clanet, A. Cambon-Thomsen. Polymorphism analysis of TNF genes regions in multiple sclerosis (Submitted)
39. K.V. Boylan, N. Takahashi, D. Paty, A.D. Sadovnick, M. Diamond, L.E. Hood and S.B. Prusiner, DNA polymorphisms to the myelin basic protein gene is associated with multiple sclerosis. Ann. Neurol. 27: 291 (1990).
40. P.J. Tienari, J. Wikstrom, A. Sajantila, J. Palo, L. Peltonen, Genetic susceptibility to multiple sclerosis linked to MBP gene, Lancet 340: 987 (1992).
41. F. Clerget-Darpoux, C. Bonaiti-Pellie, J. Feingold, Marqueurs genetiques et maladies multifactorielle, Quelle stratégie adopter ? Med. Sciences 10: 1065 (1992).

THE ROLE OF THE BLOOD-BRAIN BARRIER IN THE PATHOGENESIS OF MULTIPLE SCLEROSIS

Charles M. Poser

Department of Neurology, Harvard Medical School, and
Neurological Unit, Beth Israel Hospital, Boston MA USA
Address for correspondence: CMP: Neurological Unit,
Beth Israel Hospital, 330 Brookline Avenue, Boston MA 02215
USA FAX #617-735-5216

INTRODUCTION

A relationship between the blood-brain barrier (BBB) and the pathogenesis of multiple sclerosis (MS) has been suspected ever since Dawson[1] demonstrated that MS plaques were located close to small cerebral blood vessels. Broman[2] then proved that a loss of impermeability of the BBB did indeed exist at the site of the plaques, but it took almost 50 years to determine if this alteration of the BBB was the result of the destruction of myelin or was an important step leading to lesion formation. The availability of neuroimaging technics, first computer assisted tomography (CT) aided by iodine-containing contrast media, and more recently and more importantly, magnetic resonance imaging (MRI) with gadolinium enhancement, has added considerably to our understanding of the crucial role played by this structure in the development of MS. This means that some questions remain unanswered and must await further investigation.

In previous articles[3,4] regarding the pathogenesis of MS I proposed two new concepts: first that an MS "trait" existed as a systemic condition, characterized by a state of hyperactive immunocompetent responsiveness, and second that an alteration of the BBB was an obligatory step in the formation of the MS lesion, which consists of edema and inflammation that can but does not always lead to demyelination. I emphasized that this change in BBB permeability results in a greater vulnerability, is inherent in the MS trait, and could be increased by a number of factors including immune responses, trauma, electrical injury, and other events which can be considered as facilitators. Numerous publications have since provided further support for the view that the BBB plays a crucial role in the pathogenesis of MS. In addition, recent data have suggested that the persistent alteration of BBB permeability may play a role in the progression of the illness.

A Multidisciplinary Approach To Myelin Diseases II
Edited by S. Salvati, Plenum Press, New York, 1994

THE NATURE OF THE MS TRAIT

In order to appreciate the importance of the alteration of the BBB it is necessary to define the MS trait and to show that one of its features is an increased vulnerability of the BBB. The MS trait is defined as a permanent state of hyperactive or intensified immunocompetent responsiveness or capability, which is triggered in the genetically susceptible individual by exposure to a non-specific antigen that is almost certainly of viral origin, either an acute viral infection or a vaccination.

The presence of the MS trait simply means that a systemic condition, not itself involving the nervous system, exists that predisposes a person to MS, a disease of the central nervous system (CNS), provided a further event takes place. Classic examples of traits are acute intermittent porphyria, sickling, and glucose-6-phosphate dehydrogenase deficiency. Not all the components of the MS trait have been defined. One of them appears to be a very vigorous antibody response to a great variety of viral, and possibly other, antigens. This was originally demonstrated by Brody *et al.*[5] and confirmed by several other authors.[6, 7,8] Another probable component of the MS trait consists of the presence of immunoglobulin G (IgG) oligoclonal bands (OCB), in the CSF of unaffected siblings of MS patients, such as non-concordant twins.[9,10] B-lymphocyte and plasma cell secretion of IgG may be an expression of the increased but non-specific immunocompetent responsiveness in the MS trait.

An additional and most important aspect of this exaggerated immune response is an inflammatory, primarily lymphocytic, infiltration of small blood vessels, mostly venules and capillaries, that does not lead to the development of white matter alterations. This was clearly demonstrated by Adams *et al.*[11] and further illustrated by Gay and Esiri,[12] who showed that perivascular infiltrates found in normal white matter of MS patients which involve the blood-vessel wall itself, produce a very minor alteration of the BBB that cannot be demonstrated by gadolinium enhancement of MRI. These vascular changes may be the pathway for B-lymphocytes to penetrate into the CNS, where they produce OCB.

THE ROLE OF THE IMMUNE SYSTEM IN THE ALTERATION OF THE BBB

A major unresolved and quite controversial problem is the role played by the immune system, in particular the T-lymphocytes, in the pathogenesis of MS and more specifically their role in crossing the BBB. It has been known for a long time that activated T-cells have such ability[13] although the source and nature of this activation *prior to myelin destruction* has never been satisfactorily explained. Two possible explanations may be offered for the activation of T-lymphocytes by myelin components. The first one is that it is a manifestation of the phenomenon of molecular mimicry. Alvord[14] and Jahnke *et al.*[15] have demonstrated that several viruses share protein amino acid sequences with myelin basic protein and therefore a person who has the MS trait (as well as normal people) may respond to those antigens. Another example of possibly relevant molecular mimicry is the report of protein sequence similarity between decapeptides from rubella and measles virus and human proteolipid protein.[16] A second possible explanation is that the obligatory opening of the BBB and the destruction of myelin releases into the peripheral circulation breakdown products of myelin which then start T-cell activation. The origin of the substances from destroyed myelin was first mentioned by Oger *et al.*[17] and also by Zamvil and Steinman[18]. Olsson *et al.*[19] pointed out that they had observed autoreactive T-cells as a result of BBB breakdown in acute cerebrovascular disease.

If indeed T-cells play a role in MS pathogenesis, as is firmly believed by many investigators, then the CNS-to-blood traffic through the opened BBB would provide a constant source of antigenically activating substances for T-cells, but, obviously, only *after* some myelinoclasia had occurred.

WHAT IS THE TARGET ORGAN IN THE PATHOGENESIS OF MS?

There is consensus that MS is an inflammatory disease and that the primary change in the blood-vessel wall that leads to a possible or eventual alteration of the BBB is an inflammatory one. The exact mechanism by which this occurs remains unknown, although a number of suggestions have been made.

It is likely that the most common cause of an inflammatory reaction involving the blood vessel is a viral infection. This has been demonstrated in canine distemper virus infection of the brain, in which a perivascular reaction and inflammatory changes in the vessel wall were shown to be identical to those seen in MS[20]. This can also occur as the result of a vaccination, just as is seen in EAE[21]. Reik[22] proposed that the inflammatory reaction is set off by the deposition of immune complexes, and Coyle *et al.* [23] and others felt that this was also the causative mechanism in MS. Gay and Esiri[12] demonstrated the presence of immune complexes in the perivascular spaces of acute MS lesions, but cautioned against assuming that finding immune complexes in lesions proves that they bring about the damage. None the less, a primary pathologic role for the complexes is suggested by their exclusive association with acute plaques and their deposition with the HLA-DR+ive macrophages in the advancing borders of small hyperactive lesions.

The phenomenon of molecular mimicry [14,15] may also play an extremely important role in making the MS trait develop into the actual CNS disease. Because excess antibodies against certain viral proteins are present in the blood of the MS trait carrier, an infection by a second virus, or vaccination with viral proteins which share epitopes with the antigen(s) that originally produced the trait, may result in a further increase in either or both humoral or cellular immunoresponsiveness and the formation of immune complexes.

Hickey[24] and others have proposed that the inflammatory reaction is caused by activated lymphocytes and macrophages, but without stipulating an exact mechanism. It has already been pointed out that an inflammatory reaction of blood vessels resulting from an initial viral infection or vaccination may be considered to be part of the MS trait and not necessarily lead to an alteration of the BBB, at least one that is sufficient to affect myelin[12] or to cause MRI changes[25].

Before discussing the obligatory alteration of the BBB impermeability in the formation of the MS lesion, some consideration must be given to examining what is the target organ in MS. There are several possibilities that are not necessarily mutually exclusive. They include the oligodendroglial cell (ODC), the myelin sheath itself, and the BBB. Because the focus of this essay is the role of the BBB in MS, I will not discuss the pros and cons of the ODC and/or the myelin sheath itself as the targets of the disease. At this time neither of them has been eliminated from serious consideration.

THE BLOOD-BRAIN BARRIER AS THE TARGET ORGAN

In a recent review, Gay and Esiri[12] offered convincing evidence to support the view that the BBB, and in particular the blood vessels' endothelial lining, may be the target organ for MS. Tsukada *et al.*[26] were able to produce perivascular inflammatory, edematous, and demyelinating lesions very similar to those seen in EAE by injecting antibodies against the animals' endothelial cells. Hughes[27] said in his essay on the pathogenesis of MS: "evidence for the primary or a very early abnormality of leukocyte endothelial interaction in MS comes from three sources. Firstly, the inflammatory cell infiltrates commonly surround the blood vessels in the white matter of patients with MS even in the absence of demyelination. Secondly, retinal vasculitis has been discovered in 18% of 50 consecutive cases of MS and inflammation in the retina cannot be explained by a neighboring immune response against myelin since the retinal axons are not myelinated. Finally, the earliest radiologically detectable event in MRI studies is a lesion which enhances following the intravenous injection of gadolinium, indicating local breakdown of the BBB." Compston[28] offered a purely immunogenetic (but still unproved) alternative hypothesis: "The available evidence suggests the cell surface adhesion molecule expression alters when circulating T-lymphocytes and macrophages are activated, increasing their endothelial cell attachment; the secretion of cytokines and locally active enzymes leads to transendothelial passage, opening the barrier and bringing an array of potentially pathogenic inflammatory cells and mediators to the abluminal surface of blood vessels."

The key word in Compston's statement is "mediators," i.e., substances rather than cells. Lassman *et al.*[29] referred to a number of humoral agents such as complement, tumor necrosis factor, and gamma-interferon, which might conceivably play a role in a direct attack on myelin, or as effectors of "bystander demyelination" in MS. Gay and Esiri[12] commented that "the appearance of considerable protein leakage in normally myelinated areas around plaques suggests that leakage *per se* is not necessarily demyelinating. Our results argue against specific humoral immune-mediated damage to myelin in MS and support the idea that myelin may be damaged as an innocent bystander rather than as a specific target in the inflammatory reactions which develop around and within the walls of cerebral vessels." The concept of "bystander demyelination" is important because it is mentioned by a several experienced investigators of the question of what happens after the BBB loses its impermeability.

THE BBB IN THE PATHOGENESIS OF MS

It has been known since 1949[2] that an alteration of the BBB existed in MS plaques, but the concept that such a breakdown was a crucial and obligatory step in the pathogenesis of the CNS MS lesions was not proposed until 1986[3]. The background for this hypothesis has been reviewed in detail elsewhere[3,4]. Confirmation of the importance of this step has been made by several investigators. Compston[30] stated that "BBB penetration can be regarded as the primary disease process without which none of the events directly responsible for myelin injury would occur." Rudge[31] commented "that the disease (MS) is focal, characterized by demyelinating plaques, rather than diffuse in nature could be explained in terms of local breakdown of the BBB. It is clear that such a breakdown is an early, if not the first, step in plaque generation." Hickey[24] makes the statement that "It is obvious that any mechanism which physically destroys the components of the BBB will render the CNS open to the cellular and molecular constituents of the blood. This occurs in traumatic or surgical injury, infarction and hemorrhage. In such circumstances the required participants for inflammation are rapidly delivered to the site of injury in a gross, non-specific fashion." It is clear from these statements that breakdown of the BBB is obligatory, regardless of what mechanism then becomes activated, whether by immunocompetent cells or humoral factors or as a non-specific facilitator bystander phenomenon.

THE EFFECT OF TRAUMA ON THE BBB

Much of the relevant material has already been discussed elsewhere[3,4] and will not be repeated. The importance of trauma in some MS patients, *at some times,* is that it greatly facilitates the entry into the brain and/or spinal cord of whatever cellular or humoral substances will result in the formation of new lesions or the enlargement of already existing MS lesions. The fact that trauma even as mild as that causing concussion could cause anatomic alterations of the brain, i.e. recognizable lesions, had been pointed out as long ago as 1964 by Ommaya *et al.*[32], by Nevin[33], and again by Oppenheimer in 1968.[34] The classic review by Hardman[35] of the effect of trauma on the nervous system reinforces this point. The understanding that trauma could result in alterations of the BBB in man, which was already recognized by Bakay *et al.* in 1977[36], was supported by studies using the MRI.[37,38,39]. Several papers report that MRI abnormalities were found in both gray and white matter. Levin *et al.*[38] indicated that out of 50 of their patients who had suffered mild head injuries, lesions restricted to the subcortical white matter were present in nine. Figure 1 is an illustration that trauma can produce white matter lesions.

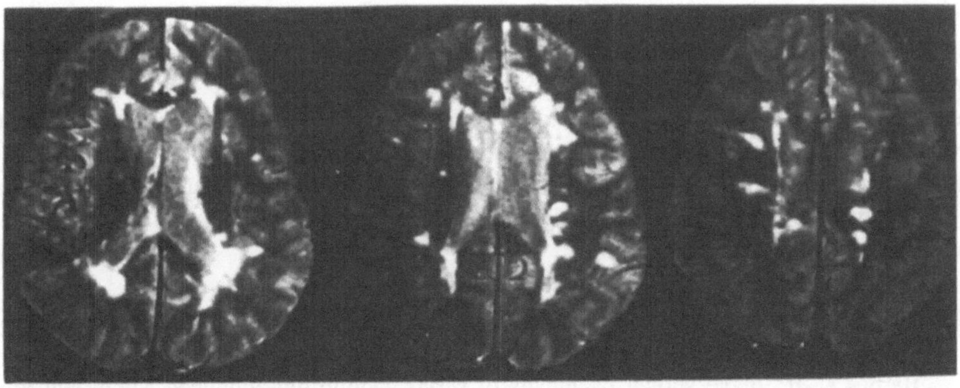

Fig. 1: T-2 weighted MRI demonstrating multiple white matter lesions in a 46-year-old man involved in a car accident with questionable loss of consciousness, complaining of anxiety attacks, and double and triple vision in the right eye only. Has bilateral Babinski signs but the rest of the neurological examination is normal. Note the enlarged left ventricle. Past history revealed another accident resulting in one week of coma 14 years previously. Reproduced from C. Poser: The unfortunate triumph of mechanodiagnosis is multiple sclerosis: a clinician's lament. Clin. Neurol. Neurosurg. 94 (suppl):S139 (1992), with permission from Prof. G. Bruyn, editor of Clinical Neurology and Neurosurgery.

Much of the animal experimental data showing that alterations of the BBB can be produced by mild concussional trauma has been reviewed[3,4]. Some additional material is worth mentioning. Maxwell *et.al.*[40] produced petechial hemorrhages, including areas of disruption of endothelial cells, in the central white matter of baboons subjected to lateral head acceleration. Ultrastructural changes resulting from concussional trauma to the spinal cord of animals were also demonstrated by Hsu *et al.*[41], Kapadia[42], and by Goodman *et al.*[43]. A particularly interesting publication is the one by Domer *et al.*[44], who used intravenously administered radioactive pertechnetate to show a significant increase in BBB permeability in animals subjected to a whiplash injury who sustained no trauma to the head.

THE EFFECT OF TRAUMA ON THE BBB IN MS

Gonsette *et al.*[45] were the first to recognize that opening the BBB by using a brain needle to perform a thalamotomy on several MS patients resulted in the formation of new plaques near the needle tract.

Oppenheimer[46] in an important post-mortem study of 18 MS cases implicated trauma as a pathogenetic factor in the formation of MS lesions in the cervical cord as a result of the normal stretching of the tethered cord during flexion and extension of the neck, especially in the presence of cervical spondylotic ridges. Rudge (1991) agreed when he stated that "although there are many potential reasons for the BBB to break, a simple model of traumatic damage could account for the commoner sites of lesions being in the highly mobile optic nerve and cervical cord, especially when tethered by the dentate ligaments, and the periventricular areas, particularly those where acute angles occur, resulting in high shear stress." The formation of new lesions, or the enlargement of old ones as a result of a major break of the BBB, may occur only in individuals who have the MS trait or asymptomatic disease. If, as many MS researchers assume, many similarities exist between EAE and MS (although they are not identical), there is no more convincing example of the crucial role that can be played by traumatic injury to the BBB in facilitating the formation of lesions than the classic experiments of Clark and Bogdanove[47]. They showed that the lesions of EAE, which are usually disseminated throughout the white matter, could be localized to an area where the BBB had been previously disrupted by an electrolytic lesion. Because the degree of vulnerability of the BBB undoubtedly varies from patient to patient as well as from time to time in individual patients, depending upon many factors such as age, the duration of disease, the state of immunological responsiveness, and other unknown factors, it is unrealistic to expect every MS patient to have a clinical exacerbation following trauma to head, neck, or back.

THE ROLE OF THE BBB IN THE PROGRESSION OF MS

Barnes *et al.*[48] have shown by means of serial gadolinium-enhanced MRIs that after alterations of the BBB, complete repair does not always occur and that the normal impermeability of the BBB may not be restored for quite some time. They felt it possible that a BBB defect may persist in long-standing lesions and be of pathogenic significance in the progression of the disease. Evidence of BBB damage was found in 70% of lesions, was less severe in old lesions than in acute lesions, and may result from repeated previous inflammatory insults. Some of the MS lesions had a slow, modest pattern of enhancement which suggested, they said, that they might "reflect a noninflammatory BBB defect resulting from incomplete repair following repeated previous inflammatory insults. If, as has been suggested, blood-borne disease mediators such as complement are important in MS, then a chronic BBB defect of the kind we have observed might allow such mediators to diffuse through open lesions and promote new inflammation at their myelinative margins, as is commonly seen both pathologically and on MRI."

The persistence of an open BBB obviously creates a two-way street, since myelin *abbau* substances such as MBP, PLP, and MOG, as well as other substances, may leave the CNS and enter the peripheral circulation, where they may activate immunocompetent cells and humoral mediators. The possibility also exists that there is a constant exchange of immunologically active or inactive substances that play a role in either maintaining the alteration of the BBB or participating in the myelinoclastic process, either directly or indirectly, by unknown means. Whatever the role of activated lymphocytes may be, this scenario was suggested by Zamvil and Steinman,[18] and Oger *et al.*[17] and Davison[49] made somewhat similar suggestions. It is extended here to include non-cellular elements. This means that as long as the BBB remains permeable, MS must be considered a self-perpetuating disease. Regardless of the specific mechanism that leads to the opening of the BBB, whether it is purely an immunological phenomenon or the effect of a facilitator such as trauma or electrical injury, progression of the disease, although not necessarily of its clinical manifestations, may continue for a very long time after the initial insult to the BBB.

CONCLUSION

In recent years, due largely to the evidence provided by MRI, it has become clear that an obligatory alteration of BBB impermeability plays an important role in the pathogenesis of the MS lesion. In addition to its initial compromise in the MS trait it may be further altered by immune mechanisms of still unknown nature and/or facilitators which include trauma. Finally, it is also apparent that the BBB defect may persist for a very long time and contribute to the progression of the disease.

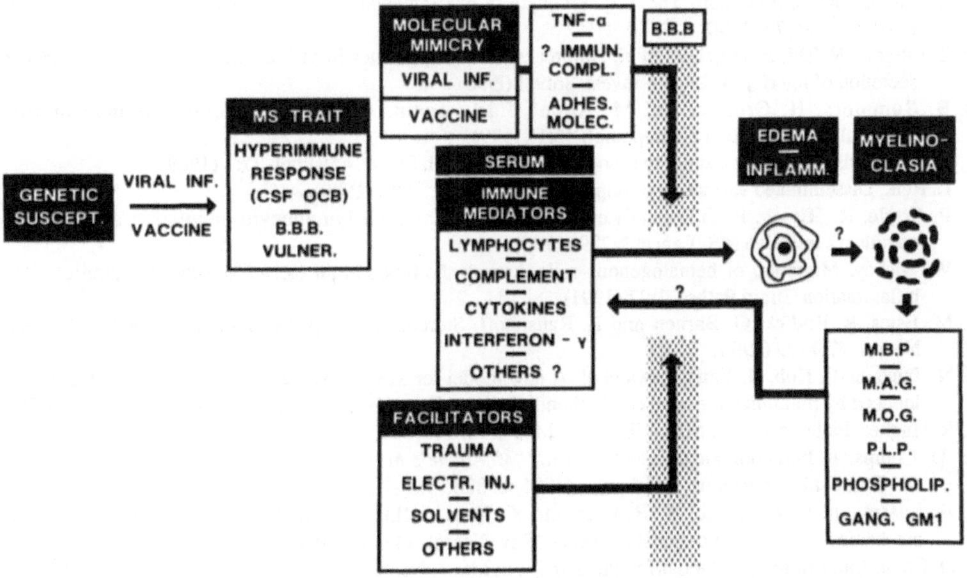

Fig. 2: Diagramatic representation of a scheme for the pathogenesis of MS. Abbreviations: OCB: oligoclonal bands; BBB vulner.: Blood-brain barrier vulnerability; Immun. Compl.: immune complexes; Adhes. Molec.: adhesive molecules; TNF-α: tumor necrosis factor α ; Electr. Inj.:electrical injury; MAG: myelin-associated glycoprotein; MOG:myelin-oligodendrocyte glycoprotein; PLP: proteolipid protein; MBP: myelin basic protein; Phospholip.: phospholipids; Gang. GM1: GM1 ganglioside.

REFERENCES

1. J. Dawson, The histology of disseminated sclerosis, Trans. Roy. Soc. Edinburgh,50:517 (1916).
2. T. Broman, *The Permeability of the Cerebrospinal Vessels in Normal and Pathological Conditions.* Munksgaard, Copenhagen (1949), p 66.
3. C. Poser, The pathogenesis of MS: a critical reappraisal. Acta Neuropathol.71:1 (1986).
4. C. Poser, Multiple Sclerosis: Observations and reflections - a personal memoir,J. Neurol. Sci. 107:127 (1992).
5. J. Brody, J. Sever and T. Henson, Virus antibody titers in MS patients, siblings and controls. J.A.M.A. 216: 1441 (1971).
6. M. Panelius, A. Salmi, P. Halonen et al. Virus antibodies in serum specimens from patients with MS, from siblings, and matched controls. Acta Neurol. Scand. 49:85(1973).
7. J. Woyciechowska, J. Dambrozia, P. Leinikki, et al. Viral antibodies in twins with MS. Neurology 35:1176 (1985).
8. E. Kinnunen, M. Valle, L. Piiranen et al. Viral antibodies in MS: a nationwide co-twin study. Arch Neurol. 47:743 (1990).
9. X. Xu and D. McFarlin, Oligoclonal bands in CSF: twins with MS. Neurology 34: 769 (1984).
10. P. Duquette, Familial subclinical MS. Neurology 41: 159 (1991).
11. C. Adams, R. Poston, S. Buk et al. Inflammatory vasculitis in MS. J. Neurol. Sci. 69; 269 (1985).
12. D. Gay and M. Esiri, Blood-brain barrier damage in acute MS plaques. Brain 114: 557 (1991).
13. R. Lisak, Immunological abnormalities. In: *Multiple Sclerosis,* p. 74. W. McDonald and D. Silberberg, eds., Butterworths, London, (1972)
14. E. Alvord, Experimental allergic encephalomyelitis and experimental allergic neuritis, in: "*Handbook of Neurology: Demyelinating Diseases.*" J. Koetsier, ed., Elsevier, Amsterdam, 3:429 (1985).
15. U. Jahnke, E. Fischer and E. Alvord, Sequence homology between certain viral proteins and proteins related to encephalomyelitis and neuritis. Science 229; 282 (1985).
16. G. Atkins, E. Daly, B. Sheahan, D. Higgins and P. Sharp, MS and molecular mimicry. Neuropathol. Appl. Neurobiol. 16:179 (1990).
17. J. Oger, L. Kastrukoff, D. Li, et al. MS: in relapsing patients, immune functions vary with disease activity as assessed by MRI. Neurology 38: 1739 (1988).
18. S. Zamvil and L. Steinman, The T lymphocyte in experimental allergic encephalomyelitis. Ann. Rev. Immunol. 89:579 (1990.
19. T. Olsson, W. Zhi, B. Hojeberg et al. Autoreactive T lymphocytes in MS determined by antigen-induced secretion of interferon. J. Clin. Invest. 86:981 (1990).
20. B. Summers, H. Greisen and M. Appel, Early events in canine distemper demyelinating encephalomyelitis. Acta Neuropathol. 46:1 (1979).
21. C. Poser, Disseminated vasculomyelinopathy. Acta Neurol. Scand. 45;suppl. 37:1 (1969).
22. L. Reik, Disseminated vasculomyelinopathy. Ann. Neurol. 7:291 (1980).
23. P. Coyle, R. Hirsch, P. O'Donnell et al. Cerebrospinal fluid lymphocyte population and immune complexes in active MS. Lancet 2:229 (1990).
24. W. Hickey, Migration of hematogenous cells through the blood-brain barrier and the initiation of CNS inflammation. Brain Pathol. 1:97 (1991).
25. M. Estes, R. Rudick, G. Barnett and R. Ransohoff, Stereotactic biopsy of an active MS lesion. Arch Neurol. 47:1299 (1990).
26. N. Tsukada, C. Koh, N. Yanagisawa et al. A new model for MS: chronic experimental encephalomyelitis induced by immunization with cerebral endothelial cell membrane. Acta Neuropathol. 73:259 (1987).
27. R. Hughes, Pathogenesis of MS. J. Roy Soc. Med 85:373 (1992).
28. D. Compston. Immunological aspects of MS. "*McAlpine's Multiple Sclerosis,*" 2nd ed., W. Matthews, ed., Churchill Livingstone, Edinburgh, (1991) p. 328.
29. H. Lassmann, F. Zimprich, K. Rossler and K. Vass, Inflammation in the nervous system: basic mechanisms and immunological concepts. Rev. Neurol. 147:763 (1991).
30. D. Compston, Limiting and repairing the damage in MS. J. Neurol. Neurosurg. Psychiat. 54:945 (1991).
31. P. Rudge, Does a retrovirally encoded superantigen cause MS? J.Neurol. Neurosurg. Psychiat.54:853 (1991).
32. A. Ommaya, D. Rockoff and M. Baldwin, Experimental concussion: a first report. J.Neurosurg. 21:249 (1964).
33. N. Nevin, Neuropathologic change in the white matter following head injury. J. Neuropathol. Exper. Neurol. 26:77 (1967).
34. D. Oppenheimer, Microscopic lesions in the brain following head injury. J. Neurol. Neurosurg. Psychiat. 31:299 (1968).

35. J. Hardman, The pathology of traumatic brain injuries. Adv. Pathol. 22:15 (1979).
36. L. Bakay, J. Lee, G. Lee et al. Experimental cerebral concussion. Part I: An electron microscopic study. J. Neurosurg. 47:346 (1977).
37. A. Jenkins, G. Teasdale, M. Howley, P. Macpherson and J. Rowan, Brain lesions detected by MRI in mild and severe head injuries. Lancet 2:445 (1986).
38. H. Levin, D. Williams, M. Crofford et al. Relationship of depth of brain lesions to consciousness and outcome of closed head injury. J. Neurosurg. 69:861 (1988).
39. H. Levin, D. Williams, H. Eisenberg, W. High and F. Guinto, Serial MRI and neurobehavioral findings after mild to moderate head injury. J. Neurol. Neurosurg. Psychiat. 55:255 (1992).
40. W. Maxwell, A. Irvine, J. Hume Adams, D. Graham and T. Gennarelli, Response of cerebral microvasculature to brain injury. J. Pathol. 155:327 (1988).
41. C. Hsu, E. Hogan, R. Gadsden, K. Spicer, M. Shi and R. Cox, Vascular permeability in experimental spinal cord injury. J. Neurol. Sci. 70:275 (1985).
42. S. Kapadia, Ultrastructural alterations in blood vessels of the white matter after experimental cord trauma. J. Neurosurg. 61:539 (1984).
43. J. Goodman, W. Bingham and W. Hunt, Ultrastructural blood-brain barrier alteration and edema formation in acute spinal cord trauma. J. Neurosurg. 44:418 (1976).
44. F. Domer, Y, Lin, K. Chandran and K. Krieger, Effect of hyperextension-hyperflexion (whiplash) on the function of the blood-brain barrier of Rhesus monkeys. Exper. neurol. 63:304 (1979).
45. R. Gonsette, G. Andre-Balisaux and P. Delmotte, La permeabilite des vaisseaux cerebraux. VI: demyelinsation experimentale provoquee par des substances agissant sur la barriere hemato-encephalique. Acta Neurol. Belg. 66:247 (1966).
46. D. Oppenheimer, The cervical cord in MS. Neuropathol. Appl. Neurobiol. 4:151 (1976).
47. G. Clark and L. Bogdanove, The induction of the lesions of allergic meningoencephalomyelitis in a predetermined location. J. Neuropathol. Exper. Neurol. 14:433 (1955).
48. D. Barnes, P. Munro, B. Youl, J. Prineas and W. McDonald. The longstanding MS lesion. Brain 114:1271 (1991).
49. A. Davison, The relevance of experimental allergic encephalomyelitis to MS. J. Roy Soc. Med. 85:425 (1992).

INTERFERON AND COPOLYMER I TREATMENT IN MULTIPLE SCLEROSIS: CLINICAL AND PATHOBIOLOGICAL IMPLICATIONS

Kenneth P. Johnson and Hillel S. Panitch

Department of Neurology
University of Maryland at Baltimore
22 S. Greene Street
Baltimore, MD 21201

INTRODUCTION

Multiple sclerosis (MS), although well-described clinically and pathologically for over 120 years, is only now poised for a new era of improved therapy brought about primarily by findings obtained from clinical research. Over the past 12 years, various investigations of new treatments have not only identified effective agents but have provided unexpected new insights into the basic mechanisms of disease activity in MS. The combination of carefully designed and conducted clinical trials, coupled with serial MRI studies of the brain, has allowed us to determine the therapeutic benefit of one and perhaps two treatments with low risk and toxicity which are likely to be available to the general MS community within 6 to 24 months.

An ideal therapy for MS would cure or prevent the disease, however, such a treatment has not yet been identified. Alternatively, agents which delay and reduce long-term neurologic disability would be a major improvement over currently available therapies. Such agents should have low toxicity and should be relatively convenient for patients to administer, possibly for long periods. Several clinical studies have shown that beta interferon (beta IFN) and perhaps Copolymer I (Cop I) display these characteristics and may soon be recognized and widely available therapies for exacerbating-remitting (E/R) MS.

The clinical usefulness of beta IFN and Cop I is based on Phase II or Phase III-controlled studies which show that they reduce the number of clinical exacerbations. This end point has been criticized with some justification because prevention of long-term neurologic disability should be the ultimate goal of any MS therapy. There is, however, a recognized relationship between the frequency of exacerbations early in MS and the amount of neurologic disability displayed years or decades later.[1] Therefore, one may assume that a reduction in the rate of exacerbations during the early stages of MS will result in less neurologic disability later on. To be useful, any therapy administered at the onset of MS, must induce little toxicity or lead to major sequelae. Both beta IFN and Cop I have been shown to produce few side effects and have been administered safely to patients for periods as long as 7 years.

The following sections will briefly review the historical developments leading to the identification of beta IFN and Cop I as potentially useful therapies for E/R MS. In the case of beta IFN, the essential Phase III clinical trials are complete and a publication is currently in press[2] which document its clinical usefulness. Some information from this positive trial is included in this paper. For Cop I, a definitive Phase III trial in E/R MS is now underway in the United States and is expected to be completed in the spring of 1994.

INTERFERON THERAPY IN MULTIPLE SCLEROSIS

Interferons are natural human proteins produced by virus-infected cells to block spread of virus to other cells. Interferon was first described in 1957 by Isaacs and Lindenman.[3] Subsequent studies have defined two types of interferon, Type 1 which includes leukocyte or *alpha interferon* and fibroblast or *beta interferon*, and Type 2 which includes the single agent, immune or *gamma interferon*. All three types have been experimentally tested in E/R MS over the past 10 years.

Type 1 interferons (alpha and beta) are closely related, being derived from genes on chromosome 9. They share the same cell receptor and produce many of the same immunological effects. Gamma interferon is an entirely different protein capable of producing numerous unique immunologic effects. All three interferons are available for clinical experimentation in both crude natural forms and in highly purified recombinant preparations. Recombinant alpha and gamma interferon have been shown to be useful in limited virus-induced neoplasms and have been licensed for general distribution. Extensive information on the basic biology and immunology, as well as the clinical uses of interferons can be found in an recent comprehensive review by Panitch.[4]

Natural Interferons in MS

In 1979 studies by Neighbour et al[5], suggested that interferon production was deficient in MS patients. Soon after, methods were developed to produce sufficient amounts of natural alpha and beta interferon for clinical experimentation. Admittedly these preparations were highly impure, containing only about 1% interferon. During the early 1980's, MS investigators were faced with the dilemma of not knowing what type or dose of interferon to use, what route to employ, what stage of disease to target or whether to provide continuous therapy or to treat only during exacerbations. Now, 12 years later, 11 major therapeutic trials with interferons have been completed which provide well-documented answers to many of these daunting questions.

The first two adequately controlled MS trials employed natural alpha IFN given subcutaneously (sc) or natural beta IFN given intrathecally and both showed positive trends in limiting new exacerbations. With sc administered alpha IFN, the side effects experienced by 24 patients treated with 5 million international units (MIU) daily for 6 months were considered unacceptable. In the case of intrathecally administered natural beta IFN, the inconvenience of intrathecal therapy and the potential risk of adhesive arachnoiditis decreased the enthusiasm for this form of therapy. Nevertheless, a double-blind, placebo-controlled trial of intrathecal natural beta IFN was undertaken. Sixty-nine patients were enrolled and those receiving interferon were shown to have fewer relapses during therapy. Two subsequent studies, one employing lymphoblastoid interferon, a combination of alpha and beta interferons and the other using poly IC-LC to induce interferon, failed to provide convincing evidence of usefulness. Please refer to reference[4] for specific literature reports.

Recombinant Interferon Therapy

By 1982, highly purified recombinant interferons were becoming available for clinical experimentation and, based on the trends toward improvement shown by the natural alpha and beta interferon studies, there was enthusiasm for testing these new agents in E/R MS. Because of the positive trend observed with sc natural alpha IFN, a low-dose study of recombinant alpha IFN (2 MIU 3 times weekly) was undertaken at the University of Maryland and at Temple University using a Schering Plough Inc., Kenilworth, New Jersey, product. Over 90 patients received 2 MIU of alpha IFN or placebo sc for one year. In this double-blind trial, there were few side effects, however, no clinical benefit was identified.

In 1984, the pivotal study of gamma IFN in E/R MS was initiated using a recombinant product supplied by the Biogen Company, Cambridge, Massachusetts, and conducted at the University of Maryland. It was already appreciated that the immunologic effects of gamma IFN were strikingly different than the Type 1 interferons and that there was a risk of stimulating rather than inhibiting MS activity. Therefore, a cautious trial was undertaken, which enrolled 18 patients who received one of three doses of gamma IFN intravenously twice weekly for one month. Seven of the 18 patients experienced new attacks of MS during the treatment. In each case, the MS symptoms appeared in a neurologic area which had previously been involved, suggesting that gamma IFN had reactivated activity within an already existing MS plaque. While this study was a therapeutic failure, it nevertheless strongly suggested that not only was MS an autoimmune disease but that gamma IFN was a major pathogenic stimulant of new disease activity. The study also suggested that agents which blocked or inhibited gamma IFN might be valuable therapies for MS.

In 1985, a specially engineered molecule of beta IFN in which a serine residue replaced a cysteine residue at position 17 and now identified as interferon beta 1b or Betaseron was provided by Berlex Laboratories, Richmond, California, for clinical evaluation in MS. Because of the recent experience with gamma interferon, it was felt prudent to conduct a pilot trial to determine potential toxicity and to identify an appropriate dose with the concept that the highest acceptable dose was likely to be the most effective. Thirty patients, divided into five groups, received 0, 4.5, 22.5, 45 or 90 MIU sc three times per week, for six-months at the University of Maryland, Thomas Jefferson University and Temple University. It soon became apparent that 90 MIU produced unacceptable side effects, whereas patients could accept 45 MIU three times weekly with little discomfort, especially after the first four to six weeks of therapy. There was a trend toward reduction of relapses with the two higher doses in this small pilot trial. After six months, all patients were reassigned to receive 45 MIU three times weekly and the majority of them have remained on therapy, some for over seven years.

Based on these encouraging preliminary results, a large, eleven university Phase III trial of Betaseron was undertaken during 1988 in the United States and Canada. A total of 372 ambulatory E/R MS patients were randomized into one of three treatment groups; 45 MIU, 9 MIU or placebo injected sc every other day. The primary experimental end points were exacerbation rates and the proportion of patients remaining free of exacerbations for two years. The double-blind protocol was maintained by using two neurologists, one designated the examining neurologist who was unaware of the patient's complaints and the other a treating neurologist who was at liberty to provide steroid therapy in the event of serious MS exacerbations. Drop-out rates were low and over 300 patients had completed the two years of study by July 1991. The side effects typical of interferon therapy, the so-called flu-like syndrome of fever, fatigue, malaise and muscle aches occurred for a few weeks at onset, after which side effects were minimal. A third year of study was added to the protocol and most patients agreed to continue.

This Phase III comprehensive trial of Betaseron has now been analyzed and a publication fully describing the results is in press[2]. Table 1 shows some of the data indicating that Betaseron has a highly significant effect on exacerbation rates after two or three years and the proportion of patients who remain free of exacerbations at two years. The time to first exacerbation is also significantly extended for the patients receiving 45 MIU. Several other measures, such as the need for steroid therapy or number of hospitalizations, were also reduced for patients in the high dose group.

Table 1. Betaseron Trial of E/R MS

Effect of IFN-ß on Exacerbations

		IFNß		
	Placebo N=112	9 MIU N=111	45 MIU N=115	p-value+
		2-Year Data		
Exacerbation Rate	1.27	1.17	0.84*	0.0001
Exacerbation Free Subjects	18	23	36	0.007
Median Time to 1st Exacerbation	153	180	295**	0.015
		3-Year Data		
Exacerbation Rate	1.21	1.05	0.84	0.0004
Exacerbation Free Subjects	17	23	27	N.S

MIU = million international units
* P< 0.01, 45 MIU vs 9 MIU
** P< 0.05, 45 MIU vs 9 MIU
+ p-value for 45 MIU vs placebo

All enrolled patients were studied by a non-enhanced MRI scan yearly during the Betaseron trial. Table 2 shows that the MS disease burden, as measured by total area of involvement, continued to expand in patients in the placebo or low dose groups, whereas the area of involvement actually declined in the group treated with 45 MIU. These results show that high dose Betaseron therapy is the only treatment that has substantially altered the natural history of MS in a properly controlled clinical trial.

Table 2. Betaseron Trial of E/R MS

MRI Lesion Area: Percent Change in Relation to Baseline

		IFN-ß	
	Placebo	9 MIU	45 MIU
1 Year	+12.4	+4.1	-1.1
2 Years	+20	+10.5	-0.1
3 Years	+17.1	+1.1	-6.2

MIU = million international units

Another recombinant form of beta interferon, Bioferon, is currently available and differs from Betaseron, in that it is produced in mammalian cells rather than in bacteria and is glycosylated. This product is also undergoing clinical evaluation in the United States employing a much lower dose, 6 MIU given sc once per week. As of February 1993, enrollment is virtually complete and results will hopefully be available sometime in the next two to three years.

Identified Mechanisms of Interferon Action in MS

Information continues to accumulate indicating that MS is a disease of immune origin[6]. The recent review by Panitch[4] has carefully defined the immunologic nature of MS and has provided extensive documentation of the mechanisms which may be operative during interferon therapy. Two specific functions have been well-described and may be of central importance: the partial inhibition of gamma IFN and its effects and the enhancement of suppressor activity.

Several laboratories have documented that beta IFN is a potent but incomplete inhibitor of gamma IFN and its effects. A variety of experimental systems have been used to document this result. Most interestingly, Panitch[7] has shown that patients treated with beta IFN continue to show the inhibitory effect on gamma IFN for prolonged periods, well over three years.

A deficiency in suppressor activity has long been recognized as an important immunologic defect in MS. Noronha et al[8] reported that recombinant beta IFN improves suppressor cell function of T cells from both MS patients and control subjects, a finding, since confirmed by Panitch[9] who has also documented that beta IFN can modify the regulatory function of T cells on gamma IFN synthesis which has been shown to be a sensitive indicator of suppressor T cell activity.

COPOLYMER I THERAPY IN MS

In the early 1970's, M. Sela, R. Arnon and co-workers at the Weizman Institute in Israel began to search for rational therapies for MS. They based their search on available information about experimental allergic encephalomyelitis (EAE) on the premise that it was a useful model of human MS. It is well-known that fragments of myelin basic protein (MBP) containing as few as 10 to 12 amino acids could either produce or inhibit EAE, however, the positive or negative effect of these small fragments differed by animal species, i.e. the same fragment would protect in one species but produce EAE in another. An effort was made to produce random polymers of the common amino acids of MBP and to test them in multiple species of animals with EAE. A synthetic polymer containing alanine, glutamic acid, lysine and tyrosine and labeled Copolymer I was found to protect several species of mammals from EAE.[10] Experimental therapy with this polymer also appeared to produce very few side effects or late sequela.

Following preliminary human experience with Cop I in Israel, Dr. Murray Bornstein initiated several highly interesting studies at the Albert Einstein School of Medicine in New York (Table 3) in which he determined exceptional safety of Cop I and also identified an acceptable dose. This led to an important Phase II trial showing benefit in E/R MS.[11] All completed clinical studies of Cop I are described in a recent review by Bornstein and Johnson.[12]

The first pilot study included 16 patients with various clinical stages of MS and showed some benefit in five of the subjects who were treated with daily sc injections for six to eighteen months. In the Phase II study, which has stimulated great interest and enthusiasm, 48 E/R patients were treated for up to 24 months with daily sc doses of 20 mg. of Cop I or placebo.[11] There was a noticeable decline in exacerbations among the Cop I treated patients and the proportion of patients on Cop I who were free of attacks was also impressive. A third study recruited patients in the chronic-progressive (CP) stage of MS at two centers, Einstein School of Medicine in New York and Baylor University School of Medicine in Houston. One hundred six CP patients were recruited into a double-blind, placebo-controlled protocol with treatment for 24 months. No significant differences in outcome were recognized between the groups, although there were trends toward stabilization or reduced decline among the Cop I treated patients.[13]

Based on these encouraging studies, especially in E/R patients, a phase III, placebo-controlled, double-blind study was initiated in October 1991 at 11 university sites in the United States. A total of 253 well-qualified patients were rapidly accrued over six months and the half-way point for the trial will occur in May 1993. The previous safety profile of few adverse reactions or late sequela has been continued in this pivotal study. Several laboratory investigations to define mechanisms of immunologic action of Cop I are also underway.

Table 3. Multiple Sclerosis: Clinical Trials with Copolymer I

Phase	# of Pts.	Clinical Type	Duration	Result
Pilot	16	All	6-18 mos.	5 improved
Phase II	48	E/R	24 mos.	Fewer Exacerbations
Phase III	106	CP	24 mos.	Trend toward benefit w/Cop I
Phase III	253	E/R	24 mos.	In Progress

The mechanism of action of Cop I in MS patients has not been investigated directly but immunologic studies of animals and cell cultures have demonstrated that Cop I can inhibit specific responses to basic neural antigens, particularly MBP. Cop I was originally synthesized to mimic EAE suppressive determinants on the MBP molecule and cross reactivity with epitopes of MBP has been demonstrated both at the B cell level with monoclonal antibodies and at the T cell level with cell mediated responses. Studies of murine EAE have shown that Cop I can induce MBP specific suppressor cells which mediate protection from EAE. Studies of proliferative responses of guinea pig and rabbit lymphocytes sensitized to MBP in *vitro* and *in vivo* showed in some cases an inhibition of responses to MBP *in vitro* by Cop I. A study of MBP specific clones showed that Cop I can directly block T cell responses to MBP in an antigen specific but not MHC restricted fashion which suggested that Cop I peptides might be capable of displacing MBP peptides from Ia and thus inhibit antigen presentation. Although studies of Cop I support a direct effect on the immune response to MBP, it is not clear that this effect is responsible for suppressing relapses in MS patients because sensitization to MBP has not been conclusively demonstrated in MS patients.

DISCUSSION

The modern era of experimental studies to identify effective therapies for MS can arguably be stated to have begun in approximately 1981. In fact, a conference sponsored by the National Multiple Sclerosis Society of the United States and held on Grand Island, New York, in the spring of 1982, led directly to the formation of the American Advisory Committee on Trials of New Agents in MS and indirectly to the creation of ECTRIMS in Western Europe. These organizations have been instrumental in stimulating new interest in clinical research in MS and in upgrading the quality of the protocols to test the effectiveness of new therapies.

Two trends in experimental MS research were evident throughout the 1980's; one, the experimental focus on major but non-specific immunosuppressive agents, such as cyclosporine, azathioprine and cyclophosphamide, and the second approach focused on the agents described in this article, various interferons and Cop I. After ten years of intense investigation, one can state that the major immunosuppressive drugs have shown relatively little value as therapeutic agents in MS or have produced unacceptably high toxicity. On the other hand, it seems almost certain that beta IFN (interferon beta 1B or Betaseron) will be recognized and licensed as an effective form of therapy for E/R MS within a very few months and that Cop I may follow a similar course within the next two to three years. This is, of course, significant for MS clinicians everywhere and their anxious and concerned patients, for Betaseron is the first treatment ever shown to convincingly improve the natural course of multiple sclerosis.[2]

A number of other immunologically active agents are currently undergoing preliminary investigation and may, in this decade, also join the list of effective therapies for MS. Hopefully these new agents can be investigated more rapidly as improved methods are recognized to measure evidence of clinical effect. The employment of serial MRI's perhaps enhanced with gadolinium is almost sure to become an important surrogate of clinical benefit, especially when the findings from the Betaseron study become more widely appreciated. In this period of increased therapeutic enthusiasm, it is important not to lose sight of the fact that carefully controlled trials must remain the single gold standard in the search for better therapies in MS. Such controlled trials will certainly become more complex and difficult if, in fact, one must compare a new therapy against a standard therapy, such as beta IFN, rather than against placebo. At the present time, there is no recognized effective therapy for CP MS, therefore, a comparison of treatment vs. placebo may continue to be acceptable.

Clinical research in MS will also increase in complexity as we approach the era of evaluating combined therapies employing several agents simultaneously in an effort to better control the disease. In this circumstance, hopefully, increased understanding of the basic immunologic defects in MS discussed at this meeting may aid in detecting the therapeutic effectiveness of multiple agents.

REFERENCES

1. B.G. Weinshenker, B. Bass, G.P.A. Rice, J. Noseworthy, W. Carriere, et al, The natural history of multiple sclerosis: a geographically based study. 2. Predictive value of the early clinical course, *Brain* 112:1419-1428 (1989).
2. The IFN- Multiple Sclerosis Study Group, Interferon-beta 1 B is effective in relapsing-remitting multiple sclerosis: Clinical results of a multicenter, randomized, double blind, placebo controlled trial, *Neurol.* 43: in press (1993).
3. A. Isaacs, J. Lindenmann, Virus interference, I. The interferon, Proceedings of the Royal Society of London (Biology) 147:258-267 (1957).
4. H.S. Panitch, Interferons in multiple sclerosis. A review of the evidence, *Drugs.* 44(6):946-962 (1992).
5. P.A. Neighbor, A.E. Miller, B.R. Bloom, Interferon responses of leucocytes in multiple sclerosis, *Neurol.* 31:561-566 (1981).
6. S. Dhib-Jalbut, D.E. McFarlin, Immunology of multiple sclerosis, *Ann Allergy.* 64:433-444, (1989).
7. H.S. Panitch, J.S. Folus, K.P. Johnson, Recombinant beta interferon inhibits gamma interferon production in multiple sclerosis, Abstract, *Ann Neurol.* 22:139 (1987a).
8. A. Noronha, A. Toscas, M.A. Jensen, Interferon beta augments suppressor cell function in multiple sclerosis, *Ann Neurol.* 27:207-210 (1990).
9. H.S. Panitch, J.S. Folus, K.P. Johnson, Activated suppressor cells inhibit synthesis of interferon-_ in patients with multiple sclerosis patients and normal subjects, Abstract, *J Neuroimmunol.* In press (1992b).
10. R. Arnon, Experimental allergic encephalomyelitis - susceptibility and suppression, *Immunol.* Rev. 55:5-30 (1981).
11. M.B. Bornstein, A. Miller, S. Slagle et al, A pilot trial of Cop 1 in exacerbating-remitting multiple sclerosis, *N Engl J Med.* 317:408-414 (1987).
12. M.B. Bornstein, K.P. Johnson, Treatment of multiple sclerosis with Copolymer I, in: "Handbook of Multiple Sclerosis," S.D. Cook, ed., Marcel Dekker, Inc., New York and Basel (1990).
13. M.B. Bornstein, A. Miller, S. Slagle et al, A placebo-controlled, double-blind, randomized, two-center, pilot trial of Cop 1 in chronic progressive multiple sclerosis, *Neurol.* 41:533-539 (1991).

THE CORRELATION BETWEEN MAGNETIC RESONANCE IMAGING ABNORMALITIES AND OTHER ASPECTS OF MULTIPLE SCLEROSIS WITH PARTICULAR EMPHASIS ON THERAPEUTIC TRIALS

Donald W. Paty

Division of Neurology
The University of British Columbia, Canada

INTRODUCTION

Magnetic resonance (MR) has had a major impact on the evaluation of patients with multiple sclerosis (MS). The most obvious area of influence has been on diagnosis. Diagnostically abnormal head scans can be seen in at least 90% of patients with clinically definite MS (CDMS). In addition, even though the spinal cord has been difficult to image in the past, techniques are improving. However, the findings on individual magnetic resonance imaging (MRI) scans do not usually correlate very well with the clinical (neurological) status of the patient, or with the prior clinical course. That lack of correlation should not be a surprise, however, because the same lack of correlation has always existed between the clinical history and pathological findings.

Serial imaging of individual patients often shows the asymptomatic accumulation of new lesions, enlargement of pre-existing lesions, and striking changes in the blood brain barrier (BBB) over time. It has been thought by many investigators that MRI detected disease activity provides an objective measure of disease activity which has been shown to occur at a greater rate than does clinical activity. In addition, MRI activity may be partially independent of changes in clinical neurological status. It is also reasonable to expect that a treatment which affects disease activity at a fundamental level should stop or slow down the development of new lesions and other MRI signs of activity.

DIAGNOSIS

MRI is about 10 times more sensitive than the unenhanced CT scan for the detection of MS lesions. A number of studies have shown (1,2) that multiple MRI abnormalities in the white matter, are present in more than 90% of patients with CDMS. Also the presence of multiple lesions on the MRI scan in patients suspected of having MS predicts the ultimate development of CDMS (3,4).

A conservative approach to the identification of MRI lesions as suggestive of MS is recommended (5) because of the non-specificity of multiple white matter lesions. It is

A Multidisciplinary Approach To Myelin Diseases II
Edited by S. Salvati, Plenum Press, New York, 1994

generally recommended that there should be at least 4 typical white matter lesions greater than 3mm in diameter disseminated in space in order for the scan to be considered strongly suggestive of MS. Since non-specific white matter abnormalities are quite common in the older age group (up to 20%) (6) a very conservative approach must be taken concerning the specificity of the MRI scan for the diagnosis for MS. However, in spite of the foregoing, most studies have shown that multiple white matter lesions on the MRI scan usually predict the onset of MS when seen in patients under the age of 50 with initial symptoms suggestive of MS such as optic neuritis (ON). Table 1 lists a number of MRI features that increase the specificity for the diagnosis of MS.

TABLE 1

LESION CHARACTERISTICS SUGGESTIVE OF MS

Size	Large (>6mm diameter)
Shape	Oval
Location	(a) Periventricular
	(b) Corpus callosum
	(c) Brainstem & cerebellum

In addition, in the MRI evaluation of patients with ON there is a very good correlation between enhancement, other lesions seen in the optic nerve and the onset of ON. Miller and his colleagues found that the length of the lesion in the optic nerves seen on the MRI scan correlated quite well with the severity and duration of visual loss (7). Resolution of enhancement is also associated with clinical improvement (8).

CORRELATION WITH PATHOLOGY

MRI alone cannot reliably identify the specific pathology of individual lesions. However, the appearance of MRI lesions in patients with MS is very reminiscent of the classical appearance of periventricular demyelination seen at autopsy. In addition, post mortem MRI studies (9,10,11) show an extensive degree of abnormality very similar to the abnormalities seen on scans done in life on severely affected patients.

In addition, there is quite a bit of variation in intensity and other signal characteristics in the lesions seen both in life and in post mortem scans. Some of the differences are obviously due to technical problems, but some differences are also probably due to variations in pathology.

As might be expected the correlation between MRI changes and pathology is not exact. Small lesions are frequently missed. The total "error" in measurement from slice to slice has varied between 13-30% However, MRI probably underestimates rather than overestimates the extent of disease. One report has also found a correlation between lesion changes and the detection of immune function abnormalities (12). There also is at least one report (13) showing that an enhancing lesion was very inflammatory on pathological examination.

QUANTITATIVE STUDIES

A quantitative approach to measuring the extent of abnormal areas (disease burden) on the MRI scan can be used to follow the evolution of the pathological process over time (14). The number, size, and distribution of lesions for each subject can be identified and followed in a systematic way.

Serial frequent studies on the same patient can then contribute greatly to the understanding the evolution of the *in-vivo* pathology in that patient. However, very careful repositioning is essential in order to avoid artifacts. Slight differences in repositioning (3 or 4 degrees) can make a dramatic difference in the appearance of the scan.

Koopmans and his colleagues (15) did a quantitative correlation study comparing 32 benign MS patients with 32 moderately disabled relapsing progressive (RP) patients paired and matched for age, sex, and duration of disease. Most benign patients had a much lower mean total area of lesion involvement than did the chronic patients, however, in 6 pairs (20%) the extent of disease appeared to be greater in the benign member of the pair as compared to the more disabled member. In addition, correlation between location of MRI lesions in the brainstem and cerebellum and clinical symptoms showed only about 50% concordance. For example, only 50% of the benign patients who had brainstem lesions seen on the MRI had a history of brainstem symptoms. The opposite was also true. Only about 50% of the benign patients who had brainstem symptoms in the past had brainstem lesions detectable on the scan.

In another quantitative correlation study Honer (16) examined 8 patients with both MS and diagnosed psychiatric disease and 8 matched patients with MS alone. He showed a greater degree of involvement of the temporal lobe in patients with both MS and psychiatric disease than in the MS patients without psychiatric disease.

The overall correlation between the extent of the MS process and disability has been statistically significant, but the correlation co-efficients have not been high. The most consistent clinical MRI correlation has been between the involvement of the corpus callosum and neuropsychological abnormalities (17,18,19,20). In addition, follow-up MRI and neuropsychological testing have suggested that there are changes in neuropsychological features over time. However, those changes are very slow to occur (21). In addition, T1 relaxation time abnormalities in apparently normal white matter also correlate with some aspect of cognitive impairment (20). Ventricular enlargement was not correlated to the score on the Beck Depression Inventory (22). Wiebe and colleagues, however, did show a correlation between MRI activity and neurological clinical activity (23).

Clinical severity in MS is to a great degree determined by the location of lesions, particularly those in the spinal cord. The general lack of high correlation between neurologically determined severity (EDSS) in MS and the extent of cranial MRI lesions should not be surprising since there are large areas of the cerebral hemispheres which are considered neurologically silent.

Perhaps the most important aspect of clinical MRI correlation has been that the mean of the extent of disease measured by MRI has been documented by several studies to slowly increase over time (24,25). This general increase in MRI detected disease burden correlates very well with what we expect to see in our chronic patients, that is, increasing neurological impairment.

EVALUATION OF THERAPEUTIC TRIALS

Serial cranial MRI studies in relapsing remitting patients and relapsing progressive patients have shown that MRI detected disease activity is much greater than is clinically detected activity. New and recurrent MRI lesions appear and disappear during periods of clinical remission (26,27,28). The lack of clinical expression of the majority of new lesions places severe constraints on the utilization of clinical assessment alone to detect disease activity, disease progression, and disease modification by therapy.

The use of quantitative and systematic MRI scanning has added an objective dimension to the evaluation of clinical trials. Prospective systematic cranial MRI studies reveal changes that are more frequent and dynamic than are clinical manifestations. MRI studies are now used in parallel with clinical assessment as an outcome measure in the evaluation of therapeutic trials. The two MRI methods should be as follows:

1. Quantitation of the extent of all cranial lesions to measure disease burden on a systematic basis at regular intervals.

2. Frequent quantitative and systematic studies with a scanning interval between 2 and 6 weeks in order to reveal the dynamics of asymptomatic disease activity.

Kappos and his colleagues (29) used MRI as an outcome measure during the last 6 months of a treatment trial with cyclosporine using a visual assessment method. They were unable to show any therapeutic effect. In a controlled trial of 100 chronic progressive MS (CPMS) patients treated with alpha-lympoblastoid interferon (24) 80 patients had quantitative MRI evaluations at entry, at 6 months and at 2 years of the trial duration. A quantitative increase in MRI detected disease burden was shown with a mean increase in extent of disease of 21% in the placebo group over 2 years. No MRI or clinical benefit was shown.

Another quantitative MRI study was done on 157 CPMS patients treated in a controlled trial with cyclosporine (30). There was also no significant MRI benefit shown from the cyclosporine. However, in both of these studies, the quantitative MRI method showed a clear cut increase in lesion area over time that provided a measure of increasing burden of disease. An increase in burden of disease over time agrees with the expected increase in clinical disease severity over time.

In addition, the frequent scanning technique has added another new dimension. Frequent scanning can detect dynamic activity by the identification of new lesions, significantly enlarging lesions, and recurrent lesions. In addition, the burden of disease can also be quantitated. A European committee looking at the issue of MRI use in clinical trials has recommended (31) that MRI scans be performed every month on all patients in therapeutic trials. Statistical power calculations show that the frequent MRI scanning method can provide a rapid and objective way of evaluating therapeutic effects using relatively small numbers of patients over relatively short periods of time. MRI evaluation will be particularly helpful in the screening of new drugs.

The Interferon Beta 1b Trial

A multi-centre randomized double blind placebo controlled trial designed to evaluate the affect of interferon beta 1b (Betaseron® - IFNB), on exacerbations has just been completed. Clinical evaluation showed a showed a 34% reduction in relapse rates over 3 years (32) due to IFNB therapy at a dose of 8 million international units (MIU) every other day. Clinical impairment (disability) measured by the EDSS (33) showed a trend to slower progression in the treated patients than in the placebo patients. However, the EDSS changes were not very great overall (only 28% of placebo patients had an increase of 1 or more

EDSS points in 3 years) and the differences between the treatment groups were not significantly different.

In contrast, the MRI studies showed quite striking treatment effects (34). The MRI activity findings in the treated group correlated quite well with the reduction in relapse rates. The MRI lesion activity in a sub-group of 52 patients who had scans every 6 weeks for 2 years showed a median decrease of 80% fewer active scans in the high dose treatment group (8 MIU of IFNB every other day) as compared to the placebo group. The high dose group also showed a median reduction of 83% in the rate of active lesions and a median reduction of 75% in the rate of new lesion formation when compared to placebo group. Seven of the high dose patients were free of MRI signs of activity during the study compared to 1 placebo patient (P=0.039). This method of MRI assessment is equivalent to clinical assessment of the rate of relapses.

The MRI evaluation of the impact of the treatment on burden of disease at 3 years showed that the placebo group had a mean increase of 17.1% compared to a mean decrease of 6.2% in the high dose treatment group (P=<0.002).

TABLE 2

DISEASE ACTIVITY DETECTED BY MRI SCANNING EVERY 6 WEEKS FOR 2 YEARS IN THE IFN BETA 1b CLINICAL TRIAL

Measurement	**TREATMENT GROUP** Placebo			**PLACEBO VS 8 MIU P VALUE**
	Placebo	**1.6 MIU**	**8 MIU**	
N (Patients) =	17	18	17	
Mean % scans with activity	34.6	17.0	15.4	0.0062
Mean active lesion rate/pt/yr	4.9	1.8	2.0	0.0089
Median active lesion rate/pt/yr	3.0	1.0	0.5	0.0289
Mean new lesion rate/pt/yr	3.2	1.1	1.2	0.0026
Median new lesion rate/pt/yr	2.0	0.5	0.5	0.0026

the MRI activity rates in the treatment groups.

TABLE 3

MRI DETECTED BURDEN OF DISEASE CHANGES OVER 3 YEARS IN THE IFN BETA 1b CLINICAL TRIAL

	TREATMENT GROUP			PLACEBO VS 8 MIU P VALUE
Measurement	**Placebo**	**1.6 MIU**	**8 MIU**	
N =	110	110	107	
Baseline MRI area (mm^2)	2611	2750	2367	NS
Mean percent change	17.1%	1.1%	- 6.2%	0.002
Median percent change	15.0%	0.2%	- 9.3%	0.002
Median change in total area (mm^2)	198	0.0	-118.9	0.002

some of the quantitative MRI data from the study.

The clinical/MRI correlations in this study were very significant, as follows:

1. Clinical evaluation showed a significant reduction in the relapse rate, and the frequent MRI studies showed an even more significant reduction in the dynamic MRI activity rate. Even though the correlation in individual patients was not exact enough to be used for clinical purposes, the correlation between treatment groups was obvious.
2. There was a trend for the increase in neurological impairment to be less in the high dose treatment group and the MRI measure of burden of disease increase was very significantly less in the same group. The EDSS both at baseline and at endpoint correlated with the MRI burden of disease (P=0.001).
3. There was a very striking relationship overall between the relapse rate and the mean change in MRI area. The patient group that had 0-.5 relapses per patient per year had a mean MRI increase of 77.6mm^2 over 3 years. However, the patient group that had more than 1.5 relapses per patient per year had a mean MRI increase in burden of disease of 516.6mm^2 (P=0.045).
4. EDSS did not change enough during the study did show a correlation between the change in MRI detected burden of disease and change in EDSS.

TABLE 4

CLINICAL/MRI CORRELATIONS IN MS

Area of Correlation		Comments
Diagnosis	(a)	90% of CDMS patients have MRI correlations of the diagnosis
	(b)	Multiple MRI lesions at onset predict the diagnosis
Optic Neuritis	(a)	Enhancement in the optic nerve correlates with symptoms
	(b)	Extent of lesions correlates with severity of symptoms
	(c)	Reduction in enhancement correlates with improvement in symptoms
Pathology	(a)	Single demyelinated lesions correlate well with the MRI appearance
	(b)	Enhancement correlates with inflammation
Neurological Findings	(a)	Benign patients usually have less extent of disease than chronic disabled patients
	(b)	Quantitation of burden of disease increases overtime
	(c)	50% of brainstem lesions and symptoms are correlated
	(d)	MRI burden of disease is correlated with clinical impairment
Neuropsychological Findings	(a)	Atrophy of the corpus callosum correlates with dementia
	(b)	Periventricular lesion load correlates with cognitive impairment
Therapeutic Trials	(a)	Relapse rates and burden of disease changes significantly correlate in groups but not in individuals
	(b)	Treatment effect on relapse rates correlates with MRI activity
	(c)	Extent of disease and neurological impairment correlated both at baseline and at end of study

summary of currently accepted MRI/clinical correlations.

SUMMARY

It is clear that MRI can contribute significantly to the understanding of the evolution of disease in MS. Quantitative and serial MRI studies have helped to reveal a new aspect of measurable activity of the pathological process. The degree of activity revealed by serial MRI studies is considerably greater than is the degree of activity determined by history and physical examination alone. Evaluation of the extent of the abnormality present (burden of disease) by a quantitative approach to lesion measurement and detection of dynamic disease activity by frequent scanning taken together add a major dimension to the tools available for all clinical studies not just therapeutic trials.

The results of the IFNB trial showed that MRI methods are much more sensitive than clinical assessment in detecting and measuring treatment effects. However, careful clinical follow-up studies must be done in order to better understand the long term prognostic implications of the MRI data. However, the conclusion must be that MRI methods are more sensitive and more objective than are clinical methods in determining both disease activity and disease burden. In the future, MRI methods must be mandatory components for therapeutic trials and disease activity studies.

ACKNOWLEDGEMENTS

Dr. Joy Wallenberg and Lynne Hannay for their help with the manuscript.

REFERENCES

1 Young IR; Hall AS; Pallis CA; Bydder GM; Legg NJ; Steiner RE. Nuclear magnetic resonance imaging of the brain in multiple sclerosis. Lancet. 1981; 2: 1063-6.

2. Robertson WD; Li DK; Mayo JR; Fache JS; Paty DW. Assessment of multiple sclerosis lesions by magnetic resonance imaging. Can Assoc Radiol J. 1987 Sep; 38(3): 177-82.

3. Miller DH; Ormerod IEC; McDonald WI; MacManus DG; Kendall BE; Kinglsey DPE; Moseley IF. The early risk of multiple sclerosis after optic neuritis. J Neurol Neurosurg Psychiatry. 1988; 51: 1569-71.

4. Lee KH; Hashimoto SA; Hooge J; Kastrukoff LF; Oger J; Li D; Paty DW. MRI in the diagnosis of MS: A prospective 3 year follow-up with comparison of clinical evaluation, evoked potentials, oligoclonal banding and CT. Neurology. 1991;41:657-660.

5. Paty DW; McFarlin DE; McDonald WI. Magnetic resonance imaging and laboratory aids in the diagnosis of multiple sclerosis [editorial]. Ann Neurol. 1991; 29: 3-5.

6. Fazekas F; Offenbacher H; Fuchs S; Schmidt R; Niederkorn K; Horner S; Lechner H. Criteria for an increased specificity of MRI interpretation in elderly subjects with suspected multiple sclerosis. Neurology. 1988; 38(12): 1822-5.

7. Miller DH; Newton MR; van der Poel JC; du Boulay EP; Halliday AM; Kendall BE; Johnson G; MacManus DG; Moseley IF; McDonald WI. Magnetic resonsance imaging of the optic nerve in optic neuritis. Neurology. 1988; 38: 175-9.

8. Youl BD; Turano G; Miller DH; Towell AD; MacMannus DG; Moore SG; Jones SJ; Barrett G; Kendall BE; Moseley IF; Tofts PS; Halliday AM; McDonald WI. The pathophysiology of acute optic neuritis. Brain 1991. 114, 2437-2450.

9. Stewart WA; Hall LD; Berry K; Paty DW. Correlation between NMR scan and brain slice: Data in multiple sclerosis. Lancet. 1984; 2: 412.

10. Ormerod IE; Miller DH; McDonald WI; du Boulay EP; Rudge P; Kendall BE; Moseley IF; Johnson G; Tofts PS; Halliday AM; Bronstein AM; Scaravilli F; Harding AE; Barnes D; Zilkha KJ. The role of NMR imaging in the assessment of multiple sclerosis and isolated neurological lesions. Brain. 1987; 110: 1579-1616.

11. Newcombe J; Hawkins CP; Henderson CL; Patel HA; Woodroofe MN; Hayes GM; Cuzner ML; MacManus D; du Boulay EP; McDonald WI. Histopathology of multiple sclerosis lesions detected by magnetic resonance imaging in unfixed post mortem central nervous system tissue. Brain. 1991; 114: 1013-23.

12. Oger J; Kastrukoff LF; Li DK; Paty DW. Multiple sclerosis: In relapsing patients, immune functions vary with disease as assessed by MRI. Neurology. 1988; 38: 1739-44.

13. Katz D; Taubenberger J; Raine C; McFarlin D; McFarland H. Gadolinium-enhancing lesions on magentic resonance imaging: Neuropathological findings. Ann Neurol 1990;28:243.

14. Paty DW. Multiple sclerosis: Assessment of disease progression and effects of treatment. Can J Neurol Sci. 1987;14:518-520.

15. Koopmans R; Li DK; Grochowski E; Cutler P; Paty DW. Benign versus chronic progressive multiple sclerosis: Magnetic resonance imaging features. Ann Neurol. 1989; 25:74-81.

16. Honer WG; Hurwitz T; Li DK; Palmer M; Paty DW. Temporal lobe involvement in multiple sclerosis patients with psychiatric disorders. Arch Neurol. 1987; 44: 187-90.

17. Huber SJ; Paulson GW; Shuttleworth EC; Chakeres D; Clapp LE; Pakalnis A; Weiss K; Rammohan K. Magnetic resonance imaging correlates of dementia in multiple sclerosis. Arch Neurol. 1987; 44: 732-6.

18. Izquierdo G; Campoy Jr F; Mir J; Gonzalez M; Martinex-Parra C. Memory and learning disturbances in multiple sclerosis. MRI lesions and neuropsychological correlation. European J Radiology. 1991; 13:220-224.

19. Pozzilli C; Passafiume D; Bernardi S; Pantano P; Incoccia C; Bastianello S; Bozzao L; Lenzi GL; Fieschi C. Spect, MRI and cognitive functions in multiple sclerosis. J Neurol Neurosurg Psychiatry. 1991; 54: 110-115.

20. Feinstein A; Kartsounis LD; Miller DH; Youl BD; Ron MA. Clinically isolated lesions of the type seen in multiple sclerosis: a cognitivie, psychiatric, and MRI follow up study. J Neurol Neurosurg Psychiatry. 1992; 55: 869-876.
21. Mariani C; Farina E; Cappa SF; Anzola GP; Fablia L; Bevilacqua L; Capra R. Neuropsychological assessment in multiple sclerosis: a follow-up study with magnetic resonance imaging. J Neurol 1991; 238: 395-400.
22. Clark CM; James G; Li D; Oger J; Paty D; Klonoff H. Ventricular size, cognitive function and depression in patients with multiple sclerosis. Can J Neurol Sci. 1992; 19: 352-356.
23. Wiebe S; Karlik SJ; Lee DH; Hopkins M; Vandervoort MK; Rice GP; Ebers GC; Noseworthy JH. Serial cranial and spinal cord quantitative MRI in multiple sclerosis: Clinical correlations. Neurology. 1990; 40(Suppl 1):377.
24. Kastrukoff LF; Oger JJ; Hashimoto SA; Sacks SL; Li DK; Palmer MR; Koopmans RA; Petkau AJ; Berkowitz J; Paty DW. Systemic lymphoblastoid interferon therapy in chronic progressive multiple sclerosis. I. Clinical and MRI evaluation. Neurology. 1990; 40: 479-86.
25. Paty DW; Bergstrom M; Palmer M; MacFadyen J; Li D. A quantitative magnetic resonance image of the multiple sclerosis brain. Am Acad Neurol. Dallas, April 1985, 137.
26. Isaac C; Li DK; Gento M; Jardine C; Grochowski E; Palmer M; Kastrukoff LF; Oger J; Paty DW. Multiple sclerosis: a serial study using MRI in relapsing patients. Neurology. 1988; 38: 1511-5.
27. Thompson AJ; Kermode AG; Wicks D; MacManus DG; Kendall BE; Kingsley DP; McDonald WI. Major differences in the dynamics of primary and secondary progressive multiple sclerosis. Ann Neurol. 1991; 29: 53-62.
28. Bastianello S; Pozzilli C; Bernardi S; Bozzao L; Fantozzi LM Buttinelli C; Fieschi C. Serial study of gadolinium-DTPA MRI enhancement in multiple sclerosis. Neurology. 1990; 40: 591-5.
29. Kappos L; Stadt D; Ratzka M; Keil W; Schneiderbanger Grygier S; Heitzer T; Poser S; Nadjmi M. Magnetic resonance imaging in the evaluation of treatment in multiple sclerosis. Neuroradiology. 1988; 30(4): 299-302.
30. Koopmans RA; Li DKB; Zhao GJ; Redekop WK; Paty DW. MRI assessment of cyclosporine therapy of MS in a multi-center trial. Neurology 42 (Suppl 3), April 1992, 210.
31. Miller DH; Barkhof F; Berry I; Kappos L; Scotti G; Thompson AJ. Magnetic resonance imaging in monitoring the treatment of multiple sclerosis: concerted action guidelines. J Neurol Neurosurg Psychiatry. 1991; 54: 683-8.
32. IFNB MS Study Group (Sibley W.). Interferon-beta 1b is effective in relapsing-remitting multiple sclerosis: Clinical results of a multicenter, randomized, double blind, placebo-controlled trial. Neurology. 1993; 43:655-661
33. Kurtzke JF. Rating neurologic impairment in multiple sclerosis: An expanded disability status scale (EDSS). Neurology. 1983. 33; 1444-52
34. Paty DW; Li DKB; MS/MRI Study Group; IFNB MS Study Group. Interferon-beta-1b is effective in relapsing-remitting multiple sclerosis: II-MRI analysis results of a multicenter randomized, double-blind placebo control trial. Neurology. 1993;43:662-667

T LYMPHOCYTE RESPONSE TO POTENTIAL AUTOANTIGENS IN PATIENTS WITH MULTIPLE SCLEROSIS

G. Ristori, M. Salvetti, C. Buttinelli, M. Falcone, A. Pisani, C. Pozzilli and C. Fieschi

Department of Neuroscience
Università "La Sapienza"
V.le Università, 30 Roma

INTRODUCTION

Myelin basic protein (MBP), a major myelin antigen, is considered a potential target of the immunopathologic process in Multiple Sclerosis (MS). MBP-specific T lymphocytes indeed mediate experimental allergic encephalomyelitis, the animal model of MS. Furthermore *in vivo*- activated T cell clones reactive to MBP were detected in the peripheral blood of patients, but not of healthy individuals (1). For these reasons the T cell response against this self-antigen is being extensively investigated in patients with MS.

We approached this matter by using new laboratory techniques (such as polymerase chain reaction and the "split-well" culture method, a highly efficient tool for the selection of antigen specific T lymphocyte lines) in order to study the pattern of reactivity of MBP specific T cells and its possible correlation with the polymorphisms of the immune response genes.

Another major aim of our work consists in investigating the role of other potential targets of the (auto)immune response in MS. In other autoimmune disorders, multiple autoantigens may contribute to the pathogenesis of the disease and other myelin determinants, different from MBP, may be involved in the immunopathogenesis of MS (2).

Among the potential autoantigens "alternative" to MBP we chose to investigate the heat shock proteins (HSP) as possible targets of autoreactive T lymphocytes. HSP are now under scrutiny in various autoimmune conditions by virtue of their conservation in phylogeny and immunogenicity (3). HSP are constitutively expressed and also stress-induced in the mammalian brain (4); moreover a preferential expression by the oligodendrocytes, among the glial cells, has been recently demonstrated (5).

T CELL REACTIVITY TO MBP

Studies on populations of MS patients showed that T-lymphocytes recognize multiple MBP epitopes (6,7), though some regions of the molecule also showed a certain degree of immunodominance (8,9,10). We studied the fine specificity of MBP reactive T lymphocyte lines (TLL) in detail in single patients, with special attention also to its stability over time.

A minimum of six MBP specific TLL was assayed from each of nine patients with MS against a panel of overlapping peptides, encompassing the whole MBP sequence.

Predominant responses could be detected in some patients with MS (fig 1a). In patient PP all TLL recognized the same peptide. Though for a different region, the same is true for patient GU and, to a lesser extent, for patient AM. The remaining patients displayed a broad reactivity to a variety of sequences (fig 1b). The major clinical characteristics did not differ between patients with homogeneous vs. non homogeneous responses to MBP; the

A Multidisciplinary Approach To Myelin Diseases II
Edited by S. Salvati, Plenum Press, New York, 1994

feature shared only by the patients with predominant responses was the DR2 phenotype. In three DR2+ non MS donors (two healthy and one with chronic inflammatory demyelinating disease) similarly dominant responses were not evident, though some slightly preferential recognitions were apparent (fig. 1c). The predominant sequence in patients GU and (namely 86-99) is encompassed by previously reported immunogenic regions of MBP such as 87-106 and 84-102 (8, 9,10). As in our study, the recognition of the latter regions was frequently associated with the DR2 phenotype. Interestingly, in DR2+ patients, predominant responses to other sequences of MBP can be detected, as demonstrated by the recognition of peptide 16-38 in patient PP.

We then decided to investigate whether the predominant responses were also stable. We therefore raised new MBP-specific T lines from patient PP eight months after the first attempt, when the disease had shifted to a phase of clinical activity. Seven new T lines were assayed (fig. 2): six confirmed the reactivity to peptide 16-38. The additional reactivities in polyclonal TLL (86-105 and 139-153) had been already detected during the first set of experiments. Peptides 7-26 and 28-48 that recalled the remaining three positive proliferative responses had sequence overlapping with the predominant region.

Since the latter result might suggest a possible associacion between patterns of T cell reactivity and DR2 subtypes, in the six DR2+ subjects (3 patients and 3 non-MS donors) we performed the genomic typing of the DRB1 and DRB5 genes. The DRB5 alleles were typed by conventional oligotyping. The DRB1 alleles were typed by applying the DNA heteroduplex analysis to the DR2+ subjects for the first time (11).
The following typings were obtained: patient AM (DRB1*1501, DRB5*0101); patient PP (DRB1*1501, DRB5*0101); patient GU (DRB1*1601, DRB5*02); donor CM (DRB1*1601, DRB5*02); donor ML (DRB1*1602, DRB5*02); donor PR (DRB1*1601, DRB5*02). Patient GU and two non MS donors had the same haplotype, hence differences in the DRB1 and DRB5 haplotypes do not account for the different patterns of response to MBP. Furthermore (table 1), the same haplotype was also shared by patients (PA and AM) with predominant recognitions of different MBP regions; conversely, patients (AM and GU) recognizing the same region of the molecule had different haplotypes.

The MHC restriction of the TLL was assessed in the DR2+ patients by using L cell transfectants as APCs. The tranfectants expressed either the DRB1*1501 or the DRB5*0101 products. As already reported by others (12), the MHC restriction was permissive; in our experiments, the variability of the restriction elements was confirmed on multiple TLL with the same peptide specificity, derived from the same individual.

T CELL REACTIVITY TO HSP

A previous study (13), concerning a possible role of HSP in MS, showed a colocalization of lymphocytes bearing gamma-delta T-cell receptor (TCR) and oligodendrocytes that expressed 65 kDa HSP (HSP65) in chronic lesions of MS brains.

We approached the problem from a "functional" point of view, by testing the response of long-term PPD-specific TLL to recombinant M. bovis HSP65 and M. tuberculosis 70 kDa (HSP70). The strategy of selecting and expanding TLL with PPD rather than with HSP was chosen in order to reduce the risk of eliciting T cell responses against contaminant antigens used to produce the recombinant proteins.

Seventy-six PPD specific TLL from 10 MS patients and 71 from healthy donors were screened. The results are summarized in table 2: 24 (31.6%) TLL from patients, and 16 (22.5%) from controls gave a positive proliferative response to HSP65 (chi-square 1.2; not signiifcant), while 19 (25.0%) and 5 (7.0%), respectively, were positive against HSP70 (chi-square 9.77; $P < 0.005$).

The phenotype of PPD-specific TLL from 6 patients and 3 healthy individuals was studied by cytofluorimetric analysis with particular attention to TCR gamma-delta determinants. Lymphocytes bearing gamma-delta TCR are indeed increased in number in immunological diseases (14) and in reumathoid arthritis synovial fluid, gamma-delta T cells recognizing M. bovis HSP65 have been found (15). In our work, with one exception, the population of gamma-delta T cells remained a minority in each line.

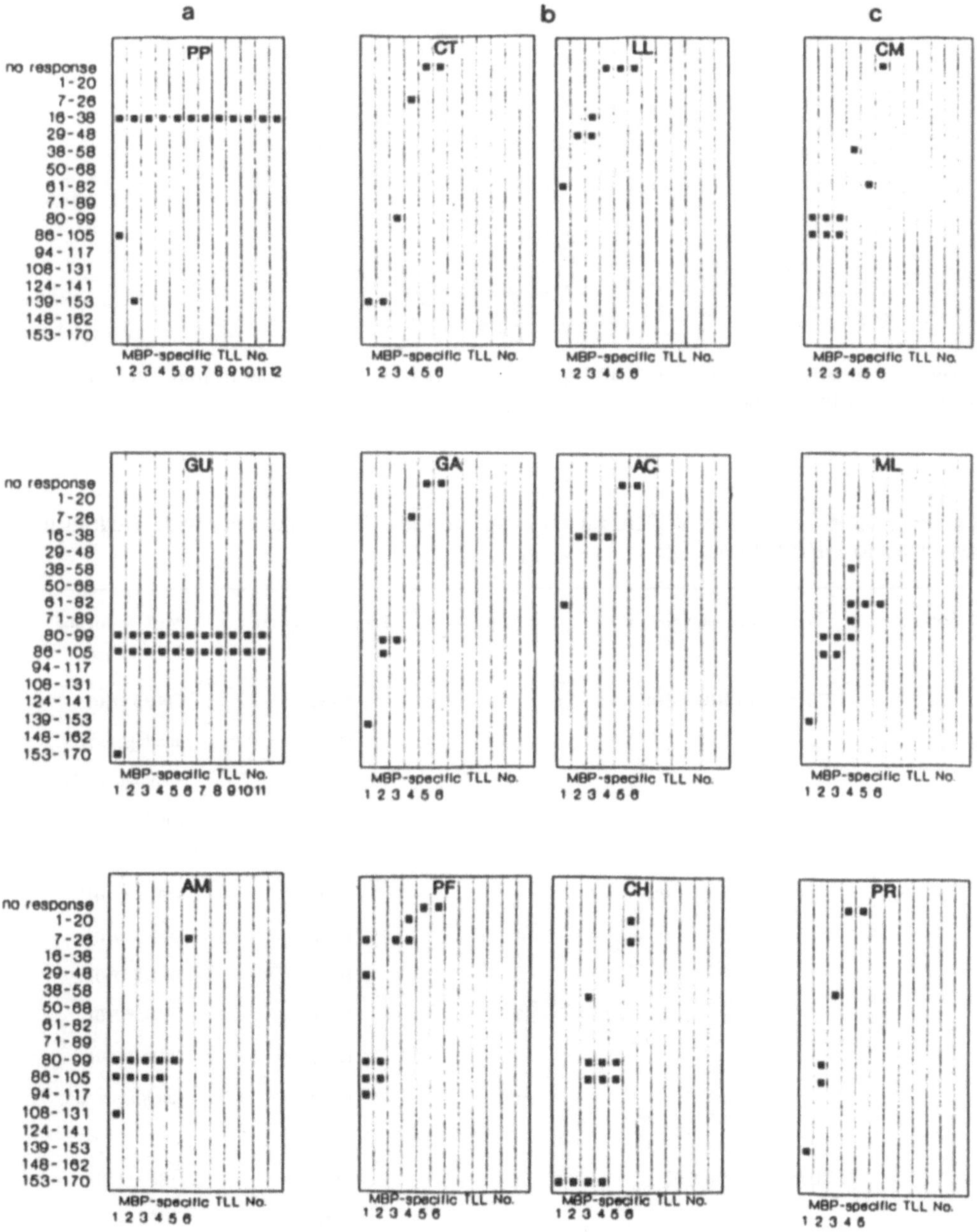

Figure 1. Proliferative responses to MBP peptide sequences of TLL from nine patients and three non-MS donors. The pattern of reactivity of MBP-specific TLL in each individual is shown. a) DR2+ MS patients; b) DR2- patients; c) DR2+ non-MS donors. "no response": TLL not recognizing any of the peptides.

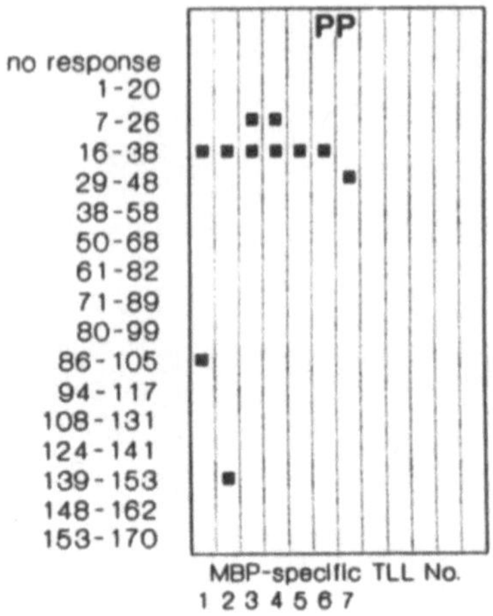

Figure 2. Proliferative responses to MBP peptide sequences of seven TLL raised from patient PP eight months after the first attempt.

Table 1. MHC restriction of TLL raised from the three DR2+ patients with predominant responses to MBP regions.

Patients	HLA-DR2 subtype	T-cell line	Target sequence	MHC restriction[b]
AM	DRB1*1501;DRB5*0101	1	86-99	DRB5*0101
		2	86-99	autologous APC
		3	86-99	autologous APC
		4	86-99	autologous APC
PP	DRB1*1501;DRB5*0101	3	16-38	DRB5*0101
		4	16-38	DRB5*0101
		5	16-38	DRB5*0101
		6	16-38	DRB5*0101
		1A [a]	16-38	DRB1*1501
		6A [a]	16-38	autologous APC
GU	DRB1*1601;DRB5*02	3	86-99	DRB1*1501
		4	86-99	DRB1*1501
		5	86-99	autologous APC
		6	86-99	autologous APC

[a] T-cell lines raised 8 months after the first attempt;
[b] MHC restriction: DRB1*1501 or DRB5*0101 indicate positive proliferative responses to MBP and to the relevant peptide in the presence of autologous APCs as well as in the presence of one of the transfectants;"autologous APC" indicates no responses with the transfectants.

Table 2. Frequencies of HSP65 and HSP70 specific T cell lines from patients with MS and controls. Results are expressed as number of HSP reactive T cell lines / number of PPD specific T cell lines tested.

Patients	HSP65	HSP70	Controls	HSP65	HSP70
SP	0/4	1/4	FP	1/2	1/2
TA	0/2	0/2	TL	2/7	1/7
CT	2/10	3/10	CV	2/3	1/3
GA	2/7	2/7	ML	2/3	0/3
GU	1/7	1/7	MC	0/2	0/2
CA	5/10	4/10	MA	0/5	0/5
AM	0/1	0/1	MM	0/5	0/5
CP	0/4	0/4	FA	0/3	0/3
SC	1/11	1/11	OA	1/5	0/5
PF	13/20	7/20	VC	0/6	1/6
			AN	1/11	0/11
			BU	7/19	1/19
	24/76	19/76		16/71	5/71

CONCLUSION

Predominant and stable T cell response to MBP regions can be detected in some patients with MS. This result, in agreement with the one of Wekerle et al. (personal comunication), may encourage an attempt to design individual-specific immunotherapies. The association between predominant response and DR2+ phenotype may suggest a genetic influence on the pattern of T cell reactivity, though there is no strict influence of the polymorphisms of the DR2 genes. A better understandiing of the mechanisms which impart dominance to portions of the molecule may come from familial studies on possible correlations between patterns of T cell reactivity and polymorphisms of immune response (new) genes or MBP genes.

A question to be addressed concerns the role of T cells recognizing autoantigens different from MBP in the pathogenesis of MS. Our results suggest that a potentially autoaggressive immune response to HSP70 may not be properly regulated in patients with this disease. Epitope mapping studies, which include sequences from the human homologs of HSP, are now required in order to ascertain cross-recognition of conserved epitopes by our HSP-specific T clones.

REFERENCES

1. M. Allegretta, J.A. Nicklas, S. Sriram, R. Albertini, T cell responsive to myelin basic protein in patients with multiple sclerosis, *Science* 247:718 (1990)
2. R. Martin, H.F. Mc Farland, D. Mc Farlin, Immunological aspects of demyelinating diseases, *Annu .Rev .Immunol.* 10:153 (1992).
3. R. A. Young and T.J. Elliot, Stress protein, infection and immune surveillance, *Cell* 59:5 (1989).
4. S. Lindquist, The heat-shock response, *Annu. Rev. Biochem.* 55: 1151. (1986).
5. M.S. Freedman, N.N. Buu, T.C.J. Ruijs, K. Williams and J.P. Antel, Differential expression of heat shock proteins by human glial cells, *J. Neuroimmunol.* 41:231 (1992).

6. M.Pette, K. Fujita, D. Wilkinson, D.M. Altmann, J. Trowsdale, G. Giegherich, H. Hinkkanen, J.T. Epplen, L. Kappos and H. Wekerle, Myeiln autoreactivity in multiple sclerosis: recognition of myelin basic protein in the context of HLA-DR2 products by T lymphocytes of multiple sclerosis patients and healthy donors *Proc. Natl. Acad. Sci.* 87: 7968 (1990).
7. T. Olsson, J. Sun, J. Hillert, B. Hojeberg, G. Andersson, O. Olerup and H. Link, Increased numbers of T cells recognizing multiple myelin basic protein epitopes in multiple sclerosis, *Eur. J. Immunol.* 22:1083 (1992).
8. R. Martin, D. Jaraquemada, M. Flerlage, J. Richert, J. Whitaker, E.O. Long, D. McFarlin and H. McFarland, Fine specificity and HLA restriction of myelin basic protein-specific cytotoxic T cell lines from multiple sclerosis patients and healthy individuals, *J. Immunol.* 145: 540 (1990).
9. K. Ota, M. Matsui, E.L. Milford, G.A. Mackin, H.L. Weiner and D.A. Hafler, T cell recognition of an immunodominant myelin basic protein epitope in multiple sclerosis, *Nature* 346:183 (1990).
10. Z. Jingwu, R. Medaer, G.A. Hashim, Y. Chin, E. Van den Berg-Loonen and J.C.M. Raus, Myelin basic protein-specific T lymphocytes in multiple sclerosis and controls: precursor frequency, fine specificity and cytotoxicity, *Ann. Neurol.*. 32:330 (1992).
11. R. Sorrentino, C. Iannicola, S. Costanzi, A. Chersi, R. Tosi, Detection of complex alleles by direct analysis of DNA heteroduplexes, *Immunogenetics* 33:118 (1991).
12. R. Martin, M.D. Howell, D. Jaraquemada, M. Flerlage, J. Richert, S. Brostoff, E.O. Long, D. McFarlin and H.F. McFarland, A myelin basic protein peptide is recognized by cytotoxic T cells in the context of four HLA-DR types associated with multiple sclerosis, *J. Exp. Med.* 173: 19 (1991).
13. K.Selmay, C. Brosnan and C. Raine, Colocalization of lymphocytes bearing gamma-delta T-cell receptor and heat shock protein hsp 65+ oligodendrocytes in multiple sclerosis. *Proc. Natl. Acad. Sci.* 88: 6452 (1991).
14. Editorial, Gamma/delta T-cell receptors, *The Lancet* 337:207 (1991).
15. J.Holoshitz, F. Koning, J.E. Coligan, J. De Bruyn and S. Strober, Isolation of CD4-CD8-mycobacteria-reactive T lymphocyte clones from rheumatoid arthritis synovial fluid, *Nature* 339: 226 (1989).

IMMUNOSUPPRESSION IN MULTIPLE SCLEROSIS: STATE OF THE ART AND FUTURE PERSPECTIVES

Luca Massacesi, Maria Pia Amato and Luigi Amaducci
Ia Clinica Neurologica Universita Florence, Italy

EVIDENCE OF IMMUNOPATHOGENESIS OF MS

It has long been suspected that multiple sclerosis (MS) is due to a primary autoimmune response to the CNS, perhaps related to an antecedent viral infection. Independent lines of evidence suggest that T-lymphocytes participate directly in the formation of MS lesions. Indeed demyelination in MS is associated with chronic inflammatory lesions in CNS white matter, consisting predominantly of T-cells that infiltrate brain white matter from blood[1-2] and activated T-cells appearing in the cerebrospinal fluid (CSF) of MS patients with active disease[3-5].

In recent years, several observations have strengthened the hypothesis that myelin protein reactive T-cells may play a role in MS: firstly, T-cells containing inflammatory infiltrates are observed only in CNS white matter, mainly composed of myelin and oligodendrocytes and not in gray matter where the myelin component is minimal[6]; secondly an increased frequency of myelin protein reactive T-cells was present in MS patients compared to disease controls, and enrichment of these cells in CSF compared to PBL was present[7,8,9]; thirdly experimental autoimmune encephalomyelitis (EAE), a disease model that resembles MS in many aspects, is mediated by T-cells of restricted specificity and signs of EAE may be lessened by immunosuppressive therapy and aggravated by immunostimulation[10,11].

EAE AS A MODEL OF MS

EAE is an autoimmune disease of the central nervous system (CNS), inducible in susceptible animals by immunization with myelin antigens or by passive transfer of sensitized T-cells to syngenic recipients[10,11]. In rodents, chronic and relapsing forms of the disease have some similarities to human inflammatory demyelinating diseases of the CNS, representing a useful model for the study of the effect of immune mechanisms of CNS damage occurring in MS[10,11]. Nonetheless, the phylogenetic differences among human and rodent immune response genes are remarkable. For this reason EAE has been extensively

A Multidisciplinary Approach To Myelin Diseases II
Edited by S. Salvati, Plenum Press, New York, 1994

studied also in non-human primates and most studies have been carried out in macaques[12]. However in most individuals of this strain, EAE takes the form of a hyperacute fulminating disease characterized by hemorragic-necrotic lesions in the CNS, hardly representing a useful model of human demyelinating lesions[12]. Recently, a chronic relapsing form of EAE characterized by mild neurological signs and a relapsing remitting course was described in the marmoset[13,14]. In this monkey strain early CNS lesions were characterized by scattered perivascular lymphomonocytic inflammatory infiltrates surrounded by demyelination and mild astrogliosis. The lesions were mainly localized in brain and spinal cord white matter. In older lesions, mononuclear cells mainly of the monocyte/macrophage lineage and also the widespread infiltration of large sharply delimited areas of demyelination and reactive astrogliosis was noted. These lesions were larger than the more acute ones and differed also in the extension of mononuclear cell infiltrates. In addition, these areas seemed to be the result of earlier and smaller lesion confluence, recalling "plaque" pathogenesis and suggesting the occurence of a sequence of events observed also in early MS lesions. However, in animals sacrificed after a long relapsing remitting course, coexistence of early lesions, morphologically similar to those observed in acute disseminated encephalomyelitis (ADE), and of older lesions similar to early MS plaques, suggests that the relationship between MS and ADE should be reconsidered. Indeed EAE, because of the morphology and distribution of CNS lesions has usually been classified as a model closer to ADE than to MS[15], but these findings suggest that differences between the CNS lesions observed in the two human diseases may be due to the different timing of usual CNS lesion observation: acutely in ADE, after many years from disease onset in MS.
In conclusion, the clinical and pathological features of marmoset EAE suggest that pathogenesis of this disease is similar to pathogenesis of MS. In addition, relationships between MS and ADE, a disease mainly sustained by hyperreactivity to MBP, should be reconsidered.

IMMUNOSUPPRESSIVE STRATEGIES IN AUTOIMMUNE DISEASES

Because of the role of the immune system in the pathogenesis of MS, the efficacy of immunosuppression has been investigated to prevent disease progression. As in other immune mediated diseases, *non-selective immunosuppressive strategies* are currently used in selected forms of the disease, but research for less toxic strategies of selective or of specific immunosuppression are currently under investigation.

Non-selective immunosuppression: azathioprine efficacy and long term hazards

Azathioprine (AZA), a nitroimidazole substitute form of 6-mercaptopurine, is the immunosuppressive agent longest in use for the treatment multiple sclerosis (MS). A critical review of the main clinical trials involving AZA therapy for MS appeared in 1988[16,17]. More recently Yudkin and associates[18], applying meta-analysis to the results of all controlled, randomized trials of AZA treatment published to date examined data from 793 patients enrolled in five double-blind and two single-blind studies. They found no significant difference in the Kurtzke disability scale scores of treated and control groups after one year of therapy, though a trend in favor of the AZA group emerged during the second and third years. Even in the first year, relapsing-remitting patients taking AZA had a significantly higher probability than controls of not suffering a relapse, a difference that doubled during the second year of treatment. Two other double masked, placebo-controlled trials have been completed more recently[19,20]. In the British and Dutch trial[19], after 3 years of follow-up the mean deterioration in the Kurtzke disability score did not significantly differ between the AZA and the placebo group. After 3 years there were slightly fewer relapses in the AZA group than in the placebo group, but the difference was also not significant. Nonetheless, Goodkin et al.[20], using a daily dosage high enough to induce white blood cell count decrease, in selected relapsing remitting cases, found a significant difference favoring AZA for observed mean exacerbation rate and time to deterioration in both the Ambulation Index and the Kurtzke scale.

Despite the widespread use of AZA over many years, however, there is surprisingly little information on the potential oncogenic risk posed by current dosages of the drug for MS patients. Studies in this field often lack controls altogether or are based on historical controls and the number of patients involved tends to be too small[21-24]. Lhermitte et al.[24] identified 10 tumors among 131 AZA-treated patients (100 mg daily; mean duration of treatment, 75 months) and four tumors in a nontreated group. These data have caused some alarm in the medical community, although the difference did not reach statistical significance and all cases involved solid tumors, not lymphomas. Data emerging from the Anglo-Dutch trial cited by Yudkin et al. are still preliminary: two AZA-treated MS patients had died of cancer by the end of the study in October 1987. New cancers were subsequently identified in two patients in the placebo group and two in the AZA group. In a longitudinal study Amato et al[25] found five malignancies in the azathioprine group (2.0 mg/kg daily; mean duration of treatment, 4.16 years) compared with seven in the control group. The age-adjusted occurrence rate was 3.62/1,000 person-years (95% CI, 1.17 to 8.43) in the treated and 4.24/1,000 person-years (95% CI 1.70 to 8.73) in the nontreated group; the age-adjusted relative risk of cancer was 0.85. In both groups tumors were typically solid, mostly carcinomas, and no lymphoma was found; most of the patients who developed malignancies were women, and the site of tumors was predominantly the uterus or the breast, common tumor sites in the female population.

Taken together these data indicate that AZA therapy is a safe treatment active in preventing MS progression when administered in relapsing remitting cases and that no alarm about an increased oncogenic risk of AZA therapy in MS is justified.

Specific immunosuppression

If MS is an autoimmune disease, identification of pathogenetic lymphocytes, may allow their inactivation and therefore a specific immunosuppressive therapy. For this reason the presence of autoimmune T-cells has been investigated in EAE and in MS.

In EAE the encephalitogenic T-cell repertoire to the myelin protein MBP and, to a lesser extent, proteolipidic protein (PLP), have been characterized in some species. The immunodominant region of MBP and the TCR repertoire utilized in this response has been shown to be strain- and species-specific, allowing development *in vivo* of specific inactivation of pathogenetic T-cells[26-33]. Nonetheless, a passive transfer model of EAE in outbred non-human primate species indicates that different immunodominant epitopes and different T-cell populations are encephalitogenic in each individual[14].

In MS the repertoire of T-cell clones reactive to MBP in DR2+ MS patients was found in one study to be specific to the aminoacid 84-102 region of MBP and to utilize Vß 17 TCR gene segments[34-35] whereas other studies indicate that a more heterogeneous T-cell response, in terms of both the fine specificity of MBP recognition and TCR usage is likely to be present[36-38].

These findings indicate that specific inactivation of potentially pathogenetic MBP specific T-cells is not currently feasible.

Selective immunosuppression

Considering that identification and specific inactivation of pathogenetic T-cell at present does not seem feasible, treatment strategies selectively active on the immune system represent probably the most realistic approach to MS therapy.

See elsewhere in this volume for data related to Beta-IFN activity in MS.

REFERENCES

1. Traugott U., Reinherz E.L., Raine C.S.: Multiple Sclerosis. Distribution of T-cells, T-cell subsets and Ia-positive macrophages in lesions of different ages. J.Neuroimmunol., 4:201-221, 1983.
2. Hayashi T., Morimoto C., Burks J.S. et al.: Dual-label immunocytochemistry of the active multiple sclerosis lesions:major histocompatibility complex and activation antigens. Ann. Neurol. 24:523-531, 1988.
3. Noronha A.B.C., Richman D.P., Arnason B.G.W.: Detection of in vivo stimulated cerebrospinal-fluid lymphocytes by flow cytometry in patients with multiple sclerosis. N. Eng. J. Med. 303:713-717, 1980.
4. Hafler D.A., Fox D.A., Manning M.E. et al.: In vivo activated T lymphocytes in the peripheral blood and cerebrospinal-fluid of patients with multiple sclerosis. N. Eng. J. Med. 312:1405-1411, 1985.
5. DeFreitas E.C., Sandberg-Wollheim S., Schonely K. et al.: regulation of interleukin-2 receptors on T cells from multiple sclerosis patients. Proc. Natl. Acad. Sci. USA 83:2637-2641, 1986.
6. McFarlin D.E., McFarland H.F.: Multiple sclerosis. N. Engl. J. Med. 307, 1246-1251, 1992.
7. Olsson T., Wang W.Z., Hojeber B., Kostulas V., Jiang Y.P., Anderson G., Ekre H.P., Link H.: Autoreactive T lymphocytes in multiple sclerosis determined by antigen induced secretion of interferon. J. Clin. Invest. 86:981-5, 1991.
8. Allegretta, M., Nicklas, J.A., Sriram, S. and Albertini R.J. T cell responsive to myelin basic protein in patients with multiple sclerosis. Science 247, 718-721, 1990.
9. Chou Y.K., Bourdette D.N., Offner H., Whitham R., Wang R.Y., Hashim G.A., Vanderbark A.A.: Frequency of T cell specific for myelin basic protein and myelin proteolipid protein in blood and cerebrospinal fluid in multiple sclerosis. J. Neuroimmunol. 38:105-14, 1992.
10. Alvord E.C., Kies M.W., Suckling A.J.: Experimental allergic encephalomyelitis; A useful model for multiple sclerosis. Alan Liss, New York, 1984.
11. Davison A.N. Cuzner M.L.: The suppression of multiple sclerosis and experimental allergic encephalomyelitis. Academic Press N.Y., 1980.
12. Massacesi L., Parritz D., Joshi N, Letvin N. and Hauser S.L.: Experimental allergic encephalomyelitis in Cynomologus Monkeys: quantitation of T-cell responses in peripheral blood. J.Clin. Invest. 90: 399-404, 1992.
13. Massacesi L., Joshi N., Parritz D.L., Seboun E., Rombos A., Letvin N.L. and Hauser S.L.: The cellular immunology of primate EAE. Neurology 41 (suppl.1): 360, 1991.
14. Massacesi L., Joshi N., Parritz D.L., Canfield D., Letvin N.L. and Hauser S.L.: Passive transfer EAE and T-cell repertoire in primates. J. Neuroimmunol. 34 (suppl.): 112, 1990.
15. Adams and Victor. Principles of Neurology. McGraw-Hill (New York), 1989.
16. Ellison GW, Myers LW, Mickey MR, Graves MC, Tourtellotte WW, and Nuwer MR. Clinical experience with azathioprine: the pros. Neurology 1988;38 (suppl 2):20-23.
17. Silberberg DH. Azathioprine in multiple sclerosis: the cons. Neurology 1988;38 (suppl2):24-27.
18. Yudkin PL, Ellison GW, Ghezzi A, et al. Overview of azathioprine treatment in multiple sclerosis. Lancet 1991;338:1051-1055.
19. British and Dutch multiple sclerosis azathioprine trial group. Double masked trial of azathioprine in multiple sclerosis. Lancet 1988;ii:179-183.
20. Goodkin DE, Bailly RC, Teeztzen ML, Hertsgaard D, and Beatty WW. The efficacy of azathioprine in relapsing-remitting multiple sclerosis. Neurology 1991;41:20-25.
21. Rosen JA. Prolonged azathioprine treatment of non-remitting multiple sclerosis. J Neurol Neurosurg Psychiatry 1979;42:338-344.
22. Aimard G, Confavreux C, Ventre JJ, Guillot M, Devic M. Etude de 213 cas de sclerose en plaques traites par l'azathioprine de 1967 a 1982. Rev Neurol (Paris) 1983;139:509-513.
23. Sabouraud O, Oger J, Darcel F, Madigand M, Merienne M. Immunosuppression au long cours dans la sclerose en plaques: evaluation des traitments commences avant 1972. Rev Neurol (Paris) 1984;140:125-130.
24. Lhermitte F, marteau R, Roullet E, De Saxce H, Loridan M. Traitement prolonge de la sclerose en plaques par l'azathioprine a doses moyennes. Bilan de quinze annees d'experience. Rev Neurol (Paris) 1984;140:553-558.
25. Amato MP, Pracucci G, Ponziani G, Siracusa G, Fratiglioni L, and L Amaducci. Long-term safety of azathioprine therapy in multiple sclerosis. Neurology 1993;43:831-833.

26. Kumar V., Kono D., Urban J. et al.: The T-cell receptor repertoire and autoimmune diseases. Ann. Rev. Immunol., 7:657, 1989.
27. Kumar V., Urban J.L., Horvath J. et al.: Aminoacid variation at single residue in an autoimmune peptide profoundly affect its properties: T-cell activation, MHC binding, and ability to block experimental allergic encephalomyelitis. Proc. Natl. Acad. Sci. USA, 87:1337, 1990.
28. Acha-Orbea H., Mitchell D.J., Timmerman L. et al.: Limited heterogeneity of T-cell receptors from lymphocytes mediating autoimmune encephalomyelitis allows specific intervention. Cell, 54:263, 1988.
29. Hashim G.A., Vandembark A.A., Galang A.B. et al.: Antibodies specific for Vß8 receptor peptide suppress experimental autoimmune encephalomyelitis. J. Immunol. 12:4621, 1990.
30. Howell M.D., Winters S.T., Olee T. et al.: Vaccination against experimental allergic encephalomyelitis with T-cell receptor peptides. Science, 246:668, 1989.
31. Sakai K., Zamvil S.S., Mitchell D.J. et al.: Prevention of experimental encephalomyelitis with peptides that block interaction of T-cell with MHC proteins. Proc. Natl. Acad. Sci. USA, 87:970, 1989.
32. Vanderbark A.A., Hashim G., Offner H.: immunization with a synthetic T-cell receptor V region peptide protects against experimental autoimmune encephalomyelitis. Nature, 341:541, 1989.
33. Zaller D.M., Osman G., Kanagawa O. et al.: Prevention and treatment of murine experimental allergic encephalomyelitis with T-cell receptor Vß specific antibodies. J. Exp. Med. 171:1943, 1990.
34. Ota K., Matsui M., Milford E.L., Mackin G.A., Weiner H.L., Hafler D.A.: T-cell recognition of an immunodominant myelin basic protein epitope in multiple sclerosis. Nature 346, 183-6, 1990.
35. Wucherpfenning K.W., Ota K., Endo N., Seidman J.G., Rosenzweig A., Weiner H.L., hafler D.A.: Shared human T cell receptor Vß usage to immunodominant region of myelin basic protein. Science, 248, 1016-9, 1990.
36. Martin, R., Howell M.D., Jaraquemada D., Flerlage M., Richert J., Brostoff S., Long E.O., Mc Farlin D.E. Mc Farland H.F.: A myelin basic protein peptide is recognized by cytotoxic T cells in the context of four HLA-DR types associated with multiple sclerosis J. Exp. Med. 173:192-8, 1991.
37. Martin R., Jaraquemada D., Flerlage M., Richert J., Whitaker J., Long E.O., Mc Farlin D.E., McFarland H.F.: Fine specificity and HLA restriction of myelin basic protein-specific cytotoxic T cell lines from multiple sclerosis patients and healthy individuals. J. Immunol. 145: 540-7, 1990.
38. Pette M., Fujita K., Wilkinson D., Altman D.M., Trowsdale J., Giegerich G., Hinkkanen A., Epplen J.T., Kappos L. and Wekerle H.: Myelin autoreactivity in multiple sclerosis: recognition of myelin basic protein in the context of HLA-DR2 products by lymphocytes of MS patients and healthy donors. Proc. Natl. Acad. Sci. 87:7968-7972, 1990.

INDEX